Kathleen,

I hope you don't find the contents too intimidating – this stuff is really pretty simple, once you get by the vocabulary –

much love

Dad 6-9-96

Tracing Biological Evolution in Protein and Gene Structures

Tracing Biological Evolution in Protein and Gene Structures

Proceedings of the 20th Taniguchi International Symposium, Division of Biophysics, held in Nagoya, Japan, 31 October – 4 November 1994

Editors:

Mitiko Gō
Department of Biology
Faculty of Science
Nagoya University
Nagoya, Japan

and

Paul Schimmel
Department of Biology
MIT
Cambridge, Massachusetts, USA

1995

ELSEVIER

Amsterdam – Lausanne – New York – Oxford – Shannon – Tokyo

ISBN 0 444 82187 2

This book is printed on acid-free paper.

Published by:
Elsevier Science B.V.
P.O. Box 211
1000 AE Amsterdam
The Netherlands

In order to ensure rapid publication this volume was prepared using a method of electronic text processing known as Optical Character Recognition (OCR). Scientific accuracy and consistency of style were handled by the author. Time did not allow for the usual extensive editing process of the Publisher.

Printed in the Netherlands

Preface

Progress in our understanding of life, particularly in research on structures and functions of proteins, DNA and RNA and on genomic information is opening a new field in the study of biological evolution and even early evolution of life. During the last 2 decades, several important and unexpected findings in molecular biology have been revealed, such as the existence of intervening sequences (introns) and the enzymatic function of RNA (ribozymes). We also have begun to appreciate how macromolecules, like proteins and RNA, can be broken down into smaller pieces, or modules, which themselves have biological activities. Twenty years ago we could not study early biological evolution on the solid basis of molecular structure, function and interactions. However, we are now opening the door to the age of understanding biological evolution with words of biophysics and chemistry. We stand at the starting point to trace the biological evolution from the beginning of life. It seemed to be timely to assemble active researchers to exchange ideas and results during the Symposium. The organizers aimed to include researchers from several disciplines who were already involved in the elucidation of "the molecular mechanism of biological evolution", yet who might not have the chance to directly discuss their results with researchers in other disciplines.

We would like to express our heartfelt thanks to the late Mr Toyosaburo Taniguchi, Mr Yuichiro Taniguchi who has succeeded in the position of chief director of the Taniguchi Foundation, and the members of the Taniguchi Foundation for their generous financial support of the Symposium. We are also indebted to Prof Akiyoshi Wada, Chairman of the Planning Committee, and to Prof Akira Tasaki, Secretary General of the Symposium, for their helpful advice and encouragement throughout this symposium, from the planning of the program to the publication of the Proceedings.

Mitiko Gō
Paul Schimmel

Contents

©1995 Elsevier Science B.V. All rights reserved.
Tracing Biological Evolution in Protein and Gene Structures.
M. Gō and P. Schimmel, editors.

Implications of an operational RNA code for amino acids

Paul Schimmel
Department of Biology, Massachusetts Institute of Technology, Cambridge, Massachusetts, USA

Abstract. The specific aminoacylation of RNA oligonucleotides having sequences based on the acceptor stems of tRNAs led to the concept of an operational RNA code, which relates specific amino acids to sequences/structures in tRNA acceptor stems. The operational RNA code probably preceded the genetic code and later was incorporated into tRNAs where it still is an important basis for determining specific aminoacylations of contemporary tRNAs. The determinants for the operational RNA code and for the genetic code (anticodons) are segregated into separate domains of the tRNA molecule. These domains are, respectively, the acceptor-TψC minihelix and the anticodon-dihyrouridine stem-biloop. These two RNA domains interact with discrete domains of the tRNA synthetase which in turn suggests that the modular structures of both molecules were developed to fit each other. We have proposed that the acceptor-TψC minihelix structure may have been the primordial substrate for aminoacylation and that complexes of these aminoacylated structures could lead to peptide-bond formation. A model for a binary complex of a hairpin oligonucleotide structure is presented here. This complex is equivalent to an intermolecular RNA pseudoknot.

Key words: aminoacyl tRNA synthetases, aminoacylation, genetic code, origin of coding systems, RNA microhelices, transfer RNAs.

Introduction

The genetic code offers no explanation for the set of rules which assigns specific nucleotide triplets to specific amino acids. The logic for UUU corresponding to phenylalanine or GCC being a codon for glycine is obscure. While specific associations between particular triplets and amino acids is a possible explanation for the historical origins of the algorithm of the genetic code, this explanation seems unlikely on physical chemical principles. More likely, the code evolved as the culmination of a process whereby amino acids were related to nucleic acid structures and sequences by aminoacylation reactions in an evolving RNA world which was making the transition to the theater of proteins.

These aminoacylation reactions were probably carried out by ribozyme catalysts and later by tRNA synthetases. The trinucleotides of the genetic code may not have been present in these RNA amino acid acceptors. Instead, the earliest substrates may have resembled the acceptor-TψC minihelix domain of the two-domain L-shaped tRNA molecule (Fig. 1) The anticodon-containing domain was then joined later to the

Address for correspondence: Paul Schimmel, Department of Biology, Massachusetts Institute of Technology, 77 Massachusetts Avenue, Cambridge, MA 02139, USA.

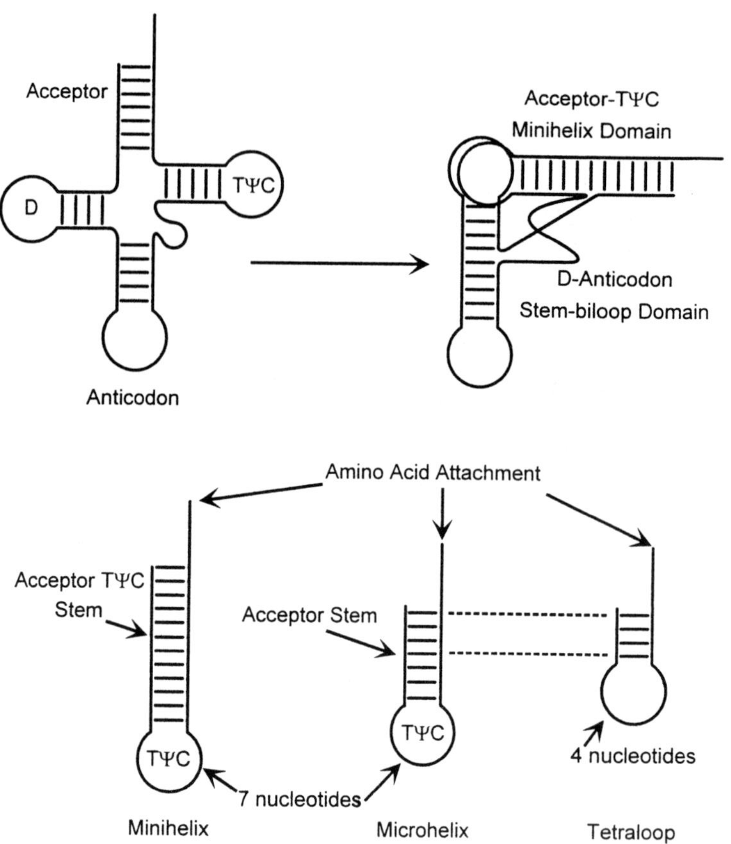

Fig. 1. Schematic illustration of the folding of the tRNA cloverleaf into two domains (top) and RNA oligonucleotide substrates for aminoacylation that are based on the acceptor-TψC domain (minihelix) or smaller fragments of that domain which contain the amino acid attachment site (bottom). In the conventional numbering of tRNA molecules, the amino acid attachment site at the 3′-end is located at A76 and is at the end of the single-stranded tetranucleotide sequence $N^{73}CCA^{76}$ which is common to all tRNAs.

acceptor helix structure to form the complete tRNA molecule.

Three independent considerations support this concept. First, RNA minihelices comprised of just the acceptor-TψC stem-loop and even smaller segments are substrates for specific aminoacylations [1–8], showing that the anticodon, per se, is not needed for aminoacylation (Fig. 1). Second, the minihelix domain on the ribosome interacts with 23S ribosomal RNA and the anticodon-containing domain with 16S rRNA [9]. This segregation of interactions suggests that the two tRNA domains had separate origins. Third, the minihelix domain is believed to be an ancient tag of RNA genomes as a structure needed for replication, long before the anticodon-containing domain was added to the tRNA molecule [10,11].

I will present here some of the background information which led to the concept of an operational RNA code for amino acids and will also explain how this operational code is associated with the evolution of tRNA synthetases. The difficult conceptual step is in making the transition from the RNA world to the theater of proteins and in connecting the operational RNA code to the genetic code. In particular, generating peptides starting with a system of noncoded synthesis and evolving to coded synthesis is a challenge. We recently proposed an explicit mechanism for noncoded peptide synthesis and how it might evolve into a system of coded synthesis using aminoacyl RNA complexes [12]. In addition to the background which led up to this proposal, I will show here a specific three-dimensional model of such a complex which is the basis for ongoing experiments.

Historical perspective on design and classification of tRNA synthetases

Early crystallographic studies of *Bacillus stearothermophilus* tyrosyl- and *Escherichia coli* (*E. coli*) methionyl-tRNA synthetases established that these two enzymes, which had little discernible sequence relationship, shared in common a nucleotide-binding fold of alternating β-strands and α-helices in their respective amino terminal domains [13]. Rossmann et al. had pointed out earlier that this design is characteristic of nucleotide-binding proteins such as the dehydrogenases and kinases [14]. These synthetases shared a "HIGH" tetrapeptide motif which is located after the first β-strand of the nucleotide-binding fold. The two conserved histidines interact with phosphates of bound ATP and the aminoacyl adenylate intermediate [15,16].

The HIGH tetrapeptide was too short to be the basis for finding statistically significant relationships between synthetase sequences. Subsequent to the structural work on tyrosyl- and methionyl-tRNA synthetases, Webster et al. obtained the sequence of *E. coli* isoleucyl-tRNA synthetase, a 939 amino acid polypeptide [17]. This sequence has an 11 amino acid match (10 identities and one conservative replacement) with a segment in *E. coli* methionyl-tRNA synthetase that ends in the HIGH tetrapeptide. This 11 amino acid peptide was designated as the "signature sequence", because it was sufficiently long to use in naive searches to find relationships between available sequences of synthetases. Later, Hountondji et al. identified the "KMSKS" motif which is common to all enzymes which have the signature sequence [18]. This pentapeptide is always on the C-terminal side of the signature sequence, where it is located in a loop after a β-strand in the second half of the nucleotide-binding fold. The second lysine of this pentapeptide has been proposed as important for stabilization of the transition state for formation of the aminoacyl adenylate [16,19].

The signature sequence and KMSKS pentapeptide together made it possible to establish that a subgroup of tRNA synthetases shared similar structures for their active site domains and were related in evolution [20,21]. These enzymes were later designated as class I tRNA synthetases and include argininyl-, cysteinyl-, glutaminyl-, glutamyl-, leucyl-, isoleucyl-, methionyl-, tryptophanyl-, tyrosyl-, and valyl-tRNA synthetases.

Not all sequenced synthetases had the signature sequence and KMSKS motif. Nor did they have a clear relationship to each other until Eriani et al. showed that they shared three highly degenerate sequence motifs [22]. These motifs are part of an eight-stranded β-structure with three α-helices which form the active site domain for adenylate synthesis and interaction with the 3′-end of the tRNA [23,24]. The motifs occur in sequential order along the polypeptide but, depending on the enzyme, may be located in the N-terminal or C-terminal half of the protein.

Motif 1 is comprised of an α-helix followed by a β-strand and is near the subunit interface in these oligomeric (mostly dimeric) proteins. Motif 2 consists of two β-strands separated by a loop which is variable in size. Residues in the strands and loops of aspartyl- and seryl-tRNA synthetases have extensive substrate interactions, including a salt bridge between a conserved arginine and the α-phosphate of ATP. There are also base-specific interactions of motif 2 residues in aspartyl-tRNA synthetase with the end of the acceptor stem of bound $tRNA^{Asp}$ [24,25]. Motif 3 is a strand-helix structure which contains the well-conserved "GLER" tetrapeptide. This motif forms part of the active site structure in the crystal structure of yeast aspartyl-tRNA synthetase, where the guanidino group of the conserved arginine of the GLER peptide bonds to oxygens of the γ-phosphate of ATP [26]. Mutagenesis of the conserved arginine in motif 3 of yeast aspartyl- and *E. coli* alanyl-tRNA synthetase (whose three-dimensional structure is not known) demonstrated the importance of this residue and of motif 3 in general for adenylate synthesis [26,27]. These enzymes, whose structures and sequences are completely unrelated to those in class I, are designated as members of class II and include alanyl-, asparaginyl, aspartyl-, glycyl-, histidyl-, lysyl-, phenylalanyl-, prolyl-, seryl-, and threonyl-tRNA synthetases.

Enzymes in both classes are in the first approximation comprised of two domains [28,29] (Fig. 2). The class-defining active site domain is believed to be the primordial synthetase. A second domain, joined to the active site structure, is idiosyncratic to the enzyme even within the same class. For example, the secondary structure of the variable domain of the class I methionyl- and glutaminyl-tRNA synthetases is predominantly made up of α-helices and β-strands, respectively [16,30,31]. Moreover, when the same synthetase is traced through evolution from bacteria to man, there is far more variability in the sequence of the second domain than in the class-defining catalytic domain [32,33].

When bound to the two-domain L-shaped tRNA molecule, the class-defining catalytic domain of enzymes in either class make contact with the acceptor-TψC minihelix [16,24]. Insertions into the nucleotide-binding fold of class I enzymes and into the multistranded β-structure of class II enzymes provide for many of these contacts. In contrast, the nonconserved second domain of synthetases makes contacts with segments outside of the acceptor-TψC minihelix, including the variable loop and anticodon [16,24,35]. This organization of synthetase-tRNA contacts may reflect a historical piece-wise assembly of a synthetase-tRNA complex, starting from the acceptor-TψC minihelix and class-defining catalytic domain of a synthetase as early functional units [36].

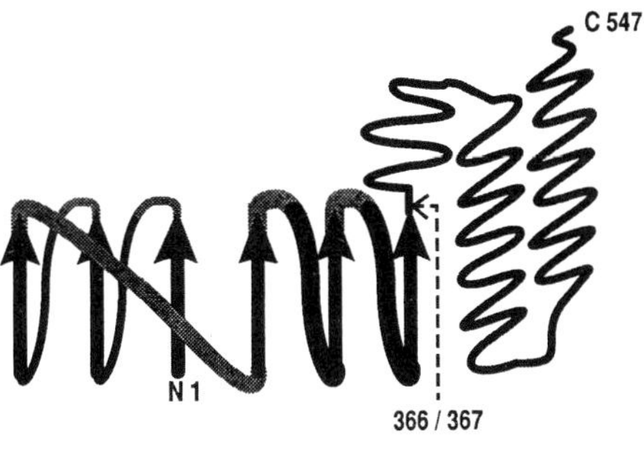

Wild-type

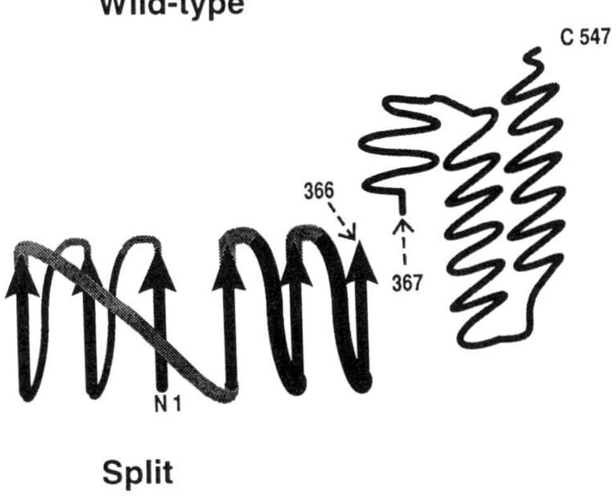

Split

Fig. 2. Schematic illustration of the active monomeric fragment of the class I *E. coli* methionyl-tRNA synthetase. The class-defining catalytic domain is located in the N-terminal region and is comprised of a nucleotide-binding fold made up of alternating β-strands (solid black arrows) and α-helices (solid black bars). This domain contains the active site and ends at approximately residue 366 where it is followed by a helical domain which provides for interactions with the anticodon of $tRNA^{Met}$. The enzyme may be split at the indicated position to create two fragments which assemble noncovalently to give an active enzyme [34].

Operational RNA code for aminoacylation

The class II *E. coli* alanyl- and seryl-tRNA synthetases make no contact with the anticodons of their respective tRNAs [35,37]. Thus, the relationship between the trinucleotides of the genetic code and alanine and serine is indirect. RNA structural elements other than the anticodon are interpreted as signals for aminoacylation with alanine or serine. Substrates comprised of the minihelix domain (acceptor-TψC stem-loop) or acceptor stem alone of the respective tRNA are aminoacylated [1,7]. These two examples are extremes, because other synthetases make anticodon contacts which are important for the efficiency of aminoacylation. In these instances, however, minihelix or acceptor stem substrates which lack anticodons are also aminoacylated specifically, although with a reduced efficiency [38].

The efficiency and specificity of aminoacylation of double-stranded oligonucleotides, which terminate in the N^{73}CCA sequence common to all tRNAs, generally depends on N^{73} (the so-called "discriminator base" at position 73 in the 76 nucleotide tRNA chain) and one or more of the first four base pairs of the acceptor stem. Specificity can be switched from one amino acid to another by the interchange of two to four nucleotides [38]. In the case of aminoacylation with alanine, a single G3:U70 base pair is the major determinant for aminoacylation. Substitution of G3:U70 with G:C or I:C or I:U abolishes aminoacylation. The unpaired exocyclic 2-amino group of G3 (in the wobble configuration) in the RNA minor groove is the required atomic group. Other functional contacts with the helix include 2′-OH groups which are within a 5 Å radius of the 2-amino group and which line the minor groove [39,40].

Thus, a specific RNA structure/sequence with a unique constellation of atoms corresponds to a particular amino acid and that relationship constitutes an operational RNA code [36]. The operational code is manifested in oligonucleotides with as few as four base pairs, corresponding to less than a half-turn of an RNA helix. The localization of signals for aminoacylation to the first four base pairs and N^{73} may reflect the historical origins of the operational RNA code. For example, if early aminoacylation catalysts were small or in any case unable to extend more than 10 to 20 Å from the site of amino acid attachment, then only those nucleotides proximal to the amino acid attachment site would be used as discriminating signals for aminoacylation.

Implications for peptide synthesis

Assuming that early aminoacylation reactions were carried out by ribozymes on RNA oligonucleotide substrates, peptides could form by bringing the aminoacyl oligonucleotides into close proximity. In the absence of a template, the charged aminoacyl RNAs could make complexes via base pairing and stacking interactions which bring the aminoacyl groups into close proximity for spontaneous peptide bond formation. The hairpin loops of minihelices are accessible for complementary hydrogen bond interactions. To assure that the aminoacyl groups are brought together, these loop-loop interactions must occur with the "sides" rather than with the "heads" of the hairpins. The resulting structure is an intermolecular pseudoknot [12].

A computer-generated model of a bimolecular pseudoknot complex shows that the activated amino acids are brought to within 11 Å of each other in one of the sterically allowed orientations (Fig. 3). In the pseudoknot a quasi-continuous helix is constructed by having the loop-loop base pairing interactions directly aligned beneath one of the hairpin helices [41]. Reaction of the aminoacyl groups results in a dipeptidyl-RNA hairpin helix and an uncharged RNA. Release of the uncharged RNA, and binding of another aminoacyl-RNA through hydrogen bond interactions with the opposite side of the loop of the dipeptidyl-RNA, sets the stage for formation of a tripeptidyl-RNA, and so on. Thus, in principle the aminoacylation of hairpin oligonucleotides can lead to peptide synthesis through the assembly of bimolecular complexes of aminoacyl-RNA pseudoknot structures [12].

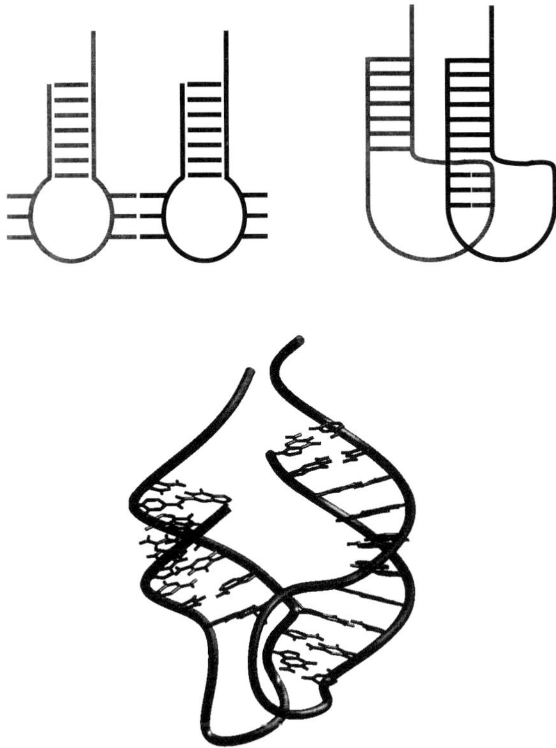

Fig. 3. Intermolecular pseudoknot dimers which bring the 3′-ends of the two microhelices into close proximity. Schematic illustration of hairpin microhelices and the pseudoknot dimers (top) with molecular model of a particular conformational isomer (bottom). Unpaired bases have been omitted for clarity. The model of the intermolecular pseudoknot was generated using MC-SYM v1.22 [42] with A-form parameters for the hairpin helices. The three-base stem forming the pseudoknot structure was also defined as an A-form helix which was stacked beneath one of the acceptor stems. Loop sizes of eight nucleotides (five unpaired and three paired) were used. A search of conformational space yielded 4,015 potential constructions. These isomers were then subjected to 10 cycles of energy minimization (conjugate gradient) using XPLOR 3.0 [43] and ranked according to total energy at the completion of the minimization process. The model depicted represents the lowest energy solution obtained by analysis of 4,015 low-energy isomers. The analysis was performed by Dr B. Henderson (MIT) who also provided the illustration.

In this scheme of noncoded peptide synthesis, advantage is taken of loop-loop interactions. If there is only one loop sequence for each hairpin helix substrate, then a given amino acid will always have the same neighbors in a peptide. However, because aminoacylation efficiency and specificity is determined by interactions with the first four base pairs of the helix and with N^{73}, loop sequences can be varied without affecting aminoacylation. These variations could include a set of different loop sequences for lateral pairing interactions fixed to each amino acid specific helix. In this way, the neighbors of any given amino acid are expanded and a diversified array of different peptide sequences could be synthesized without the use of a template. A system of this sort can also make the transition to a coded system of protein synthesis by taking advantage of interactions of the "end" or "head" of the loop sequence with a template. Lateral loop-loop interactions in this instance could still be used to bring aminoacyl groups into close proximity for peptide-bond formation [12].

Implications for assembly of tRNA and synthetase structures

The transition from the RNA world to the theater of proteins would replace ribozyme catalysts for aminoacylation with the earliest tRNA synthetases. The conserved domain of tRNA synthetases is required for amino acid activation and this function may have been the first activity associated with these enzymes. Transfer of the activated amino acid to the 3′-end of a specific RNA would require acquisition of motifs for RNA interactions. In contemporary tRNA synthetases, these motifs consist of insertions into the catalytic domain which provide contacts with the acceptor helix and bring the 3′-end of the bound RNA next to the activated amino acid. These insertions are idiosyncratic to the synthetase. Thus, the primordial synthetase is imagined to consist of the catalytic domain with motifs to interact with the primordial tRNA-like structure consisting of the minihelix domain (Fig. 4). The aminoacylated RNAs could participate in noncoded or coded peptide synthesis as outlined above.

A separate "template reading head" domain containing the anticodon was eventually added to complete the tRNA structure. This addition segregated template reading from the site of aminoacylation and the determinants of the operational RNA code. It no doubt facilitated a more efficient and accurate system of polypeptide synthesis. At the same time, addition of a second, variable synthetase domain facilitated interactions with the newly acquired tRNA domain and, in some instances, with its anticodon trinucleotide. The acquisition of the second synthetase domain provided strong selective pressure to retain the anticodon-containing tRNA domain [36].

With this scheme for assembly of synthetase-tRNA complexes, the operational RNA code imbedded in acceptor stems is integrated with the genetic code and

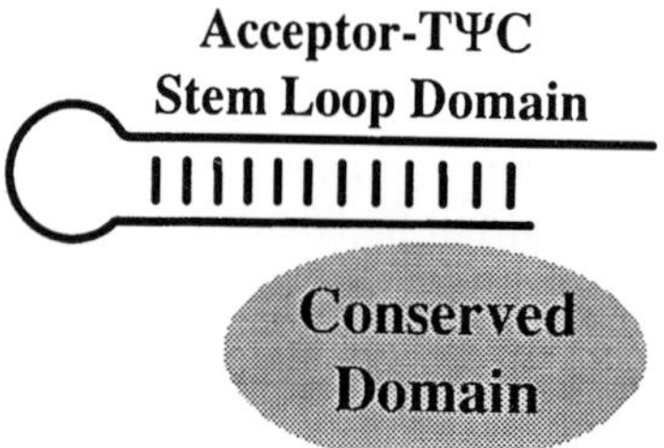

Fig. 4. Schematic illustration of the class-defining domain of a primordial tRNA synthetase as a complex with a minihelix, as an example of an early form of a synthetase-tRNA complex. The later addition to the minihelix of a template-reading head in the form of a second domain would result in a full tRNA molecule. This addition could be coincident with the acquisition of a second domain for the synthetase.

anticodon trinucleotides. The relationship between particular trinucleotides and specific amino acids would be stochastic. The advantage of a template-reading head tRNA domain is so great that once it is joined to a minihelix structure it remains fixed. The amino acid associated with a particular anticodon would be a reflection of happenstance and not of a fundamental chemical or physical relationship between the amino acid and the anticodon.

Thus, the synthetase-tRNA complex may reflect the modular, step-wise development of the genetic code, with separate modules for the operational RNA code and for the template-reading head that interacts with mRNA sequences. A synthetase fragment containing the conserved domain and missing most of the variable domain is active on minihelix substrates [44]. Noncovalent assembly of active synthetase molecules from the separated catalytic and variable domains has also been demonstrated [32,34,45]. These and other experimental approaches will provide more understanding of individual domains of tRNAs and tRNA synthetases, and the communication between them that is needed for elucidating the genetic code.

Acknowledgements

This work was supported by grant number GM15539 from the National Institutes of Health and by a grant from the National Foundation for Cancer Research.

References

1. Francklyn C, Schimmel P. Nature 1989;337:478–481.
2. Francklyn C, Schimmel P. Proc Natl Acad Sci USA 1990;87:8655–8659.
3. Martinis SA, Schimmel P. Proc Natl Acad Sci USA 1992;89:65–69.
4. Francklyn C, Shi J-P, Schimmel P. Science 1992;255:1121–1125.
5. Frugier M, Florentz C, Giegé R. Proc Natl Acad Sci USA 1992;89:3990–3994.
6. Nureki O, Niimi T, Muto Y, Kanno H, Kohno T, Muramatsu T, Kawai G, Miyazawa T, Giegé R, Florentz C, Yokoyama S. In: Nierhaus KH, Franceschi F, Subramanian AR, Erdmann VA, Wittmann-Liebold B (eds) The Translation Apparatus. New York: Plenum Press, 1993;59–66.
7. Sampson JR, Saks ME. Nucl Acid Res 1993;21:4467–4475.
8. Frugier M, Florentz C, Giegé R. EMBO J 1994;13:2218–2226.
9. Noller HF. In: Gesteland RF, Atkins JF (eds) The RNA World. Plainview, NY: Cold Spring Harbor Laboratory Press, 1993;137–156.
10. Weiner AM, Maizels N. Proc Natl Acad Sci USA 1987;84:7383–7387.
11. Maizels N, Weiner AM. In: Gesteland RF, Atkins JF (eds) The RNA World. Plainview, NY: Cold Spring Harbor Laboratory Press, 1993;577–602.
12. Schimmel P, Henderson B. Proc Natl Acad Sci USA 1994;91:11283–11286.
13. Blow DM, Bhat TN, Metcalfe A, Risler JL, Brunie S, Zelwer C. J Molec Biol 1983;171:571–576.
14. Rossmann MG, Liljas A, Branden C-I, Banazak LJ. In: Boyer PD (ed) The Enzymes, Edn. 3, Vol. 11. Orlando, FL: Academic Press, 1975;61–102.
15. Brick P, Bhat TN, Blow DM. J Molec Biol 1988;208:83–98.
16. Rould MA, Perona JJ, Söll D, Steitz TA. Science 1989;246:1135–1142.
17. Webster TA, Tsai H, Kula M, Mackie GA, Schimmel P. Science 1984;226:1315–1317.
18. Hountondji C, Dessen P, Blanquet S. Biochim 1986;68:1071–1078.

19. Fersht AR. Biochemistry 1987;26:8031–8037.
20. Ludmerer SW, Schimmel P. J Biol Chem 1987;262:10801–10806.
21. Schimmel P. Ann Rev Biochem 1987;56:125–158.
22. Eriani G, Delarue M, Poch O, Gangloff J, Moras D. Nature 1990;347:203–206.
23. Cusack S, Berthet-Colominas C, Härtlein M, Nassar N, Leberman R. Nature 1990;347:249–255.
24. Ruff M, Krishnaswamy S, Boeglin M, Poterszman A, Mitschler A, Podjarny A, Rees B, Thierry JC, Moras D. Science 1991;252:1682–1689.
25. Cavarelli J, Rees B, Ruff M, Thierry J-C, Moras D. Nature 1993;362:181–184.
26. Cavarelli J, Eriani G, Rees B, Ruff M, Boeglin M, Mitschler A, Martin F, Gangloff J, Thierry JC, Moras D. EMBO J 1994;13:327–337.
27. Shi J-P, Musier-Forsyth K, Schimmel P. Biochemistry 1994;33:5312–5318.
28. Moras D. Trends Biochem Sci 1992;17:159–164.
29. Buechter DD, Schimmel P. Crit Rev Biochem Molec Biol 1993;28:309–322.
30. Rould MA, Perona JJ, Söll D, Steitz TA. Nature 1991;352:213–218.
31. Brunie S, Zelwer C, Risler J. J Molec Biol 1990;216:411–424.
32. Shiba K, Schimmel P. Proc Natl Acad Sci USA 1992;89:1880–1884.
33. Shiba K, Suzuki N, Shigesada K, Namba Y, Schimmel P, Noda T. Proc Natl Acad Sci USA 1994;91:7435–7439.
34. Burbaum JJ, Schimmel P. Biochemistry 1991;30:319–324.
35. Biou V, Yaremchuk A, Tukalo M, Cusack S. Science 1994;263:1404–1410.
36. Schimmel P, Giegé R, Moras D, Yokoyama S. Proc Natl Acad Sci USA 1993;90:8763–8768.
37. Park SJ, Schimmel P. J Biol Chem 1988;263:16527–16530.
38. Martinis SA, Schimmel P. In: Söll D, RajBhandary UL (eds) tRNA: Structure, Biosynthesis and Function. Washington, DC: American Society for Microbiology, 1995;349–370.
39. Musier-Forsyth K, Schimmel P. Nature 1992;357:513–515.
40. Musier-Forsyth K, Usman N, Scaringe S, Doudna J, Green R, Schimmel P. Science 1991;253:784–786.
41. Wyatt JR, Tinoco I Jr. In: Gesteland RF, Atkins JF (eds) The RNA World. Plainview, NY: Cold Spring Harbor Laboratory Press, 1993;465–496.
42. Major F, Turcotte M, Bautheret D, LaPalme G, Fillion E, Cedergren R. Science 1991;253:1255–1260.
43. Brünger A, Kuriyan J, Karplus M. Science 1987;235:458–460.
44. Buechter DD, Schimmel P. Biochemistry 1993;32:5267–5272.
45. Shiba K, Schimmel P. J Biol Chem 1992;267:22703–22706.

Tracing Biological Evolution in Protein and Gene Structures.
M. Gō and P. Schimmel, editors.

Dissection of an enzyme into two fragments at intron-exon boundaries

Kiyotaka Shiba
Inheritance and Variation, Precursory Research for Embryonic Science and Technology (PRESTO), Research Development Corporation of Japan (JRDC); and Department of Cell Biology, Cancer Institute, Kami-Ikebukuro, Toshima, Tokyo, Japan

Abstract. *Background.* It has been shown that *Escherichia coli* (*E. coli*) isoleucyl-tRNA synthetase can be dissected into two enzyme fragments at various sites throughout the whole enzyme structure without losing its biological activity. Complementary enzyme fragment pairs assembled noncovalently in vivo to reconstitute activity. Reconstitution has also been achieved from assembly of three enzyme fragments. These observations support the notion that the ancient enzymes could function as assemblages of short peptides that were translated from minigenes. In the exon theory of genes proposed by Gilbert, the exon is considered to serve as a genetic unit in the evolution of a gene. Ancient enzymes could function as assemblages of peptides encoded by exons.

Results. Here, reconstitution has been tested with fragments that divide *E. coli* isoleucyl-tRNA synthetase at locations corresponding to intron-exon boundaries of the *Tetrahymena thermophila* (*T. thermophila*) isoleucyl-tRNA synthetase gene. Two out of four pairs of fragments are active in this assay.

Conclusions. The data obtained here support the notion that the ancient enzymes could function as clusters of exon-encoded polypeptides and also suggest that a covalent linkage between exons is not required for the development of genes in all cases.

Key words: aminoacyl-tRNA synthetase, evolution, fragment complementation, minigene, Rossmann fold, the exon theory of genes.

Introduction

Many present-day genes encode large proteins. For instance, the *Escherichia coli* (*E. coli*) *ileS* gene encodes the isoleucyl-tRNA synthetase which comprises 939 amino acids [1]. Isoleucyl-tRNA synthetase belongs to a family of aminoacyl-tRNA synthetases which have an important role in the translation process of genetic information. Thus, the origin of aminoacyl-tRNA synthetases should be very ancient [2]. Indeed, sequence similarity observed among prokaryotic, eukaryotic and archaebacterial isoleucyl-tRNA synthetases indicates that the origin of the gene for this enzyme predates the divergence of bacteria, archaebacteria and eukaryotes [3–5].

How did such large proteins arise? The idea that modern genes coding for larger

Address for correspondence: Kiyotaka Shiba, Department of Cell Biology, Cancer Institute, Japanese Foundation for Cancer Research, Kami-Ikebukuro, Toshima-ku, Tokyo 170, Japan. Tel.: +81-3-3918-0111. Fax: +81-3-5394-3903.

proteins evolved by fusion of a number of ancestral smaller units of genes has come to be accepted [6]. Although we cannot experimentally determine the nature of such small ancestral gene units which served as building blocks in the evolution of a gene, Shiba and Schimmel [4] were interested in the possibility that modern large genes could be dissected into several smaller genes (minigenes) without losing their biological activity. We have chosen the *E. coli* isoleucyl-tRNA synthetase as a model enzyme because: 1) it has an ancient origin (our interest is the rule that governed the birth of ancient genes); 2) it is a monomeric enzyme (results for a multimeric enzyme could be difficult to interpret [7]); and 3) it is essential for cell growth (an in vivo assay system can be easily constructed).

Isoleucyl-tRNA synthetase is encoded by the *ileS* gene in *E. coli*. The enzyme belongs to the methionine subfamily which constitutes five members of 10 class I aminoacyl-tRNA synthetases [8], i.e., methionyl-, cysteinyl-, leucyl-, valyl- and isoleucyl-tRNA synthetases [9,10]. Although the tertiary structure of isoleucyl-tRNA synthetase is not yet known, multiple sequence alignment with members of the methionine subfamily and the tertiary structure of *E. coli* methionyl-tRNA synthetase [11] allowed us to build a structural model for isoleucyl-tRNA synthetase [4,5,9,12]. Figure 1 shows a schematic diagram of the predicted structure of isoleucyl-tRNA synthetases. It has a nucleotide-binding fold (Rossmann fold) at its N-terminal region, which comprises four α-helices and six β-strands (shown by rectangles and pentagons, respectively, in Fig. 1). The HIGH [1,13] and KMSKS [14] elements are common features of class I enzymes and form the ATP-binding site in the Rossmann fold [15–18]. The Rossmann fold domain is also believed to be responsible for tRNA-acceptor helix interactions which are facilitated by the connective polypeptide 1 (CP1), placed between the first and second halves of the Rossmann fold [12,15]. For these reasons, the N-terminal domain is designated the "class-defining catalytic domain" or "core domain" [19,20]. To the class-defining N-terminal domain, the C-terminal second domain is joined. The second domain is rich in α-helices [11] and contains amino acid residues needed for interactions with the anticodon of tRNA [21]. Multiple sequence alignment analysis with five isoleucyl-tRNA synthetases showed

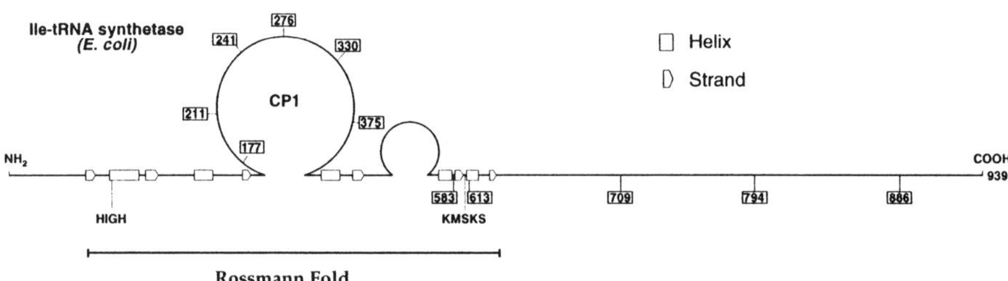

Fig. 1. Schematic representations of the organization of a secondary structure for *E. coli* isoleucyl-tRNA synthetase. The model is based on [4]. Five α-helices and six β-strands that form the Rossmann fold are shown by rectangles and pentagons, respectively. The third helix is not an element of the canonical nucleotide-binding fold, βαβαβ-βαβαβ. The HIGH- and KMSKS-conserved elements are indicated. Numbers represent codons at which *ileS* was dissected into two minigenes without losing the biological activity [4].

a much greater selective pressure on the core domain than on the second domain [5].

The coding sequence for *ileS* was divided into two parts at 18 places and the resulting N-terminal and C-terminal portions were separately cloned on ColE1-derivative (amp^r) and p15A-derivative (tet^r) expression vectors, respectively (Fig. 2) [4]. These two replicons are compatible so that these plasmids can coexist in the same *E. coli* cell. The reconstitution of active enzyme in vivo was tested by introduction of each set of plasmids into *E. coli* tester strain IQ843/pRMS711 [4]. This tester strain contains a Δ*ileS* null allele on the chromosome, and cell viability is maintained by a wild-type copy of *ileS* on the plasmid pRMS711 that has a temperature-sensitive pSC101 replicon [22]. As a result, growth of IQ843 is temperature sensitive. Complementation of the temperature-sensitive phenotype of this tester strain was used to test for reconstitution of active enzyme from the fragment assembly (Fig. 2). Among 18 pairs tested, 11 fragment pairs restored active enzyme and supported cell growth of the tester strain at 42°C [4]. In Fig. 1 the locations of the break points for the active bipartite enzymes are shown along with the predicted structure of isoleucyl-tRNA synthetases. It is noteworthy that reconstitution was achieved with fragments that divide the enzyme within or proximal to critical

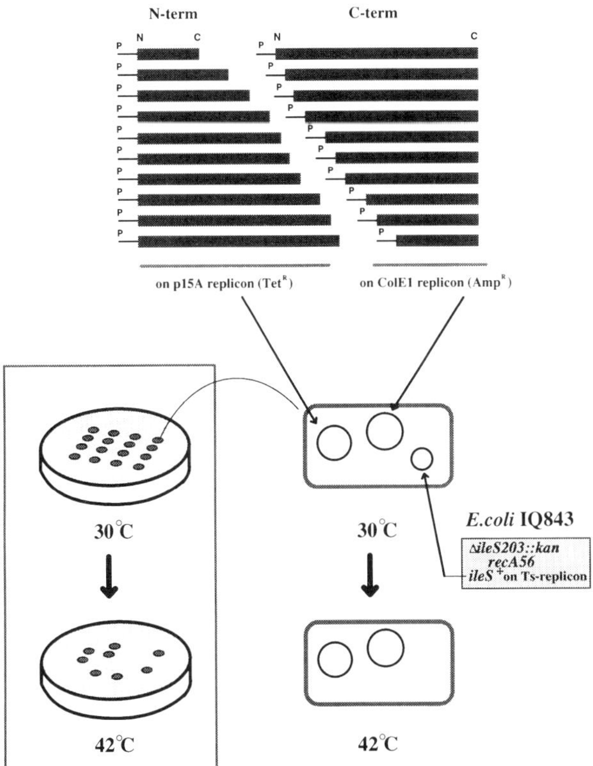

Fig. 2. Strategy of an in vivo test system for fragment complementation with isoleucyl-tRNA synthetase. See text for details.

elements of the secondary structure. For instance, the split at 613 is located in an α-helix that is adjacent to the critical KMSKS motif (Fig. 1).

We have further dissected isoleucyl-tRNA synthetase into three pieces, N-terminal, C-terminal, and a middle portion, based on the result of bipartite complementation [23]. Of 10 three-fragment combinations tested, three (1-330/329-583/585-939, 1-330/329-613/616-939 and 1-330/329-794/795-939 in Fig. 3) supported cell growth of the tester strain at high temperatures [23]. More recently, a new three-fragment construction which comprises 1-118/119-330/329-939 has been produced (Fig. 3, see below). Functional complexes containing the three polypeptides were detected at elution positions corresponding to the native enzyme in a gel filtration experiment, demonstrating that the three fragments noncovalently form a native-like structure [23].

The observations described above support the possibility that assemblages of polypeptides were intermediates in the development of large proteins. Ancient enzymes could function as clusters of short peptides that were translated from several mini-genes [4,23–25]. In the exon theory of genes proposed by Gilbert [26] the exon is considered to serve as a genetic unit in the evolution of a gene. Ancient enzymes could function as assemblages of peptides encoded by exons [25–27]. We cannot exclude, however, the possibility that a covalent linkage between exon-encoded peptides, with the help of the splicing mechanism, was absolutely required for the generation of a new gene. Here, this question has been investigated using a test system for functional assembly of isoleucyl-tRNA synthetase. Reconstitution has been tested with fragments that divide the *E. coli* isoleucyl-tRNA synthetase at locations corresponding to intron-exon boundaries of the *Tetrahymena thermophila* (*T. thermophila*) isoleucyl-tRNA synthetase gene [28].

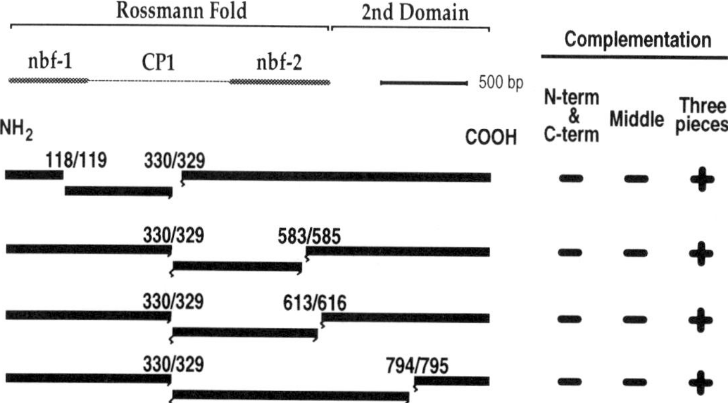

Fig. 3. Reconstitution of isoleucyl-tRNA synthetase activity from three enzyme fragments. The enzyme fragments are represented by thick lines. Numbers for the first and the last *ileS* codons of each fragment are given. The results of the complementation tests are shown in the right columns.

Material and Methods

Construction of plasmids

Plasmids that produce N-terminal or C-terminal fragments of isoleucyl-tRNA synthetase were constructed as follows. For N-terminal fragment-producing plasmids, PCR was performed on plasmid pKS21 (which has an intact *ileS* on pTZ19R [29]), using primers KY-38 (5′-CGG TAC CTT ATT TCA GAA TCT TGT TAA CCG A-3′, for "1-75"), KY-40 (CGG TAC CTT AGG TGA ATT TCT CAC CCG GCT, for "1-118"), KY-42 (CGG TAC CTT ACA CGC CCA GAC GGA TAA AGT, for "1-146") and KY-46 (CGG TAC CTT ACC ACT GCA CGC CTT TGA TCT C, for "1-443") which have sequences from the 3′ portions of the corresponding *ileS* gene fragments (plus an *Asp*718 recognition sequence (GGTACC) and a termination codon at their 5′ ends), and KUNO-22 (AGC GGA TAA CAA TTT CAC ACA GGA) which has the sequence corresponding to that of the 5′ portion of *lacZ′* on the vector portion of pKS21. Amplified fragments were digested with *Afl*II and *Asp*718, and recloned into the *Afl*II and *Asp*718 sites of pKS152 which have an intact *ileS*, p15A replicon and tet^r gene from pACYC184 [30], and fi *ori* from pTZ19U [29]. The *Afl*II site is located at the beginning of *ileS* and the *Asp*718 site is located at the vector portion. The resulting plasmids were designated pKS280 ("1-75"), pKS282 ("1-118"), pKS284 ("1-146") and pKS287 ("1-443"), respectively. For C-terminal fragment-producing plasmids, PCR was performed on pKS21 using primers KY-39 (5′-GGC CAT GGA CAT TAT CGT GAA GTC CAA-3′, for "76-939"), KY-41 (GGC CAT GGC CGC CGA GTT CCG CGC CAA, for "119-939") or KY-43 (GGC CAT GGT GGG CGA CTG GTC GCA CCC G, for "148-939") which have sequences from the 5′ portions of the corresponding *ileS* gene fragments (plus a *Nco*I recognition sequence (CCATGG) at their 5′ ends), and KUNO-21 (CGC CAG GGT TTT CCC AGT CAC GAC) which has a sequence corresponding to that of the 3′ portion of *lacZ′* on the template. Amplified fragments were digested with *Nco*I and *Eco*RI, and recloned onto the *Nco*I-*Eco*RI sites of pTV119N vector. This plasmid is a derivative of pUC119 [31] and has a unique *Nco*I site at the initiation codon of *lacZ′*, so that a cloned gene at the *Nco*I site can be translated from this initiation codon (Takara Shuzo). The resulting plasmids were designated pKS281 ("76-939"), pKS283 ("119-939") and pKS285 ("148-939"), respectively. For the C-terminal fragment of "444-939", the previously constructed pKS153 [4] was used. The N-terminal fragment of isoleucyl-tRNA synthetase from this plasmid has an amino-terminal fusion of 22 amino acids that is derived from the beginning of the β-galactosidase protein coded by pTZ19R [29].

Complementation test

Complementation was tested by introducing a pair of plasmids into the *E. coli* strain IQ844 (MC4100 [32] Δ*ileS203*::*kan recA56*/F′ *lacI^q pro$^+$ lac$^+$*)/pRMS711 (*ileS$^+$*Cm^r) [23] at 30°C by selecting for transformants on Luria-Bertani (LB) plates [33] containing ampicillin and tetracycline. Transformants were then tested for growth at 42 or 37°C on LB plates or M9 plates [33] containing 0.2% casamino acids. Elimination of the pRMS711 (*ileS*$^+$Cm^r) plasmid was confirmed by testing for sensitivity to chloramphenicol.

Results

Mapping of intron-exon boundaries on E. coli isoleucyl-tRNA synthetase

Csank and Martindale [28] reported the gene structure of the ciliate *T. thermophila* isoleucyl-tRNA synthetase. The gene has eight introns that separate the coding frame into nine exons. Schimmel et al. [34] have mapped the locations of these eight introns on the structural model of *E. coli* isoleucyl-tRNA synthetase that was made from multiple sequence alignment of 21 sequences of the methionyl-tRNA synthetase subfamily and the tertiary structure of *E. coli* methionyl-tRNA synthetase. In this map, introns 1–5 are located within the N-terminal Rossmann fold and intron-6 is located at the boundary between the N-terminal core region and the C-terminal second domain [34]. Figure 4A shows a schematic representation for locations of introns 2–6 in a Rossmann fold in isoleucyl-tRNA synthetase. Among them, three introns, intron-2, 3 and 4, are located in the first half of the Rossmann fold. Intron-2 is mapped on the first helix of the Rossmann fold and intron-3 and 4 are located on loops that connect the second β-strand and the second helix, and the second helix and the third β-strand, respectively. Intron-5 is located in the second half of the Rossmann fold, and is predicted to be within or proximal to the helix that precedes the fourth β-strand of the Rossmann fold [11]. Although this helix is not an element for the canonical nucleotide-binding fold (βαβαβ-βαβαβ), it is a critical element for forming the Rossmann fold in *E. coli* methionyl-tRNA synthetase [11].

The gene for *T. thermophila* isoleucyl-tRNA synthetase is the only one that has been reported to have multiple introns in the methionyl-tRNA synthetase subfamily. In this subclass of enzyme, genes for *Neurospora crassa* (*N. crassa*) cytoplasmic [35] and mitochondrial [36] leucyl-tRNA synthetases have been reported to have single introns, and the location of the intron in the mitochondrial leucyl-tRNA synthetase is mapped close to intron-4 of *T. thermophila* isoleucyl-tRNA synthetase [34]. Among other class I enzymes, genes for a human tryptophanyl-tRNA synthetase [37] and for a human glutamyl-tRNA synthetase [38] have been reported to contain multiple introns. The tryptophanyl-tRNA synthetase gene has at least 12 exons and more than four introns are located in the N-terminal core region. Glutamyl-tRNA synthetase exists as a fusion enzyme with prolyl-tRNA synthetase in human [39] and *Drosophila* [40]. The conserved core region for human glutamyl-tRNA synthetase is encoded by eight exons, exons IV-XI [39]. The human glutamyl-tRNA synthetase shows higher homology with *E. coli* glutaminyl-tRNA synthetase than with *E. coli* glutamyl-tRNA synthetase [41]. This high sequence similarity and the tertiary structure of *E. coli* glutaminyl-tRNA synthetase [15] allowed investigators to locate positions of introns

(A) Ile-tRNA synthetase

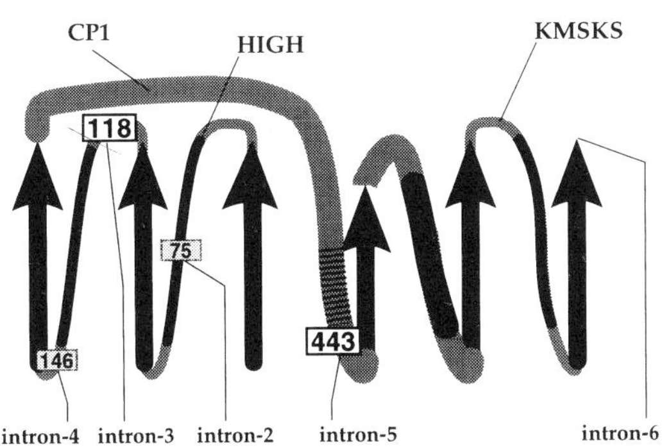

(B) Glu-tRNA synthetase

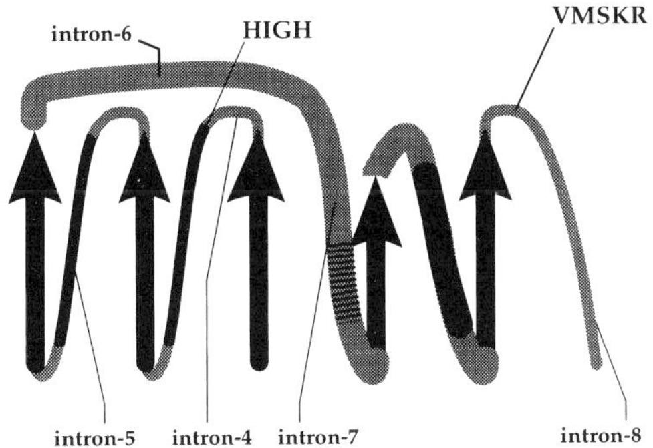

Fig. 4. Schematic representations of the secondary structure and intron locations for the N-terminal Rossmann folds from a *T. thermophila* isoleucyl-tRNA synthetase (A) and a human glutamyl-tRNA synthetase (B). Black arrows represent the β-strands and solid black rods the α-helices that form Rossmann fold. The striped black rods that precede the fourth β-strand of both enzymes represent the noncanonical α-helix that also forms a Rossmann fold [11,15]. The HIGH- and KMSKS- (VMSKR for a human enzyme) conserved elements are indicated. CP1 represents a polypeptide insertion that separates the first and second subdomains of the Rossmann fold. The numbers in (A) represent codons of the *E. coli* isoleucyl-tRNA synthetase that correspond to intron-exon boundaries of *T. thermophila* enzyme. The model for a human glutamyl-tRNA synthetase was created based on an *E. coli* glutaminyl-tRNA synthetase [15].

on the structural model for a human glutamyl-tRNA synthetase [39]. Figure 4B shows a schematic drawing for locations of introns in human glutamyl-tRNA synthetase. Intron-4 is located on a loop that connects the first β-strand and the first α-helix, and intron-5 is mapped on the second helix of the Rossmann fold. Intron-7 is proximal to the helix that precedes the fourth β-strand of Rossmann fold. This helix also forms the second subdomain of the Rossmann fold in *E. coli* glutaminyl-tRNA synthetase

[15]. Intron-8 is located near the boundary between the N-terminal Rossmann fold and the following helical subdomain [15].

It is noteworthy that in both *T. thermophila* isoleucyl-tRNA synthetase and human glutamyl-tRNA synthetase some introns separate subdomains of the Rossmann fold: introns-2, 3 and 4, of the *T. thermophila* isoleucyl-tRNA synthetase, and introns-4 and 5 in the human glutamyl-tRNA synthetase. Furthermore, these intron positions are not necessarily in loop regions, but rather some introns (intron-2 in isoleucyl-tRNA synthetase and intron-5 in glutamyl-tRNA synthetase) separate α-helices. Similar observations have been reported for other genes that encode proteins with Rossmann folds [42].

Dissection of an enzyme into two fragments at intron-exon boundaries

This study focused on four introns, intron-2, 3, 4 and 5 of *T. thermophila* isoleucyl-tRNA synthetase, that are located within the N-terminal of the Rossmann fold (Fig. 4A). The gene for *E. coli* isoleucyl-tRNA synthetase was divided into two pieces at these four intron-exon boundaries and the resulting pairs of fragments, 1-75/76-939, 1-118/119-939, 1-146/148-939 and 1-443/444-939, were separately cloned into two compatible plasmid vectors (Table 1). Reconstitution of the active enzyme was tested by introducing each pair of plasmids into the *ileS* null strain at 30°C and scoring growth of transformants at high temperatures (42 or 37°C). As shown in Table 1, among four splits, the pairs 1-118/119-939 and 1-443/444-939 restored the enzyme activity that supported cell growth at 42°C. Elimination of the wild-type copy of *ileS* on the temperature-sensitive plasmid was confirmed as described in Material and Methods. These two split points correspond to intron-3 and intron-5 of *T. thermophila* isoleucyl-tRNA synthetase, respectively. Intron-3 is on the loop that connects the second β-strand and the second α-helix of a Rossmann fold. The enzyme was further dissected into three active fragments using this intron-3 site as one of two split points (Fig. 3, data not shown). Intron-5 is predicted to be within or proximal to the helix that precedes the fourth β-strand of the Rossmann fold.

Discussion

In the exon theory of genes proposed by Gilbert [26], the exon is considered to serve

Table 1. Complementation test with two fragments that divide isoleucyl-tRNA synthetase at intron-exon boundaries.

Plasmids[a]	Enzyme fragments	Complementation
pKS280/pKS281	1-75/76-939	No
pKS282/pKS283	1-118/119-939	Yes
pKS284/pKS285	1-146/148-939	No
pKS287/pKS153	1-443/444-939	Yes

[a]p15a-derivative plasmid producing N-fragment/ColE1-derivative plasmid producing C-fragment.

as a building block in the evolution of a gene. Ancient enzymes could function as assemblages of peptides encoded by exons. According to this view, exons could have been covalently linked with the help of RNA splicing in later stages of gene evolution. If this were the case, a covalent linkage between exons might only be required for improving the efficiency of gene expression or protein folding. Previous work by Craik et al. [43] has shown that the isolated clostripain-digested fragment of human β-globin, on "31-104", which is identical to the exon-2 product of a human gene, can maintain a ferrous haemdioxygen complex when the side exon products, "1-30" and "105-146", were present in a reconstitution experiment in vitro. This observation demonstrated that β-globin executes its biochemical function even when covalent linkages at exon-intron boundaries are broken.

An alternative view is that a covalent linkage between exons was a prerequisite to the generation of a new gene. In this case, the activity of an enzyme should be dependent on the covalent bond between fragments coded by exons. It has been pointed out by Brändén [44] that many intron positions are located near amino acids that are involved in catalysis. This observation may suggest the importance of the role of covalent linkage between exons in the evolution of a gene.

Here, reconstitution has been tested with fragments that divide the *E. coli* isoleucyl-tRNA synthetase at locations corresponding to intron-exon boundaries of *T. thermophila* isoleucyl-tRNA synthetase. If a covalent linkage between exons were an absolute requirement for the generation of a new catalyst, breakage of polypeptide bonds in a present-day enzyme at intron-exon boundaries should disrupt enzyme activity. The results obtained here show that a covalent linkage between exons is not required for enzyme function in all cases.

Our previous works [4,23] and the experiments done here show that a protein can be dissected within a region of a secondary structure without losing its activity. This fact challenges the notion by Stoltzfus et al. [45] that exons or modules do not represent functional units because they are not always located between secondary elements. Our data suggest that genetic units do not necessarily correspond to structural units.

The experiment performed here assumed that four introns of *T. thermophila* isoleucyl-tRNA synthetase have an ancient origin and that their positions have not moved in the course of evolution. It is known that an intron can be inserted into a pre-existing exon in later stages of evolution [46], and also that an intron can slide in the course of evolution [47]. To find exons with an ancient origin in isoleucyl-tRNA synthetases, investigation of the pattern of intron distribution across taxa will be required [48].

Acknowledgements

I would like to express my thanks to Prof Paul Schimmel and Dr Tetsuo Noda for their encouragement and support, Dr W. Todd Miller for critical reading of the manuscript, and Ms Noriko Suzuki for technical assistance. This work was supported in part by a grant from the Human Frontiers Science Organization.

References

1. Webster TA, Tsai H, Kula M, Mackie GA, Schimmel P. Specific sequence homology and three-dimensional structure of an aminoacyl transfer RNA synthetase. Science 1984;226:1315–1317.
2. Schimmel P. Aminoacyl tRNA synthetases: general scheme of structure-function relationships in the polypeptides and recognition of transfer RNAs. Ann Rev Biochem 1987;56:125–158.
3. Nagel GM, Doolittle RF. Evolution and relatedness in two aminoacyl-tRNA synthetase families. Proc Natl Acad Sci USA 1991;88:8121–8125.
4. Shiba K, Schimmel P. Functional assembly of a randomly cleaved protein. Proc Natl Acad Sci USA 1992;89:1880–1884.
5. Shiba K, Suzuki N, Shigesada K, Namba Y, Schimmel P, Noda T. Human cytoplasmic isoleucyl-tRNA synthetase: selective divergence of the anticodon-binding domain and acquisition of a new structural unit. Proc Natl Acad Sci USA 1994;91:7435–7439.
6. Hardie DG, Coggins JR (eds). In: Multidomain Proteins - Structure and Evolution. Amsterdam: Elsevier Science Publishers B.V., 1986.
7. Crick FH, Orgel LE. The theory of inter-allelic complementation. J Molec Biol 1964;8:161–165.
8. Eriani G, Delarue M, Poch O, Gangloff J, Moras D. Partition of tRNA synthetases into two classes based on mutually exclusive sets of sequence motifs. Nature 1990;347:203–206.
9. Hou Y-M, Shiba K, Mottes C, Schimmel P. Sequence determination and modeling of structural motifs for the smallest monomeric aminoacyl-tRNA synthetase. Proc Natl Acad Sci USA 1991;88:976–980.
10. Eriani G, Dirheimer G, Gangloff J. Cysteinyl-tRNA synthetase: determination of the last *E. coli* aminoacyl-tRNA synthetase primary structure. Nucl Acid Res 1991;19:265–269.
11. Brunie S, Zelwer C, Risler J-L. Crystallographic study at 2.5 Å resolution of the interaction of methionyl-tRNA synthetase from *Escherichia coli* with ATP. J Molec Biol 1990;216:411–424.
12. Starzyk RM, Webster TA, Schimmel P. Evidence for dispensable sequences inserted into a nucleotide fold. Science 1987;237:1614–1618.
13. Ludmerer SW, Schimmel P. Gene for yeast glutamine tRNA synthetase encodes a large amino-terminal extension and provides a strong confirmation of the signature sequence for a group of the aminoacyl-tRNA synthetase. J Biol Chem 1987;262:10801–10806.
14. Hountondji C, Blanquet S. Methionyl-tRNA synthetase from *Escherichia coli*: primary structure at the binding site for the 3′-end of $tRNA_f^{Met}$. Biochemistry 1985;24:1175–1180.
15. Rould MA, Perona JJ, Söll D, Steitz TA. Structure of *E. coli* glutaminyl-tRNA synthetase complexed with $tRNA^{Gln}$ and ATP at 2.8 Å resolution. Science 1989;246:1135–1142.
16. Mechulam Y, Dardel F, Le Corre D, Blanquet S, Fayat G. Lysine 335, part of the KMSKS signature sequence, plays a crucial role in the amino acid activation catalyzed by the methionyl-tRNA synthetase from *Escherichia coli*. J Molec Biol 1991;217:465–475.
17. Schmitt E, Meinnel T, Blanquet S, Mechulam Y. Methionyl-tRNA synthetase needs an intact and mobile $_{332}KMSKS_{336}$ motif in catalysis of methionyl adenylate formation. J Molec Biol 1994;242:566–577.
18. Doublié S, Bricogne G, Gilmore C, Carter JCW. Tryptophanyl-tRNA synthetase crystal structure reveals an unexpected homology to tyrosyl-tRNA synthetase. Structure 1995;3:17–31.
19. Burbaum JJ, Schimmel P. Structural relationships and the classification of aminoacyl-tRNA synthetases. J Biol Chem 1991;266:16965–16968.
20. Schimmel P, Giegé R, Moras D, Yokoyama S. An operational RNA code for amino acids and possible relationship to genetic code. Proc Natl Acad Sci USA 1993;90:8763–8768.
21. Shepard A, Shiba K, Schimmel P. RNA binding determinant in some class I tRNA synthetases identified by alignment-guided mutagenesis. Proc Natl Acad Sci USA 1992;89:9964–9968.
22. Toth MJ, Schimmel P. Internal structural features of *E. coli* glycyl-tRNA synthetase examined by subunit polypeptide chain fusions. J Biol Chem 1986;261:6643–6646.
23. Shiba K, Schimmel P. Tripartite functional assembly of a large class I aminoacyl tRNA synthetase. J Biol Chem 1992;267:22703–22706.

24. Go M, Nosaka M. Protein architecture and the origin of introns. Cold Spring Harbor Symp Quant Biol 1987;52:915–924.
25. Seidel HM, Pompliano DL, Knowles JR. Exons as microgenes? Science 1992;257:1489–1490.
26. Gilbert W. The exon theory of genes. Cold Spring Harbor Symp Quant Biol 1987;52:901–905.
27. Gilbert W, Glynias M. On the ancient nature of introns. Gene 1994;135:137–144.
28. Csank C, Martindale DW. Isoleucyl-tRNA synthetase from the ciliated protozoan *Tetrahymena thermophila*-DNA sequence, gene regulation, and leucine zipper motifs. J Biol Chem 1992;267: 4592–4599.
29. Mead DA, Szczesna-Skorupa E, Kemper B. Single-stranded DNA "blue" T7 promoter plasmids: a versatile tandem promoter system for cloning and protein engineering. Protein Eng 1986;1:67–74.
30. Chang AC, Cohen SN. Construction and characterization of amplifiable multicopy DNA cloning vehicle derived from the P15A cryptic miniplasmid. J Bacteriol 1978;134:1141–1156.
31. Vieira J, Messing J. Production of single-stranded plasmid DNA. In: Wu R, Grossman L, Moldave K (eds) Recombinant DNA Methodology. San Diego, CA: Academic Press, Inc., 1989;225–233.
32. Silhavy TJ, Berman ML, Enquist LW. Experiments with gene fusions. Cold Spring Harbor, NY: Cold Spring Harbor Laboratory, 1984.
33. Miller JH. Experiments in molecular genetics. Cold Spring Harbor, NY: Cold Spring Harbor Laboratory, 1972.
34. Schimmel P, Shepard A, Shiba K. Intron locations and functional deletions in relation to the design and evolution of a subgroup of class I tRNA synthetases. Protein Sci 1992;1:1387–1391.
35. Benarous R, Chow CM, RajBhandary UL. Cytoplasmic leucyl-tRNA synthetase of *Neurospora crassa* is not specified by the leu-5 locus. Genetics 1988;119:805–814.
36. Chow CM, Metzenberg RL, RajBhandary UL. Nuclear gene for mitochondrial leucyl-tRNA synthetase of *Neurospora crassa*: isolation, sequence, chromosomal mapping and evidence that the leu-5 locus specifies structural information. Molec Cell Biol 1989;9:4631–4644.
37. Frolova LY, Grigorieva AY, Sudomoina MA, Kisselev LL. The human gene encoding tryptophanyl-tRNA synthetase - interferon-response elements and exon-intron organization. Gene 1993;128:237–245.
38. Kaiser E, Eberhard D, Knippers R. Exons encoding the highly conserved part of human glutaminyl-transfer RNA synthetase. J Molec Evol 1992;34:45–53.
39. Kaiser E, Hu B, Becher S, Eberhard D, Schray B, Baack M, Hameister H, Knippers R. The human eprs locus (formerly the qars locus) - a gene encoding a class I and a class II aminoacyl-tRNA synthetase. Genomics 1994;19:280–290.
40. Cerini C, Kerjan P, Astier M, Gratecos D, Mirande M, Sémériva M. A component of the multisynthetase complex is a multifunctional aminoacyl-tRNA synthetase. EMBO J 1991;10:4267–4277.
41. Fett R, Knippers R. The primary structure of human glutaminyl-tRNA synthetase: a highly conserved core, amino acid repeat regions, and homologies with translation elongation factors. J Biol Chem 1991;266:1448–1455.
42. Michelson AM, Blake CCF, Evans ST, Orkin SH. Structure of the human phosphoglycerate kinase gene and the intron-mediated evolution and dispersal of the nucleotide-binding domain. Proc Natl Acad Sci USA 1985;82:6965–6969.
43. Craik CS, Buchman SR, Beychok S. O_2 binding properties of the product of the central exon of β-globin gene. Nature 1981;291:87–90.
44. Brändén C. Anatomy of α/β proteins. In: Fletterick R, Zoller M (eds) Computer Graphics and Molecular Modeling. Cold Spring Harbor, NY: Cold Spring Harbor Laboratory, 1986;45–51.
45. Stoltzfus A, Spencer DF, Zuker M, Logsdon JM Jr, Doolittle WF. Testing the exon theory of genes: the evidence from protein structure. Science 1994;265:202–207.
46. Palmer JD, Logsdon JJM. The recent origins of introns. Curr Opin Genet Devel 1991;1:470–477.
47. Gilbert W, Marchionni M, McKnight G. On the antiquity of introns. Cell 1986;46:151–154.
48. Kersanach R, Brinkmann H, Liaud MF, Zhang DX, Martin W, Cerff R. Five identical intron positions in ancient duplicated genes of eubacterial origin. Nature 1994;367:387–389.

Tracing Biological Evolution in Protein and Gene Structures.
M. Gō and P. Schimmel, editors.

Architectures of class I aminoacyl-tRNA synthetases

Osamu Nureki and Shigeyuki Yokoyama
Department of Biophysics and Biochemistry, School of Science, University of Tokyo, Tokyo, Japan

Abstract. Aminoacyl-tRNA synthetases (aaRSs) specific for glutamate (GluRS) and glutamine (GlnRS), belonging to class I, are the most related proteins among the 20-member family. The crystal structure of GluRS from *Thermus thermophilus* (*T. thermophilus*) has been solved and refined at 2.5 Å resolution [1], showing an elongated shape composed of four tandemly arranged domains. As compared with the structure of *Escherichia coli* (*E. coli*) GlnRS, this GluRS structure exhibits a significant geometrical similarity in the N-terminal half, comprising the class I-specific Rossmann fold and an intervening subclass-specific α/β domain, in which two secondary structures are inserted. By contrast, the GluRS COOH-terminal half uniquely folds into a "quasi" four-helix bundle and an "α-helix cage" consisting of six α-helices packed by leucine-leucine interactions, which delineates an outstanding contrast to the β-barrel architecture of GlnRS. Mutagenesis studies, based on a docking examination of GluRS and tRNA, revealed that the two structural insertions in the N-terminal half and a basic amino acid cluster in the C-terminal domain play key roles in the specific recognition of the D and acceptor stems, and the anticodon region of $tRNA^{Glu}$. It is possible that the two related aminoacyl-tRNA synthetases have diverged from each other through a combination of insertions and deletions of modular structures in the class/subclass-defining domains and a total exchange of the nonconserved anticodon-binding domains.

Key words: evolution, tRNA identity, X-ray crystallography.

Introduction

In all living cells, aminoacyl-tRNA synthetases (aaRSs) catalyze docking of an amino acid to the acceptor end of the corresponding tRNA, which initiates protein biosynthesis. Based on this fundamental function, aaRSs are believed to be one of the most evolutionally early proteins, having linked the RNA world to the DNA world. As fidelity of translation of genetic information is guaranteed by the specificity of aaRSs for an amino acid and tRNA, 20 aaRSs must be designed strictly to recognize both the cognate substrates, discriminating against all others. However, aaRSs exhibit a wide structural divergence in polypeptide size, subunit organization, primary structure and three-dimensional structure, which made it difficult to comprehend how the specific recognitions for substrates are achieved.

In 1990, Eriani et al. proposed to divide 20 members of the aaRS family into two classes on the basis of ATP-binding mode (Table 1) [2]. The known crystal structures

---

Address for correspondence: Osamu Nureki, Department of Biophysics and Biochemistry, Faculty of Science, University of Tokyo, 7-3-1 Hongo, Bunkyo-ku, Tokyo 113, Japan. Tel.: +81-3-3812-1805. Fax: +81-3-5689-5609.

Table 1. Partition of 20 aminoacyl-tRNA synthetases.

Class I				Class II						
HIGH + KMSK				Motif 1,2,3				Motif 3		
MetRS	(α_2)	2′OH	RF	SerRS	(α_2)	3′OH	AP	AlaRS	(α_4)	3′OH
IleRS	(α)	2′OH		ThrRS	(α_2)	3′OH		GlyRS	($\alpha_2\beta_2$)	3′OH
ValRS	(α)	2′OH		ProRS	(α_2)	3′OH				
LeuRS	(α)	2′OH								
CysRS	(α)			AspRS	(α_2)		AP			
				AsnRS	(α_2)	3′OH				
GluRS	(α)		RF	LysRS	(α_2)	3′OH				
GlnRS	(α)	2′OH								
ArgRS	(α)	2′OH		HisRS	(α_2)	3′OH				
TyrRS	(α_2)		RF	PheRS	($\alpha_2\beta_2$)	3′OH				
TrpRs	(α_2)		RF							

Note: RF and AP represent Rossman fold and antiparallel β-sheet, respectively. The specificity toward either the 2′OH or the 3′OH of the tRNA terminal adenosine as the amino acid attachment site are shown.

of five aaRSs (TyrRS, MetRS, GlnRS, SerRS and AspRS) [3—7] provide a structural basis for this classification; class I aaRSs possess a canonical Rossmann fold (five or six parallel β-sheets flanked by α-helices) providing ATP-binding HIGH (His-Ile-Gly-His)/KMSK (Lys-Met-Ser-Lys) motifs, whereas class II aaRSs exhibit central seven-stranded antiparallel β-sheets with three motifs for ATP binding. Catalytic features of aaRSs, such as their capacity to charge the amino acid to the 2′OH or 3′OH group of the terminal adenosine, also appear to be linked to this classification [2]. Extended sequence similarities allowed 10 members of class I or II aaRSs to be subdivided into three or four subgroups, respectively, which also correlates well to the chemical properties of the amino acid substrate [2].

Conversely, aaRSs features toward tRNA recognition suggest their idiosyncratic evolution. Identity nucleotides, which specify the cognate recognition by aaRS, are diversely located in each tRNA sequence [8] in a manner not correlated with the aaRS classification. The aaRSs structures responsible for tRNA recognition which are not correlated with this classification, those of GlnRS [5] (class I) and AspRS [7] (class II), exhibit antiparallel β-barrel architecture, whereas those of TyrRS [3], MetRS [4] (class II) and SerRS [6] (class II) all show α-helical topology. Furthermore, answers to the question of how each aaRS distinguished its cognate tRNAs from all others remains elusive, partly because the above five aaRSs belong to different subgroups which makes it difficult to compare the structures for tRNA recognition.

Glutamyl- and glutaminyl-tRNA synthetases (GluRS and GlnRS, respectively) are closely related aaRSs with extended sequence similarities [9]. They are believed to have diverged most recently in the aaRS evolution (Fig. 1). In organisms lacking

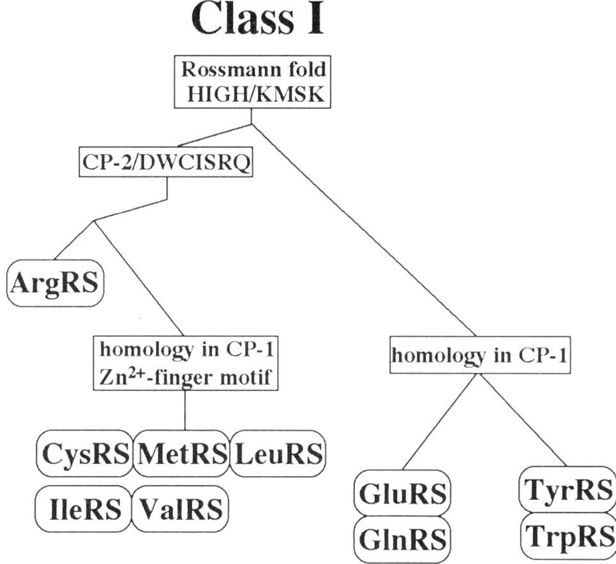

Fig. 1. Presumed evolutionary descent of class I aminoacyl-tRNA synthetases. CP-1 and CP-2 denotes connective polypeptide 1 and 2, respectively, which intervenes into the Rossmann fold.

GlnRS, such as chloroplasts, plant and animal mitochondria and several species of Gram-positive bacteria, GluRS substitutes for this homologue [10,11]. GluRS and GlnRS, with arginyl-tRNA synthetase from a subgroup of class I aaRS, initiate adenylation of amino acid on the trigger of binding with the cognate tRNA [12]. Here, we describe the three-dimensional structure of GluRS from thermophilic bacteria *Thermus thermophilus* (*T. thermophilus*) [1]. Docking model with the cognate tRNA, which coincides with biochemical studies on the GluRS (presented here) and on $tRNA^{Glu}$ (manuscript in preparation), delineates how the ingeniously designed structural modules confer the specific RNA recognition, and how a signal of tRNA-binding is transmitted to the enzyme catalytic center. This GluRS structure, when compared with that of GlnRS [5], visualizes a piece of the scenario of aaRS evolution.

Results and Discussion

Domain architecture and the active site

T. thermophilus GluRS is a bent and twisted cylinder, roughly banana shaped, with overall dimensions of 80 × 40 × 30 Å (Fig. 2). The backbone structure consists of four domains, which are tandemly arranged to form an elongated and curved shape with an axial ratio of 3.4:1.

The N-terminal central domain (domain 1, residues 1—70 and 187—322) contains a typical Rossmann fold, which folds into a five-stranded parallel β-sheet (β 1—3, 10

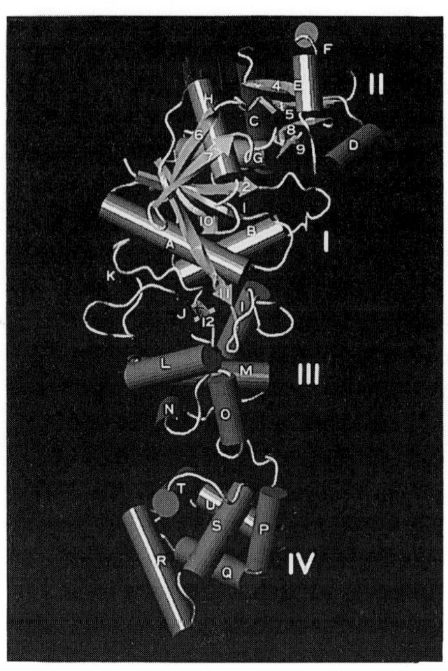

Fig. 2. Schematic diagram of the three-dimensional structure of *T. thermophilus* GluRS. The α-helices are represented by red cylinders (labeled A through U), the strands in the β sheets by cyan arrows (1 through 12), and the coils by narrow yellow rods. Helix E is a 3_{10} helix. The overall structure consists of four tandemly arranged domains labeled I through IV. Two class-I-specific signature motifs are shown.

and 11) surrounded by four α-helices (α A, B, G and H), and an additional subdomain consisting of four α-helices (α I–L) (Fig. 1). As in the case of dehydrogenases, kinases and other class I aaRSs sharing the common dinucleotide-binding fold, that of the GluRS is folded into two symmetrically related halves. The first half comprises N-terminal 70 residues in three β-strands (β 10 and 11) flanked by two α-helices (α G and H), and appears to bind ATP, while the second half, residues 187–237, consists of two β-strands (β 10 and 11) flanked by two α-helices (α G and H), and appears to provide a binding pocket for glutamate substrate. All these structural features have been identified in other class I aaRSs [3–5], indicating that their N-terminal domains have evolved from a common ancestor [2]. The active site of the GluRS lies at the bottom of a very deep pocket of the Rossmann fold; the characteristic His^{15}-Val^{16}-Gly^{17}-Thr^{18}/Lys^{243}-Ile^{244}-Ser^{245}-Lys^{246} motifs are located on two annealed loops between the β1 strand and the αA helix, and between the β11 strand and the αI helix, respectively (arrows in Fig. 1). These two loops are close to each other (Fig. 3A) in a deep cleft of the N-terminal half of GluRS (Fig. 2). Actually, through a preliminary crystal soaking experiment, the ATP substrate is found to bind to these signature motifs where His^{15} makes a stacking interaction with the adenine ring and hydrogen bonds to β phosphate, Lys^{243} to N-6 and N-7 positions of the base moiety and Ser^{245} to β and γ phosphates (manuscript in preparation).

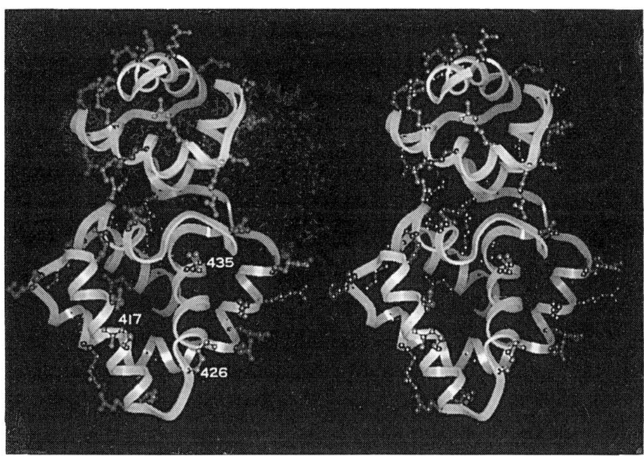

Fig. 3. Stereoviews of the GluRS C-terminal half consisting of "quasi" helix bundle and "α-helix cage" This view shows intra- and inter-helical ion pairings. Arg^{417}, Lys^{426} and Arg^{435} forming a positively charged pocket are labeled.

Between the two halves of this domain I intervenes a domain (II) of 116 residues, which folds into a four-stranded antiparallel β-sheet (β 4, 5, 8 and 9) flanked by pseudo 3-fold-symmetric distribution of α-helices (α D—F) including one distinct 3_{10} helix (α E) (Fig. 2). Similar insertions into the dinucleotide fold appear to occur at all the class I aaRSs [13], which exhibits a wide divergence in its length and sequence [14]. Its primary role is suggested to be the recognition of amino acid substrate and tRNA acceptor [5]. In the present structure, the second and third β-strands (β 5 and 8) extend in a long antiparallel β-sheet (β 6 and 7) into the central domain I (Fig. 2). This β-sheet runs approximately perpendicularly across the Rossmann fold and forms a β-hairpin (Fig. 2).

Unusual folding of the C-terminal half

The three α-helices (α M—O) of domain III (residues 323—370), together with a helix L of domain I, fold into a "quasi" four-helix bundle with the axis almost perpendicular to the largest dimension of the molecule (Fig. 1). Although a similar compact packing of α-helices is seen in histon H5, homeodomain proteins and other transcriptional factors [15,16], the GluRS domain III is folded contrariwise to them.

The C-terminal domain IV (residues 371—468) consists of six α-helices (α P—U), which fold into a hemispheral shape like an "α-helix cage" (Fig. 2). This folding is topologically quite unique among structures in any other aaRSs analyzed so far. Although the overall architecture resembles those of λ repressor DNA-binding domain [17] or POU-type homeodomain [18], the direction of folding is reversed. The two helical components (α R and S) form a "helix-turn-helix" structure showing geometrical similarity with those of repressor proteins; 22 C_{α}-carbons can be superimposed on those of λ repressor with an rms deviation of 0.93 Å, and 24 can be superimposed on those of *trp* repressor with an rms deviation of 0.88 Å.

Surprisingly, 21.4% of residues in the domain IV are Leu, which occur at 3- or 4-residue intervals in the sequence. This arrangement allows the branched Leu side chains to be aligned along the helix axis and facilitates the inter-helical contacts. Compared to a regular Leu zipper [19], the slightly irregular occurrence of the Leu residues result in the formation of an extensive internal hydrophobic core surrounded by six helices (Fig. 2). Contacts between helices also include ion pairs; many Arg, Lys and Glu with a long side chain are arranged on the helical surface, which permits both intra- and inter-helical ion pairings (Fig. 3). Conversely, Arg^{417}, Lys^{426} and Arg^{435} on the "helix-turn-helix" structure are not involved in these ion pairings (Fig. 3), which provide a positively charged pocket possibly participating in tRNA binding. The seemingly rigid "α-helix cage" (domain IV) is linked by a mobile hinge to the domain III and appears to make a loose contact. Two Leu-Leu interactions (between Leu^{354} and Leu^{453}, and Leu^{372} and Leu^{454}) and two ion pairings (between Arg^{350} and Glu^{449}, and Glu^{375} and Arg^{461}) hitch this sequestered domain to the rest of the molecule.

Structural comparison of GluRS and GlnRS

The folding of the N-terminal half of *T. thermophilus* GluRS bears a striking resemblance to that of *Escherichia coli* (*E. coli*) GlnRS, whose crystal structure was solved in the complex with $tRNA^{Gln}$ [5] (Fig. 4). Polypeptide backbone chains for the GluRS and GlnRS were superimposed to establish the geometrical correspondence between residues in the two molecules which is necessary to align the amino acid sequence.

Then, in total, a striking geometrical similarity (rms deviation of 1.14 Å on average) occurs for as many as 198 residues, even though only 25% of amino acids are identical, which may account for the common ancestor of the two protein halves. The resulting sequence alignment is shown in Fig. 5. This structural similarity is remarkably high, as compared with the superposition of the GluRS structure on that of TyrRS (class I) from *Bacillus stearothermophilus* [3] (Table 2); out of the 206 α-carbons in the Rossmann fold, 96 can be superimposed with rms deviation of 2.18 Å, and the intervening domain of TyrRS folds in a completely different manner.

In contrast to the significant similarity in the N-terminal halves, the GluRS C-terminal half, unexpectedly, bears a completely different structure from that of GlnRS. As mentioned above, the C-terminal half of GluRS folds into the α-helical bundle and the unique "α-helix cage", exhibiting striking contrast to the β-barrel architecture of GlnRS [5] (Fig. 4). In addition to these topological differences, the spatial arrangement of the third and fourth structural domains is different between GluRS and GlnRS (Fig. 4). Thus, this particular contrast between GluRS and GlnRS clearly indicates that the two C-terminal halves evolved from different ancestors.

Altogether, there are three sets of structural differences between GluRS and GlnRS: one is a switch of domain architecture in the C-terminal half as mentioned

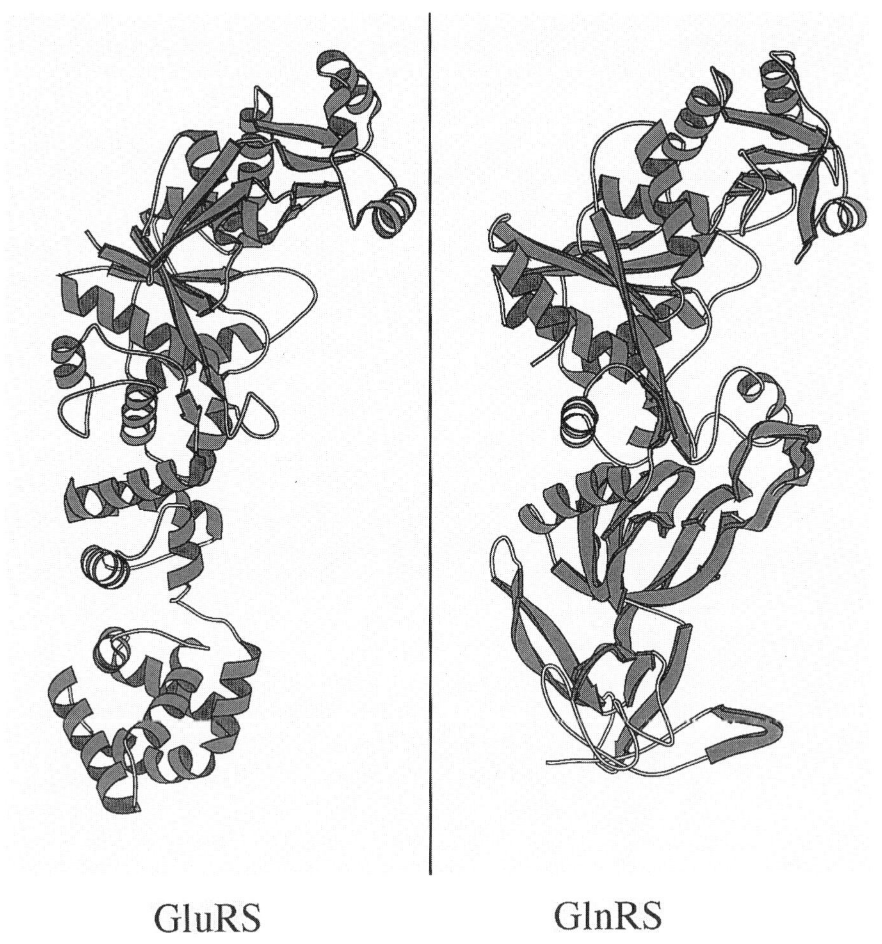

Fig. 4. Comparison of GluRS and GlnRS structures, which highlights a significant similarity in the N-terminal halves and a striking contrast in the C-terminal halves. This view is in a similar orientation to Fig. 2.

above; the other two occur as sequence deletions and insertions in the N-terminal half. In the GluRS structure, three major deletions occur compared with that of GlnRS. Firstly, GlnRS segments of residues 177—183 and 196—207 are deleted in the GluRS structure (designated as deletion 1 and 2, respectively, in Fig. 5). These segments form β loops to make specific contact with major-groove base pairs G2·C71 and G3·C70 (dominant identity elements) on $tRNA^{Gln}$ [5]. The 16-residue segment specific for GlnRS, referred to as deletion 3 (Fig. 5), folds into a flanking α-helix and presumably serves as a transmitter to the active center of the signal to which the correct tRNA anticodon is bound [5,20]. These deletions of the three

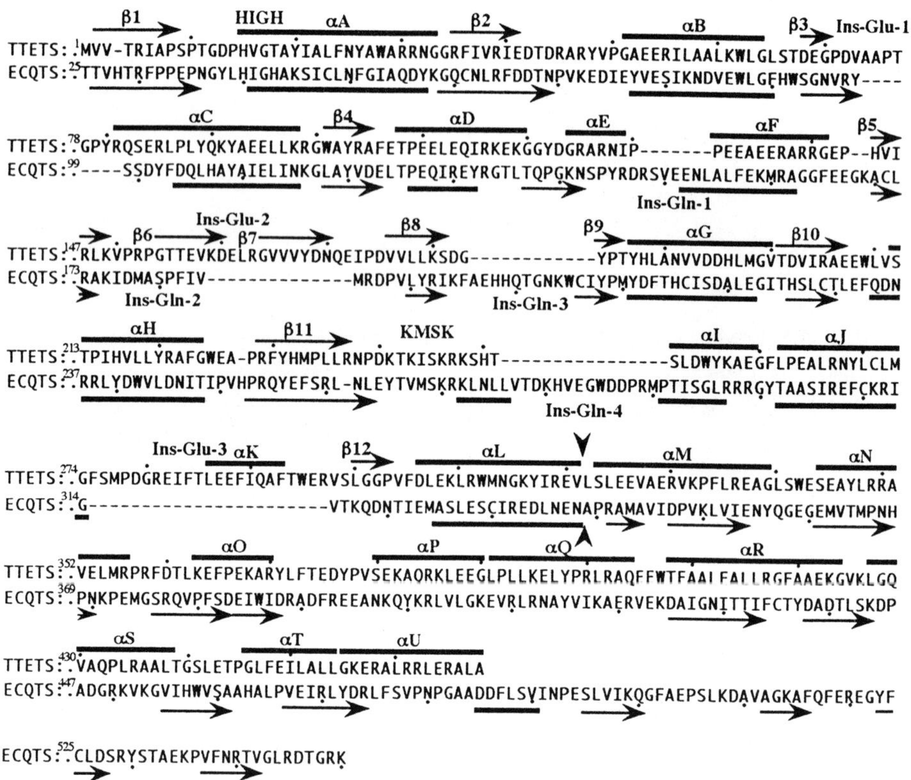

Fig. 5. Primary sequence alignment of *T. thermophilus* GluRS (TTETS) and *E. coli* GlnRS (ECQTS), based on the structural alignment.

functional segments may account for differences in the mechanisms of tRNA recognition and of enzyme activation between the two related synthetases.

Table 2. Superimposition of *T. thermophilus* GluRS structure on that of *E. coli* GlnRS. The superimposition is calculated by minimizing the distance between corresponding Cα positions for four segments as followed in the NH_2-terminal halves.

Domain	Residues in GluRS	Residues in GLnRS	Superimposed Cα atoms	Secondary structures	rms deviation (Å)
I	1–68	26–93	68	α-A,B β-1,2,3	1.06
II	85–107	104–124	23	α-C β-4	0.92
II–I	144–237[a]	170–262[a]	65	α-G,H β-5,8,9,10,11	1.07
I	252–319[b]	292–336[b]	42	α-I,J,L	1.53

[a]Except for residues 151–173, 181–183, 224–226 in GluRS and for residues 177–185, 193–207, 248–251 in GlnRS. [b]Except for residues 274–299 in GluRS and for residues 314–316 in GlnRS.

T. thermophilus GluRS, instead, exhibits three major insertions. Seven-residue segments from 74–80 (referred to a insertion 1 in Fig. 5) is looped-out at the junction between the domains I and II. The second insertion (insertion 2, Fig. 5) is the 27 Å-long antiparallel β sheet (β 6 and 7) extending from the intervening α/β domain II, as previously mentioned. The last insertion (insertion 3, Fig. 5) spans residues 275–297 and folds into α-helix (α K) with surrounding loops, which emanate from the central domain I.

tRNA recognition

Plausible superposition of the GluRS structure on that of GlnRS complexed with $tRNA^{Gln}$ allowed docking of $tRNA^{Glu}$ to GluRS (Fig. 6). Crystal structure of *E. coli* GlnRS·$tRNA^{Gln}$ complex [3,20] as well as biochemical studies [21] revealed that the first three base pairs of the acceptor stem and the three anticodon bases of $tRNA^{Gln}$ dominantly specify the cognate recognition by GlnRS. In the present GluRS structure, however, two β-loop regions of GlnRS which specifically interact with the $tRNA^{Gln}$ acceptor stem are deleted, as described (deletion 1 and 2, Fig. 5). We prepared numbers of *E. coli* $tRNA^{Glu}$ variants with T7 RNA polymerase and tested their charging of glutamate by *E. coli* GluRS (data not shown). The kinetics show that U2·A71 and C4·G69 in the $tRNA^{Glu}$ acceptor stem appreciably affect its charging activity. These nucleotides are conserved in *T. thermophilus* $tRNA^{Glu}$ (data not shown), and consequently, this translocation of the identity elements may account for the structural differences between GluRS and GlnRS. In the $tRNA^{Glu}$-docking model, the antiparallel β-sheet specific for GluRS (insertion 2, Fig. 5) extends along the tRNA acceptor stem and possibly provides some specific interactions (Fig. 6). In fact, mutation of Asp^{160} in this structure to Ala significantly reduces the GluRS activity.

Genetic and biochemical studies by Söll et al. [22] and by our group (data not shown) implicated three anticodon nucleotides as identity elements of *E. coli* $tRNA^{Glu}$. Because the structure of the GluRS C-terminal domains are distinct from those of GlnRS, the mechanism of the anticodon recognition must be different. In the tRNA-docking model, the anticodon are situated close to the cleft between the two helical domains III and IV. As mentioned above, the GluRS C-terminal domains present a cluster of positive charges at this inter-domain cleft. These positive charges are derived from Arg^{349}, Arg^{350} and Arg^{358}, which are located on the second helix (α N) of domain III. We replaced each Arg residue with Gln to examine their activities towards the anticodon recognition. All the mutations appreciably reduced the specific activity of the GluRS, suggesting interaction between the Arg cluster and the anticodon bases. Furthermore, the positively charged pocket expands to the "α-helix cage" domain IV. This exposed surface cavity with positive charge involves Arg^{417}, Lys^{426} and Arg^{435} on the helix-turn-helix structure as mentioned. Thus, it is possible that the "α-helix cage" domain IV with a rigid internal structure moves in a lump for fitting the positively charged cavity onto the tRNA anticodon.

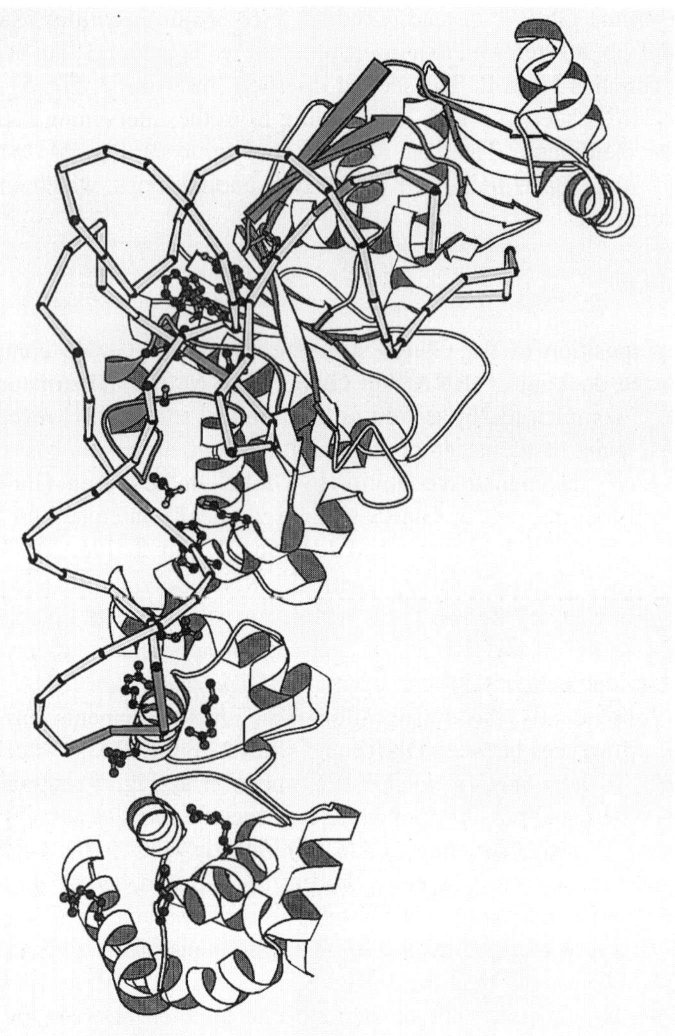

Fig. 6. Docking model between GluRS and $tRNA^{Glu}$. Identity determinants of $tRNA^{Glu}$ are colored magenta. The GluRS-specific helical and β-stranded insertions are colored in red. Amino acid residues responsible for tRNA recognition as described in the text are represented.

Modular structure participating in tRNA discrimination

The $tRNA^{Glu}$ has a distinct nature in that the most dominant identity elements are concentrated in the D-stem: U 11·A24, Ψ13·G22 and A21 (data not shown), which are highly specific for $tRNA^{Glu}$. The docking model of GluRS and $tRNA^{Glu}$ shows that the GluRS-specific modular structure including α-K (insertion 3, colored red in Fig. 6) is in close contact with this key portion of $tRNA^{Glu}$. This helical structure, emanating from the central subdomain, approaches the middle of the tRNA D-stem

Table 3. Three-dimensional–one-dimensional-compatibility analysis by three-dimensional profile method [23].

Domain (1/2) I + II 1–233			Domain I + II 1–322			Domain III + IV 323–468			Domain IV 370–468		
ARS	Source[a]	S[b]	ARS	Source	S	ARS	Source	S	ARS	Source	S
GluRS	*T.th.*	1.00	GluRS	*T.th.*	1.00	GluRS	*T.th.*	1.00	GluRS	*T.th.*	1.00
GluRS	*R.me.*	0.71	GluRS	*R.me.*	0.69	GluRS	*B.st.*	0.41	GluRS	*B.st.*	0.50
GluRS	*B.st.*	0.59	GluRS	*B.st.*	0.62	GluRS	*E.co.*	0.38	GluRS	*E.co.*	0.46
GluRS	*B.su.*	0.48	GluRS	*B.su.*	0.51	GluRS	*B.su*	0.38	GluRS	*B.su*	0.37
GluRS	*E.co.*	0.44	GluRS	*E.co.*	0.49	GluRS	*R.me.*	0.24	AlaRS	*E.co.*	0.34
ValRS	*B.st.*	0.24	ValRS	*B.st.*	0.17	ValRS	human	0.22	TrpRS	Y(mt)	0.33
LeuRS	*S.ce.*	0.21	GlnRS	*S.ce.*	0.15	LeuRS	*S.ce.*	0.21	LeuRS	*S.ce.*	0.32
ProRS	*E.co.*	0.20	Pro/Glu	f.f.	0.13	MetRS	*S.ce.*	0.21	TyrRS	Y(mt)	0.32
Pro/Glu[c]	f.f.	0.19				MetRS	*B.st.*	0.21	MetRS	*T.th.*	0.31
						IleRS	*S.ce.*	0.21	MetRS	*S.ce.*	0.31
						IleRS	*M.th.*	0.20			
						MetRS	*E.co.*	0.20			

[a]Abbreviations: *T.th.* — *T. thermophilus*; *R.me.* — *Rhizobium meliloti*; *B.st.* — *B. stearothermophilus*; *B.su.* — *B. subtilis*; *E.co.* — *E. coli*; *S.ce* — *Saccharomyces cerevisiae*; f.f — fruit fly; *M.th.* — *Methanococcus thermophilia*; Y(mt) — yeast mitochondria. [b]Relative score for compatibility. [c]*Drosophila* multifunctional synthetase with N-terminal sequence homologous to that of GluRS and C-terminal sequence homologous to that of ProRS.

to provide specific interactions with the identity nucleotides (colored magenta in Fig. 6). We replaced amino acid residues probably in specific contact with the $tRNA^{Glu}$ identity elements with Ala to investigate the function of their side chains. Mutation of Ser^{299} situated in the vicinity of Ψ13, and that of Glu^{282} in the vicinity of A24 dramatically reduced the GluRS activity. Further kinetics revealed that Lys^{309} and Trp^{312}, which meet close together with Glu^{282} on the molecular surface, are crucial for the enzymatic activity. These four residues are invariant among seven GluRSs analyzed so far, and play a pivotal role in specific recognition of $tRNA^{Glu}$. Thus, the present three-dimensional structure of GluRS, when compared with that of GlnRS, offers a novel insight that the specific cognition of tRNA is achieved by the insertion of the ingeniously designed modular structures into the aaRSs' structure.

Implication for aaRS evolution

The crystal structure of GluRS presented here permits a structural comparison of two closely related aaRSs belonging to the same subgroup, which can explain how the specific recognition of tRNA is achieved; insertions of ingeniously designed modular structures confer the recognition specificity. Moreover, from the view point of aaRS evolution, our structure shed a novel light on this process; architectural switch of the molecular half between GluRS and GlnRS strongly suggests that the two molecular parts of aaRS, the catalytic half and tRNA-recognition half, have evolved independently.

So, have the aaRS domains participating in the tRNA recognition idiosyncratically evolved? We performed a three-dimensional-profile search using the GluRS C-terminal domain IV as the test structure, which detected high structural similarity between GluRS (class I) and AlaRS (class II) (Table 3). AlaRS bears heptad leucine repeat in its sequence [24], which may permit the helical packing similar to the GluRS domain IV. Similar leucine-rich repeats occur at both class I (IleRS, 25) and class II (SerRS and PheRS, 6 and 26, respectively), and may cause a similar helical folding which partly resembles those of DNA-binding proteins. On the other hand, β-barrel architecture of the tRNA-recognition domains of AspRS and GlnRS exhibits the nature of RNA-binding proteins [5,7]. Therefore, it appears plausible to suggest that aaRS domains participating in the tRNA recognition have evolved from at least two origins with the nature of DNA-binding or RNA-binding, and have joined together with the catalytic half to establish the present-day 20 aaRSs.

Acknowledgements

We thank Drs K.Y.J. Zhang and D. Eisenberg for providing the program SQUASH and the three-dimensional-profile-search program. We are grateful to Drs N. Sakabe, A. Nakagawa and M. Konno for their generous support in data collection at KEK, Japan. Dr H. Nakamura is acknowledged for calculating the electrostatic potential and Dr K. Nishikawa is also acknowledged for three-dimensional-profile analysis.

References

1. Nureki O, Vassylyev DG, Katayanagi K, Shimizu T, Sekine S, Kigawa T, Miyazawa T, Yokoyama S, Morikawa K. Science 1995;267:1958–1965.
2. Eriani G, Delarue M, Poch O, Gangloff J, Moras D. Nature 1990;347:203–206.
3. Brick P, Bhat TN, Blow DM. J Molec Biol 1989;208:83–98.
4. Brunie S, Zelwer C, Risler JL. J Molec Biol 1990;216:411–424.
5. Rould MA, Perona JJ, Söll D, Steitz TA. Science 1989;246:1135–1142.
6. Cusack S, Berthet-Colominas C, Hartlein M, Nassar N, Leberman R. Nature 1990;347:249–255; Fujinaga M, Berthet-Colominas C, Yaremchuck AD, Tukalo M, Cusack S. J Molec Biol 1993;234:222–233; Biou V, Yaremchuk A, Tukalo M, Cusack S. Science 1994;263:1404–1410.
7. Ruff M, Krishnawamy S, Boeglin M, Poterszman A, Rees B, Thierry JC, Moras D. Science 1991;252:1682–1689; Cavarelli J, Rees B, Ruff M, Thierry JC, Moras D. Nature 1993;362:181–184; Delarue M, Poterszman A, Nikonov S, Garber M, Moras D, Thierry JC. EMBO J 1994;13:3219–3229.
8. Schulman LH. Prog Nucl Acid Res Molec Biol 1991;41:23–87.
9. Breton R, Sanfacon H, Papayannopoulos I, Biemann K, Lapointe J. J Biol Chem 1986;261:10610–10717.
10. Schön A, Kannangara CG, Gough S, Söll D. Nature 1988;331:187–190.
11. Wilcox M, Nirenberg M. Proc Natl Acad Sci USA 1968;61:229–236.
12. Kern D, Lapointe J. Biochemistry 1979;18:5809–5818.
13. Starzyk RM, Webster TA, Schimmel P. Science 1987;237:1614–1618.
14. Shiba K, Schimmel P. Proc Natl Acad Sci USA 1992;89:1880–1884.

15. Ramakrishnan V, Finch JT, Graziano V, Lee PL, Sweet RM. Nature 1993;362:219–223.
16. Schutz SC, Shields GC, Steitz TA. Science 1991;253:1001–1007; Clark KL, Halay ED, Lai E, Burley SK. Nature 1993;364:412–420.
17. Jordan SR, Pabo CO. Science 1988;242:893–899.
18. Nuria A-M, Mortishire-Smith RJ, Aurora R, Herr W, Wright PE. Cell 1993;73:193–205.
19. O'Shea EK, Klemm JD, Kim PS, Alber T. Science 1991;254:539–544; Ellenberger TE, Brandl CJ, Struhl K, Harrison SC. Cell 1992;71:1223–1237.
20. Rould MA, Perona JJ, Steitz TA. Nature 1991;352:213–218.
21. Jahn M, English S, Söll D. Nature 1991;352:258–260.
22. Sylvers LA, Rogers KC, Shimizu M, Ohtsuka E, Söll D. Biochemistry 1993;32:3836–3841.
23. Leslie AGW. In: Bailey S, Dodson E, Phillips S (eds) Proceedings of the Daresbury Study Weekend, 5–6 February 1988. SERC Daresbury Laboratory, 1988;25–31.
24. Putney SD, Royal NJ, Newman de Vegvar H, Herlihy WC, Biemann K, Schimmel P. Science 1981;213:1497–1501; Lupas A, Dyke MV, Stock J. Science 1991;252:1162–1164.
25. Csank C, Martindale DW. J Biol Chem 1992;267:4592–4599.
26 Kreutzer R, Kruft V, Bobkova EV, Lavrik OI, Sprinzl M. Nucl Acid Res 1992;20:4173–4178.

Tracing Biological Evolution in Protein and Gene Structures.
M. Gō and P. Schimmel, editors.

Structure, function and evolution of aminoacyl-tRNA synthetases

Stephen Cusack
European Molecular Biology Laboratory, Grenoble Outstation, Grenoble, France

Abstract. The structural basis for the activation of serine and the synthesis of Ap_4A by seryl-tRNA synthetase is described using crystal structures of the enzyme and its substrate complexes. Secondly, biochemical and structural evidence concerning the specific recognition of $tRNA^{ser}$ by seryl-tRNA synthetase is reviewed. These results are discussed in the context of current knowledge on structure-function relationships in other aminoacyl-tRNA synthetases of both class I and II. Using sequence alignments and phylogenetic trees, the early evolution of class II aminoacyl-tRNA synthetases is traced. Arguments are presented to suggest that synthetases are not the oldest of protein enzymes, but survived as RNA enzymes during the early period of the evolution of protein catalysts.

Key words: Ap_4A, protein-RNA interactions, seryl-tRNA synthetase, X-ray crystallography.

Introduction

The fidelity of modern protein biosynthesis rests on two key interactions, the specific attachment of amino acids to their cognate tRNA species and the accurate codon-anticodon interaction between mRNA and tRNA. The former process is catalysed by the aminoacyl-tRNA synthetases which discriminate with remarkable selectivity amongst many structurally similar tRNAs and amino acids. The accuracy of the aminoacylation reaction is based on the existence of positive and negative elements in both tRNAs ("identity" elements) and synthetases which permit both recognition and productive binding of cognate pairs and discrimination against or nonproductive binding of noncognate pairs [1]. In spite of their common catalytic function, the synthetases have long been known to be diverse in subunit structure, polypeptide size and amino acid sequence (Table 1). Quaternary structures include α, α_2, α_4 and $\alpha_2\beta_2$, and subunit polypeptide sizes range in *Escherichia coli* (*E. coli*) from 334 residues (tryptophanyl-tRNA synthetase) to 951 (valyl-tRNA synthetase). Synthetases in eukaryotes are generally larger than their prokaryotic counterparts due to additional sequences dispensable for aminoacylation [2]. In higher eukaryotes the occurrence of nine of the aminoacyl-tRNA synthetases in a 1.2 MDa multisynthetase complex (with, in a separate complex, the valyl-tRNA synthetase associated with the heavy form of elongation factor 1, see Table 1) is suggestive of a higher level of structural organisation of the biosynthetic machinery in these cells [2].

Address for correspondence: Stephen Cusack, European Molecular Biology Laboratory, Grenoble Outstation, c/o ILL, 156X, 38042 Grenoble Cedex 9, France.

Table 1. Subunit composition and size of aminoacyl-tRNA synthetases from prokaryotes (*E. coli*), lower eukaroyotes (*S. cerevisiae*) and higher eukaryotes (various). The table is an updated version of that found in [2]. For higher eukaryotes, CX indicates that the synthetase is part of the multisynthetase complex. In this complex the prolyl- and glutamyl-tRNA synthetases are fused into a single polypeptide chain [2]. The valyl-tRNA synthetase of higher eukaryotes forms a separate complex (CX′) with the heavy form of elongation factor 1 (eEf-1H).

Enzyme	Prokaryotes		Lower eukaryotes		Higher eukaryotes	
Cys	α	461	—	—	—	—
Glu	α	471	—	—	α	1714 CX+PRO
Gln	α	550	—	809	—	– CX
Arg	α	577	α	—	α	661 CX
Leu	α	860	α	1090	α	– CX
Ile	α	939	α	1073	α	1266 CX
Val	α	951	α	1058	α	1265 CX′+Ef–1H
Trp	α_2	334	α_2	—	α_2	475
His	α_2	424	α_2	526	α_2	509
Tyr	α_2	424	α_2	—	α_2	—
Ser	α_2	430	α_2	462	α_2	505
Asn	α_2	465	α_2	492	α_2	552
Lys	α_2	505	α_2	591	α_2	597 CX
Pro	α_2	572	α_2	—	α_2	1714 CX+GLU
Asp	α_2	590	α_2	557	α_2	501 CX
Thr	α_2	642	α_2	734	α_2	724
Met	α_2	677	α	751	α_2	– CX
Ala	α_4	876	α	—	α	967
Gly	$\alpha_2\beta_2$	689+303	α_2	687	α_2	687
Phe	$\alpha_2\beta_2$	795+327	$\alpha_2\beta_2$	595+503	$\alpha_2\beta_2$	—
			α (mito)	474		

In recent years, amino acid sequence analysis and crystal structures of, to date, 10 different synthetases have contributed to a deeper understanding of this apparent diversity. A major breakthrough came in 1990 when sequence analysis [3] and the crystal structure of seryl-tRNA synthetase [4] showed that the 20 aminoacyl-tRNA synthetases are partitioned into two distinct structural classes (I and II) of 10 members each. Each class is characterised by different short-sequence motifs (with in fact extremely few absolutely conserved residues) and a distinctive topology of the catalytic domain (Table 2). In addition, these analyses revealed the modular domain structure of the synthetases, which allows the identification of more closely related subclasses and explains the apparent wide diversity of synthetase subunit size [5—7]. These developments have revived speculation about the position of the synthetases in the evolution of the protein-synthesis machinery and their potential role in the origin and evolution of the genetic code.

In the following paper I would like to review what is known about seryl-tRNA synthetase structure and function, highlighting its peculiarities compared to other synthetase systems. Particular attention will also be paid to differences between class I and class II synthetases. Finally, I will comment on what sequence alignments tell

Table 2. Classification of aminoacyl-tRNA synthetases.

	Class I			Class II			
Aminoacylation site on terminal	2′OH of terminal ribose			3′OH of terminal ribose (exc. Phe)			
Sequence motifs	...HI**G**H... ...KMS**K**S...			[1]....P.. [2]..F**R**XE/D.....R/HXXXFXXXE/D [3]..GXGXGXX**R**...			
Subclass members	(a)	(b)	(c)	(a)	(b)	(c)	(d)
	Leu	Trp[a]	Glu[a]	Ser[a]	Asp[a]	Phe[a]	Ala
	Ile	Tyr[a]	Gln[a]	Thr	Asn	Gly[b]	
	Val			Gly[a,c]	Lys[a]		
	Cys			Pro			
	Met[a]			His			
	Arg						

Note: bold type in motifs indicates absolutely conserved residue. [a]Crystal structure of synthetase known; [b]GlyRS from *E. coli* ($\alpha_2\beta_2$); and [c]GlyRS (α_2).

us about the evolution of aminoacyl-tRNA synthetases and the possible implications for the evolution of the genetic code.

The structure of seryl-tRNA synthetase

The crystal structures of seryl-tRNA synthetase (SerRS) from *E. coli* (SerRSEC) and *Thermus thermophilus* (*T. thermophilus*) (SerRSTT) have both been determined at 2.5 Å resolution [4,8]. The *T. thermophilus* enzyme comprises 421 residues per subunit with a primary sequence identity of 34% to the *E. coli* enzyme (430 residues per subunit) and the three-dimensional structures are very similar (Fig. 1). The catalytic domain is based on a seven-stranded antiparallel β-sheet with two connecting helices. This antiparallel fold is characteristic of class II synthetases and has now been found in the aspartyl-tRNA synthetase [9], phenylalanyl-tRNA synthetase [10] and lysyl-tRNA synthetase [11]. In addition, seryl-tRNA synthetase has a remarkable additional N-terminal domain comprising a 60-Å-long, solvent-exposed, antiparallel-coiled coil (the “helical arm”).

Structural basis of serine activation and AP_4A synthesis by seryl-tRNA synthetase

The first step of the overall aminoacylation reaction catalysed by SerRS is activation of serine by ATP (in the presence of Mg^{2+}) leading to the stable enzyme-bound intermediate, seryl-adenylate (Ser-AMP). In addition, like many other synthetases [12], SerRS can synthesize bisadenosine tetraphosphate (Ap_4A) by attack of the

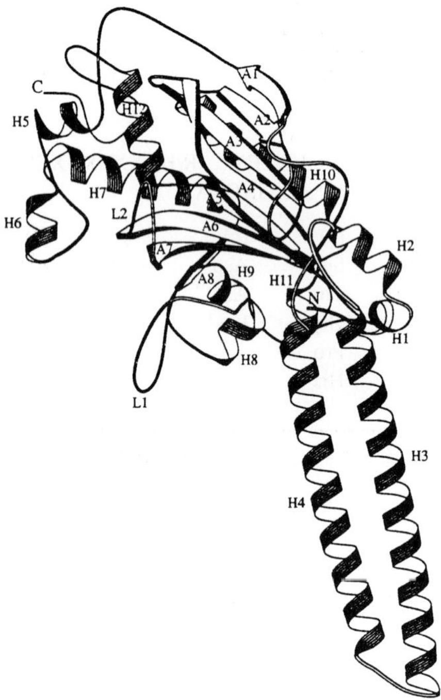

Fig. 1. Ribbon diagram of one subunit of *E. coli* seryl-tRNA synthetase showing the secondary structure. The strands of the major antiparallel sheet are labelled A1 to A7. See [4,5].

intermediate by a second ATP molecule with release of serine. A series of crystal structures at 2.3–2.6 Å resolution of binary and tertiary complexes involving SerRSTT, ATP, Mn^{2+} or Mg^{2+}, seryl-adenylate (natural and analogues) and Ap_4A have been determined which provide the structural basis required to explain the specificity and mechanism of these reactions [13,14]. Figure 2 shows a superposition of the conformations of ATP and Ser-AMP (enzymatically produced in the crystal) complexed with SerRSTT both in the presence of Mn^{2+}. These small substrates are bound by a network of hydrogen bond interactions in a deep hydrophilic cleft formed by the antiparallel β-sheet and surrounding loops of the synthetase catalytic domain. Four regions are involved in the interactions including the motifs-2 and -3 regions of class II synthetases. Apart from the specific recognition of the amino acid side chain, sequence alignments suggest that the interactions of the ATP and the adenylate are likely to be similar in all class II synthetases [13].

In the presence of a divalent cation (Mg^{2+} or Mn^{2+}) the ATP is found to be in an unusual "bent" conformation which is stabilised by the binding of the γ- and α-phosphates to the conserved arginines of motif 3 (Arg-386 in SerRSTT) and motif 2 (Arg-256), respectively, and also by the divalent cation (Fig. 2). The major divalent cation site (for Mn^{2+} or Mg^{2+}) bridges the α- and β-phosphates and also has two ligands in the protein, Glu-345 and Ser-348. In addition, two other Mn^{2+} sites

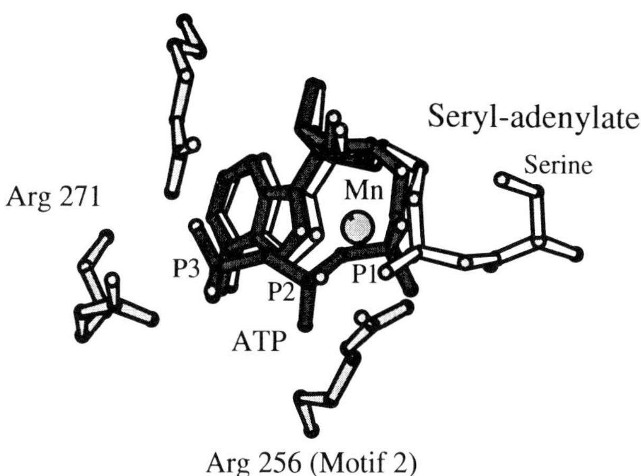

Fig. 2. Superposition of the conformations of ATP (black) and Ser-AMP (white) in the active site of SerRSTT. The α, β, and γ phosphates of the ATP are labelled P1, P2 and P3, respectively. The position of the divalent cation Mn^{2+} is also shown together with the two class II conserved arginines, Arg-256 from motif 2 and Arg-386 from motif 3. Arg-271, which is, however, not strictly conserved, also interacts with the γ-phosphate of the ATP.

(confirmed by their anomalous X-ray scattering signal) are found on either side of the β-γ bridging oxygen. Previous reports [15] that the ATP has an extended conformation can now be explained by the absence of Mg^{2+} binding in the crystallisation medium, which contains high concentrations of ammonium sulphate. If the ammonium sulphate is replaced by sodium citrate the same bent conformation of ATP is also observed in the presence of Mg^{2+}. ATP specificity is guaranteed by hydrogen bond interactions with three of the purine ring nitrogens. The Ser-AMP molecule is found in the same conformation as that already reported for two seryl-adenylate analogues [13]. Serine specificity is ensured firstly by the interaction of the side-chain hydroxyl group with Thr-380 (in motif 3) of the enzyme and secondly by the small size of the side-chain pocket. The superposition of the enzyme-bound ATP and Ser-AMP structures (Fig. 2) provide strong support for an in-line displacement mechanism for serine activation following nucleophilic attack by the serine carboxyl group on the α-phosphate of the ATP. The pentavalent transition state is presumably stabilised by the class II conserved motif 2 arginine (Arg-256) and possibly also by the major divalent cation, while class II conserved motif 3 arginine (Arg-386) assists the release of the pyrophosphate-leaving group with its associated divalent cations.

Figure 3 shows the conformation of the enzyme-bound Ap_4A-Mn^{2+} complex superimposed on that of Ser-AMP. The tetraphosphate of the Ap_4A is stabilised by interactions with the Mn^{2+} (which is at the same position as in the ATP and Ser-AMP complexes) with the P1, P2 and P4 phosphates. The superposition shown in Fig. 3, strongly suggests that Ap_4A is synthesised by a reversal of the same mechanism as

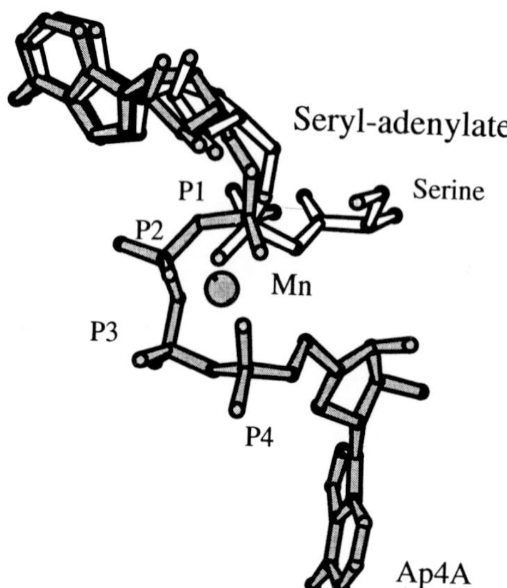

Fig. 3. Superposition of the conformations of Ser-AMP (white) and Ap_4A (grey) in the active site of SerRSTT. The α, β, γ and δ phosphates of the ATP are labelled P1, P2, P3 and P4, respectively. The position of the divalent cation Mn^{2+} is also shown. Note that the ribose conformation of the top adenosine is 3′ endo whereas that of the second adenosine is 2′ endo.

for the formation of Ser-AMP, since the attacking P_γ of the second ATP (equivalent to P2 in Ap_4A) is in the same position as the P_β of the first ATP molecule. Again the possible involvement of the divalent cation in catalysing this displacement reaction is raised. A final point concerns the site for the second adenosine of the Ap_4A. This must be very close to that of the terminal adenosine-76 of $tRNA^{ser}$, but not identical, since otherwise, aminoacylation of the 3′OH of the second ATP might occur rather than Ap_4A production, which is not the case. These findings are described in more detail elsewhere [14].

The mechanism of amino acid activation described here is similar to that proposed by Cavarelli et al. [16] for aspartyl-tRNA synthetase, apart from the role of the divalent cation. The position of the major Mn^{2+} site in SerRSTT differs from that attributed to the Mg^{2+} in the case of aspartyl-tRNA synthetase which is claimed to be between P_β and P_γ [16]. A structurally based mechanism for amino acid activation by the class I glutaminyl-tRNA synthetase (GlnRS) has also been proposed [17]. In the GlnRS complex crystal structure the two class I conserved motifs (HIGH and KMSKS) are directly implicated in binding the ATP. Furthermore, the second histidine of the HIGH motif and the second lysine of the KMSKS motif are postulated to participate in the stabilisation of the pentavalent transition state [17]. Thus recent structural studies on both class I and class II synthetases are consistent with early work concerning the mechanism of amino acid activation in which the inversion of configuration at the P_α was observed and an in-line mechanism proposed [18]. In both classes, the role of the enzyme is to enhance the reaction rate by binding the ATP and amino acid in the correct configuration for this reaction to occur and to stabilise the transition state, but two completely different structural frameworks have been used in evolution to achieve this.

tRNA^{ser} identity

E. coli possesses five iso-accepting tRNA^{ser}s (with four distinct anticodons) in order to cope with the six codons for serine which are from two distinct codon groups. In addition, the tRNA^{selcys} (an opal suppressor tRNA) necessary for the co-translational incorporation of selenocysteine into proteins [19] and an amber suppressor tRNA (*supD* gene product) are also specifically aminoacylated by SerRS. As a result there is no consistency in the anticodon bases and indeed, it turns out that the anticodon is not recognized by seryl-tRNA synthetase, as was originally shown by Normanly et al. [20,21]. Amongst the 20 systems, this exceptional feature is shared only by leucyl- and alanyl-tRNA synthetase [1]. In vivo identity switch experiments [20,21] have shown the importance for tRNA^{ser} identity of the discriminator base and bases from the first three pairs of the acceptor stem as well as the D-stem base pair C11-G24. The latter, however, is not directly in contact with the synthetase [15] and seems to be a negative determinant to avoid mischarging with leucine and glutamate. Another special feature of serine iso-accepting tRNAs is that, with exception of those from animal mitochondria, they possess a long variable arm of 14-20 nucleotides rather than the usual 4-5. This feature is shared in procaryotes by tRNA^{tyr} and tRNA^{leu} isoacceptors. Experiments by various groups [21,23] have highlighted the importance of the long variable arm as an essential recognition element and this is also clearly apparent from the crystal structure of the complex (see below). A conserved feature in prokaryotic tRNA^{ser} isoacceptors is the absence of unpaired nucleotides at the base of the variable arm stem. The tRNA^{leu} isoacceptors have one unpaired nucleotide at the 3′ end of the variable stem and tRNA^{tyr} have two. These differences, which are correlated with the number of insertions in the D-loop, are thought to influence the orientation of the variable arm. Himeno et al. [22] have argued that these factors are crucial for the discrimination between long variable arm tRNAs and have demonstrated this for tRNA^{tyr}, tRNA^{ser} [22] and tRNA^{leu} [24].

The structure of the seryl-tRNA synthetase-tRNA^{ser} complex

The crystal structure of the complex between seryl-tRNA synthetase and tRNA^{ser}(CGA) from *E. coli* has been determined at 4 Å resolution [25] and between seryl-tRNA synthetase and tRNA^{ser}(GGA) from *T. thermophilus* at 2.9 Å resolution [15]. Interestingly, the stoichiometry of this complex is one tRNA to one synthetase dimer (see Fig. 4), although it is known from both solution studies and from the *E. coli* complex crystal structure that the synthetase can simultaneously bind two tRNA molecules. The main conclusions from these crystal structures can be summarised as

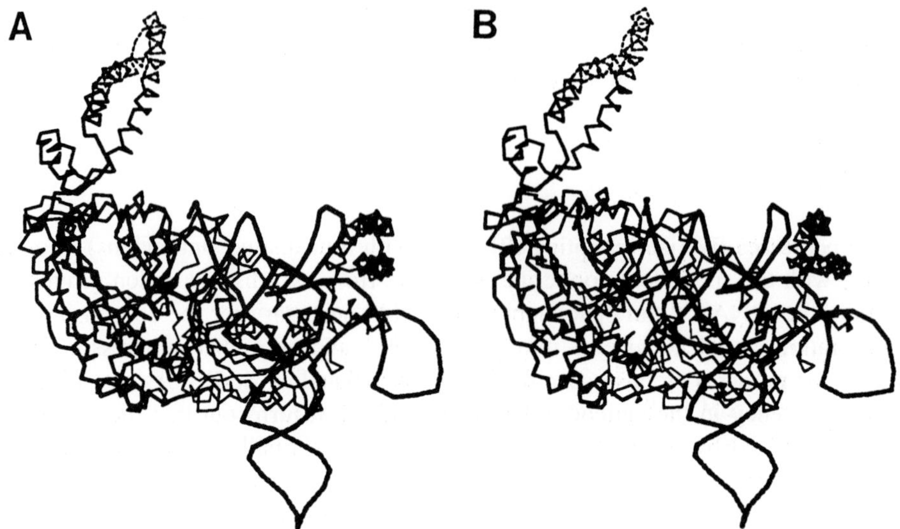

Fig. 4. Stereo diagram of the seryl-tRNA synthetase-tRNA^{ser} complex from *T. thermophilus* showing the C_α trace of the protein (A) and the phosphate backbone of the tRNA (B). Disordered regions are dotted.

follows: 1) the tRNA binds across the two subunits of the dimer; 2) the anticodon stem loop is not in contact with the synthetase; 3) upon tRNA binding the helical arm of the synthetase, which is otherwise flexible, is stabilised in a new orientation and curves between the TΨC loop and the long variable arm of the tRNA; 4) the synthetase makes several backbone contacts, particularly with the tRNA long variable arm, but few base-specific interactions; 5) contacts with the tRNA long variable arm backbone and extends until the sixth base pair, explaining the need for a minimum length of the arm, but allowing longer arms (as in the SelC gene product tRNA^{selcys}) to be accommodated; 6) bases 20A and 20B, inserted into the D-loop in prokaryotic tRNA^{ser}, both play novel roles in tertiary interactions in the core of the tRNA. In particular, the base of Gua-20B is stacked against the first base pair of the long variable arm and thus defines the direction of the latter (Fig. 5); and 7) disorder of the 3′ end of the tRNA^{ser} in this crystal form prevents detailed description of the interactions of the acceptor stem and CCA_{76} end of the tRNA with the active site [15]. However, it is clear that the 3′ end of the tRNA can enter the active site without distortion and that there are acceptor stem major groove interactions with the motif 2 loop, similar to the case of aspartyl-tRNA synthetase, also class II [26]. This contrasts with the situation in the glutaminyl-tRNA synthetase complex (class I), where the synthetase interacts with the acceptor stem minor groove making a hairpin turn of the 3′ end of the tRNA necessary in order that it can enter the active site [27].

These crystallographic results, which are in good agreement with the biochemical results on tRNA^{ser} identity (see above and reference [15]), show that both distinctive features of the serine system, the synthetase helical arm domain and the tRNA^{ser} long

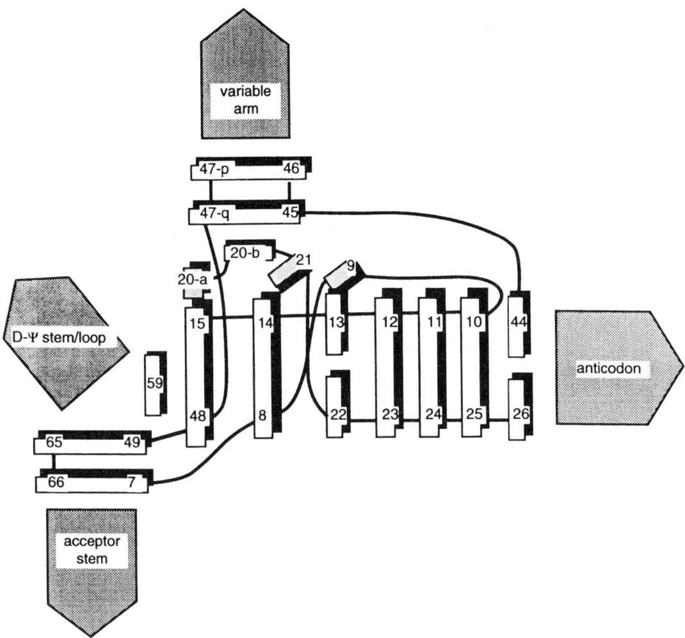

Fig. 5. Schematic diagram of the core structure of tRNA^{ser} [15]. The D-loop base Gua-20b stacks with the first base pair of the long variable arm (Ade-45:Uri-47q) and leads to the tilting of bases Ade-21 and Gua-9 which, however, still manage to make base triplets.

variable arm play the major roles in the mutual recognition of these two macromolecules. Furthermore this recognition is based on shape rather than specific nucleotide sequence. A similar conclusion was drawn by Asahara et al. [28] who exhaustively studied the in vitro effect on the aminoacylation reaction of sequence variations in the acceptor stem and variable arm of tRNA^{ser}. To test the exact role of the synthetase helical arm in aminoacylation, two mutants of the *E. coli* synthetase have been constructed with a partial or full deletion of the helical arm [29]. These mutants are not affected in the serine activation function, but have dramatically reduced aminoacylation activity. Particularly interesting is the fact that the mutant synthetases appear to have lost their specificity for tRNA^{ser} and also charge noncognate type 1 tRNA(s), although which isoacceptors are (mis)charged remains to be determined. This is the first evidence that the SerRS helical arm acts not only as a determinant for cognate tRNA recognition but also as an antideterminant for noncognate tRNAs [29].

The mode of recognition of the cognate tRNA in the serine system is very different from that observed in the two other known synthetase-tRNA complex structures, the class I *E. coli* glutaminyl system [27] and the class II yeast aspartyl system [9]. In both these cases, the anticodon is an important tRNA identity element and is recognised by special anticodon-binding domains of the synthetase, although the details of these interactions are very different in the two cases [26,30,31].

The evolution of aminoacyl-tRNA synthetases

Clearly it is of great interest to use the growing wealth of structural and sequence information on aminoacyl-tRNA synthetases to try to reconstruct phylogenetic trees for these enzymes, despite the difficulty of defining a time axis in such problems [32]. This should be done using structurally based sequence alignments as far as possible rather than relying on automatic "blind" methods of multiple sequence alignment which can lead to avoidable errors. Structurally based sequence alignments for the 10 class II synthetases have been presented previously [5,6,13] and will soon be published for class I synthetases [33]. Here we will focus on class II synthetases.

As described in Cusack [6], the three class II motifs can now be identified in all class II synthetases following recent results on the glycyl-, alanyl- and phenylalanyl-systems. It should be noted that some errors have come to light in the alignments of motif 1 presented in Table 1 of [6], but these do not affect the conclusions.

Figure 6 shows schematically the domain structure of class II aminoacyl-tRNA synthetases based on the multiple alignments. Identification of homologous domains N-terminal or C-terminal to the catalytic domain permit separation of the homo-dimeric class II synthetases into class IIa (the enzymes for serine, threonine, glycine (except *E. coli*), and histidine) and class IIb (the enzymes for aspartic acid, asparagine and lysine) [5—7]. These putative tRNA-binding domains are apparently conserved throughout evolution for a given system. In the case of aspartyl-tRNA synthetase, the N-terminal domain is involved in specific anticodon recognition [26] and a similar

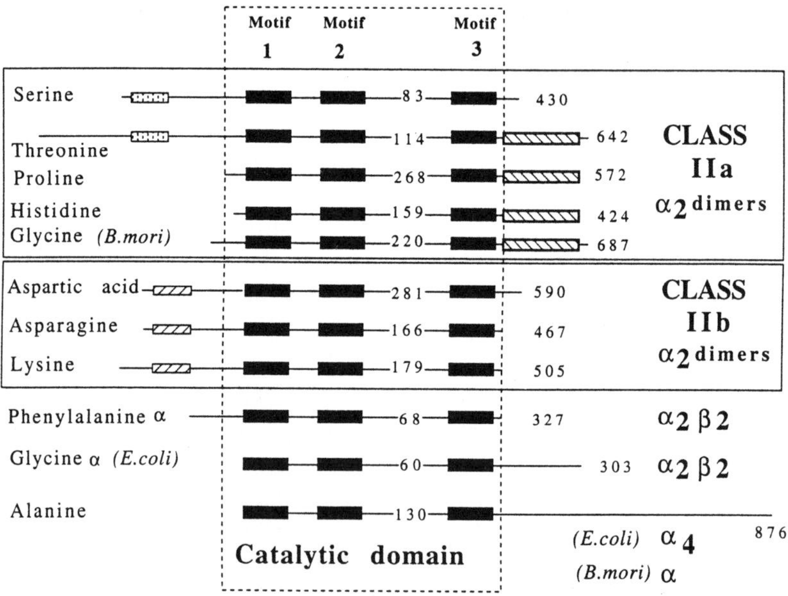

Fig. 6. Domain structure of class II aminoacyl-tRNA synthetases. Homologous domains attached to the N- or C-terminal of the catalytic domain have similar shading. The sequence numbering refers to the *E. coli* enzyme unless otherwise stated.

domain topology is found in other oligonucleotide-binding proteins [34]. As expected, a very similar fold has been found for the N-terminal domain in lysyl-tRNA synthetase [11]. The C-terminal domain characteristic of class IIa has not yet been proved to be an anticodon-binding domain, but is a very strong candidate; its absence from only seryl-tRNA synthetase is consistent with the fact that only this class IIa synthetase does not recognise the anticodon. The region between motifs 2 and 3 is also a site for the insertion of extra domains that are even more idiosyncratic and thus probably more recent acquisitions; the example of aspartyl-tRNA synthetase shows that this insertion domain is not conserved even between the yeast and bacterial enzymes [35]. Glycyl-tRNA synthetase from *E. coli*, and the phenylalanyl-tRNA synthetases (except that from yeast mitochondria, which is monomeric) have an $\alpha_2\beta_2$ composition, the α-subunits having the characteristic class II motifs (Table 1 and Fig. 6). The remarkable structure of the phenylalanyl-tRNA synthetase from *T. thermophilus* shows that both the α- and β-subunits contain a class II antiparallel fold, although only that of the α-subunit is apparently catalytically active [10].

In the above paragraph the modular nature of the class II synthetases has been emphasised. Consideration of this leads to the almost inevitable conclusion that the common ancestor to these enzymes consisted of a class II catalytic domain onto which were subsequently added N- or C-terminal anticodon-binding domains and more recently (since the evolution of different cell types) insertions between motif 2 and motif 3 (and elsewhere). If we consider only the extended sequences around the three motifs, i.e., the only regions that can be aligned amongst all class II synthetases, we can draw an unrooted phylogenetic tree (Fig. 7). This shows that even

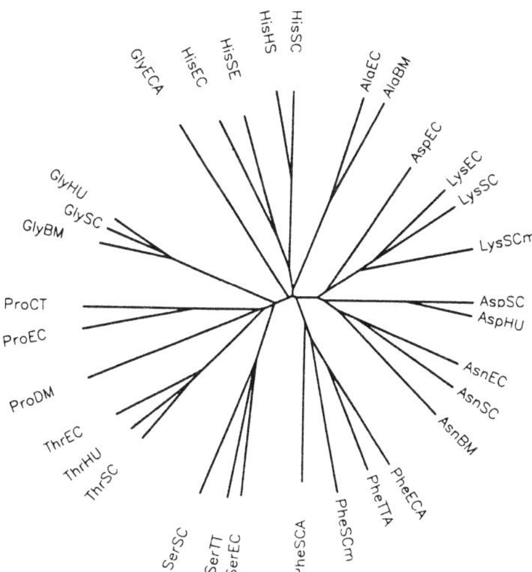

Fig. 7. Unrooted phylogenetic tree of class II aminoacyl-tRNA synthetases based on alignment of 140 residues from extended regions around the three class II motifs. Abbreviations used: GlyECA = α-subunit of glycyl-tRNA synthetase from *E. coli*; TT = *T. thermophilus*; SC(m) = *S. cerevisiae* (mitochondrial); HU = human; DM = *Drosophila melanogaster*; BM = *Bombyx mori*; SE = *Streptococcus equisimilis*; CT = *Chlamydia trachomatis*. Figure prepared using the programme CLUSTALV [50].

without putting any information in about the extra domains, apart from the catalytic domain, the subclass divisions into class IIa (except HisRS), class IIb and others (GlyRSEC, PheRS, AlaRS) is clearly reproduced.

This phylogenetic tree poses many questions. Amongst these are the different origins of the $\alpha_2\beta_2$ GlyRS from *E. coli* and the α_2 GlyRS from other organisms [6]. Similarly, why did yeast mitochondrial PheRS simplify to a monomeric enzyme? Or alternatively, why did the $\alpha_2\beta_2$ PheRS become more complex with the addition of a second catalytic-like domain [6,10]? Another question concerns the significance of the subclasses. The tree suggests that at some earlier time, there were significantly fewer synthetases. For instance, class IIb synthetases AspRS and LysRS have apparently diverged from a common ancestor by gene duplication and adaptation of the amino acid site and the tRNA specificity. (NB: Without loss of generality we do not consider AsnRS in the following argument). What was the function of this common ancestor? Was it such an inaccurate synthetase that it could not tell the difference between aspartic acid and lysine and charged a tRNA with either of them? Or was it on the contrary such a good enzyme that it could act as a double synthetase with specific LysRS and AspRS activity? Both possibilities seem unlikely. Did the common ancestor have just one of the two activities then? This possibility begs the question of what performed the synthetase function for the other amino acid? Obviously the problem is compounded if one considers the common ancestor of all the 10 class II synthetases. Very recently an interesting piece of information emerged which may shed light on the nature of the common ancestor of class II synthetases. Whereas many nonsynthetase enzymes with the same catalytic fold as class I aminoacyl-tRNA synthetases are known (kinases, dehydrogenases, etc.), to date no other enzyme has been found with the antiparallel ATP-binding fold characteristic of class II synthetases. However, it has recently been shown [36] that biotin synthetase (BirA) from *E. coli*, whose crystal structure is known [37], has such a fold, although there is no obvious sequence homology. Biotin synthetase activates biotin with ATP to form an intermediate biotin-AMP. The same enzyme then transfers the biotin to lysines of certain target proteins (e.g., acetyl-CoA carboxylase); BirA thus has a close functional similarity to an aminoacyl-tRNA synthetase. Although convergent evolution cannot be ruled out, this finding permits one to hypothesise that BirA and class II synthetases had a common ancestor which was capable of activating small molecules with ATP. Class II synthetases would then have evolved when an activation domain, specialised to an amino acid, also became able to transfer the amino acid to the 3′ end of a tRNA molecules. This is clearly consistent with the view, based on sequence alignments [6] and on experiments with mini-RNA helices [38], that early synthetases acquired tRNA-binding specificity firstly through acceptor-stem interactions with the activation domain (e.g., via the motif 2 loop in the case of class II synthetases) and later by addition of extra tRNA-binding domains (notably anticodon recognition domains).

Implications for the evolution of the genetic code

It has long been recognised that synthetases represent an evolutionary chicken and egg paradox [39], since proteins cannot be made accurately without highly specific synthetases, but synthetases are themselves accurately made proteins. The arguments presented in the previous paragraphs suggest that the hypothesis that the first aminoacyl-tRNA synthetases were protein enzymes runs into great difficulties. The fact that class II synthetases may have evolved from an existing protein with ATP-mediated activation activity also suggests that synthetases were not necessarily amongst the first protein enzymes; the same argument can be made for class I synthetases.

Fortunately, the RNA world hypothesis [40] provides a way out of this impasse, despite its difficulties [41]. A series of discoveries in the natural world as well as the "artificial" RNA world (i.e., that accessible by the in vitro selection of RNA molecules) go some way towards demonstrating the chemical plausibility of template-directed polypeptide synthesis by RNA catalysts exclusively. These discoveries include: 1) the demonstration that extensively deproteinized large ribosomal subunits retain peptidyl transferase activity, suggesting that the catalytic activity resides in the RNA component [42]; 2) the demonstration of the aminoacyl-esterase activity of the *Tetrahymena* ribozyme [43]; 3) the in vitro selection of RNAs with specific ATP [44] or polar [45] and nonpolar [46] amino acid binding sites; and 4) the in vitro selection of an RNA capable of aminoacylating itself [47]. It is beyond the scope of this article to discuss how, why and to what sophistication polypeptide synthesis may have evolved in the RNA world (see for example [48,49]). However, in relation to the question as to whether the present-day protein synthetases have anything to tell us about the origin of the genetic code, it is important to ask what the genetic code looked like when this replacement process occurred? Had the 20 amino acid genetic code already evolved at that time and functioned in a more or less rigorous way or did a simpler code with fewer amino acids exist? We are not yet in a position to answer these questions, although we will speculate briefly about how RNA synthetases might have been replaced by protein synthetases. The partition of modern synthetases into two mutually exclusive and structurally different classes might, if the modern genetic code had already evolved at the time, only reflect two different pathways in the replacement process. Using the arguments developed above, the key to the replacement process would have been the acquisition by existing amino acid activation domains (which must have been useful in their own right) of tRNA 3′-end binding. It is then relatively easy to imagine that: 1) the development of an efficient aminoacylation enzyme would soon make the corresponding RNA synthetase redundant, particularly after the acquisition of additional (e.g., anticodon-binding) domains which would make tRNA discrimination more accurate; 2) depending on the amino acid specificity (e.g., polar or nonpolar), different activation domains might have existed, thus explaining the occurrence of the two classes of synthetase catalytic domain; and 3) gene duplication and evolution of the amino acid binding site and tRNA specificity would give rise to the observed evolutionary tree of protein synthetases.

If the evolutionary relationship between protein synthetases only reflects the process of replacing existing RNA synthetases and not the co-evolution of the synthetases with the genetic code, then we have to look elsewhere for the origins of the genetic code.

Acknowledgements

I would like to thank all my colleagues at the Grenoble Outstation of EMBL who have contributed to the work on seryl-tRNA synthetase through which many of the results and ideas reviewed in this paper originated. I would also like to thank Peter Artymiuk and Silvia Onesti for communication of results prior to publication.

References

1. Saks ME, Sampson JR, Abelson JN. The transfer RNA identity problem: a search for rules. Science 1994;263:191–197.
2. Mirande M, Lazard M, Kerjan P, Bec G, Agou F, Quevilo S, Waller J-P. The aminoacyl-tRNA synthetase family: an evolutionary view of their structural organisation. In: Nierhaus KH, Franceschi F, Subramanian AR, Erdmann VA, Wittmann-Liebold B (eds) The Translation Apparatus: Structure, Function, Regulation, Evolution. New York: Plenum Press, 1993;657–668.
3. Eriani G, Delarue M, Poch O, Gangloff J, Moras D. Partition of tRNA synthetases into two classes based on mutually exclusive sets of sequences motif. Nature 1990;347:203–206.
4. Cusack S, Berthet-Colominas C, Härtlein M, Nassar N, Leberman R. A second class of synthetase structure revealed by X-ray analysis of *Escherichia coli* seryl-tRNA synthetase at 2.5 Å. Nature 1990;347:249–255.
5. Cusack S, Härtlein M, Leberman R. Sequence, structure and evolutionary relationships between class 2 aminoacyl-tRNA synthetases. Nucl Acid Res 1991;19:3489–3498.
6. Cusack S. Sequence, structure and evolutionary relationships between class 2 aminoacyl-tRNA synthetases: an update. Biochimie 1993;75:1077–1081.
7. Delarue M, Moras D. The aminoacyl-tRNA synthetase family: modules at work. BioEssays 1993;15:675–687.
8. Fujinaga M, Berthet-Colominas C, Yaremchuk AD, Tukalo MA, Cusack S. Refined crystal structure of the seryl-tRNA synthetase from *Thermus thermophilus* at 2.5 Å resolution. J Molec Biol 1993;234:222–323.
9. Ruff M, Krishnaswamy S, Boeglin M, Poterszman A, Mitschler A, Podjarny A, Rees B, Thierry J-C, Moras D. Class II aminoacyl-tRNA synthetases: crystal structure of yeast aspartyl-tRNA synthetase complexed with $tRNA^{asp}$. Science 1991;252:1682–1689.
10. Mosyak L, Reshetnikova L, Goldgur Y, Delarue M, Safro M. Structure of phenylalanyl-tRNA synthetase from *Thermus thermophilus*. Nat Struc Biol 1995;2:537–547.
11. Onesti S, Miller AD, Brick P. The crystal structure of lysyl-tRNA synthetase (LysU) from *Escherichia coli*. Structure 1995;3:163–176.
12. Goerlich O, Foeckler R, Holler E. Mechanism of synthesis of adenosine (5′) tetraphospho (5′) adenosine (AppppA) by aminoacyl-tRNA synthetases. Eur J Biochem 1982;126:135–142.
13. Belrhali H, Yaremchuk A, Tukalo M, Larsen K, Berthet-Colominas C, Leberman R, Beijer B, Sproat B, Als-Nielsen J, Grübel G, Cusack S. Crystal structures at 2.5 Å resolution of seryl-tRNA synthetase complexed with two analogs of seryl adenylate. Science 1994;263:1432–1436.

14. Belrhali H, Yaremchuk A, Tukalo M, Berthet-Colominas C, Rasmussen B, Bösecke P, Diat O, Cusack S. The structural basis for seryl-adenylate and Ap_4A synthesis by seryl-tRNA synthetase. Structure 1995;3:341–352.
15. Biou V, Yaremchuk A, Tukalo M, Cusack S. The 2.9 Å crystal structure of *T. thermophilus* seryl-tRNA synthetase complexed with $tRNA^{ser}$. Science 1994;263:1404–1410.
16. Cavarelli J, Eriani G, Rees B, Ruff M, Boeglin M, Mitscler A, Martin F, Gangloff J, Thierry J-C, Moras D. The active site of yeast aspartyl-tRNA synthetase: structural and functional aspects of the aminoacylation reaction. EMBO J 1994;13:327–337.
17. Perona JJ, Rould MA, Steitz TA. The structural basis of transfer RNA aminoacylation by *E. coli* glutaminyl-tRNA synthetase. Biochemistry 1993;32:8758–8771.
18. Langdon SP, Lowe G. The stereochemical course of amino acid activation by methionyl and tyrosyl-tRNA synthetases. Nature 1979;281:320–321.
19. Böck A, Forchhammer K, Heider J, Leinfelder W, Sawers G, Veprek B, Zinoni F. Selenocysteine: the 21st amino acid. Molec Microbiol 1991;5:515–520.
20. Normanly J, Ogden RC, Horvath SJ, Abelson J. Changing the identity of a transfer RNA. Nature 1986;321:213–219.
21. Normanly J, Ollick T, Abelson J. Eight base changes are sufficient to convert a leucine-inserting tRNA into a serine-inserting tRNA. Proc Natl Acad Sci USA 1992;89:5680–5684.
22. Himeno H, Hasegawa T, Ueda T, Watanabe K, Shimuzu M. Conversion of aminoacylation specificity from $tRNA^{Tyr}$ to $tRNA^{Ser}$ in vitro. Nucl Acid Res 1990;18:6815–6819.
23. Wu XQ, Gross HJ. The long extra arms of human tRNA((Ser)Sec) and tRNA(Ser) function as major identify elements for serylation in an orientation-dependent, but not sequence-specific manner. Nucl Acid Res 1993;21:5589–5594.
24. Asahara H, Himeno H, Tamura K, Hasegawa T, Watenabe K, Shimizu M. Recognition nucleotides of *E. coli* $tRNA^{leu}$ and its elements facilitating discrimination from $tRNA^{ser}$ and $tRNA^{tyr}$. J Molec Biol 1993;231:219–229.
25. Price S, Cusack S, Borel F, Berthet-Colominas C, Leberman R. Crystallisation of the seryl-tRNA synthetase $tRNA^{ser}$ complex of *E. coli*. FEBS Lett 1993;324:167–170.
26. Cavarelli J, Rees B, Ruff M, Thierry J-C, Moras D. Yeast $tRNA^{asp}$ recognition by its cognate class II aminoacyl-tRNA synthetase. Nature 1993;362:181–184.
27. Rould MA, Perona JJ, Söll D, Steitz TA. Structure of *E. coli* glutaminyl-tRNA synthetase complexed with $tRNA^{gln}$ and ATP at 2.8 Å resolution. Science 1989;246:1135–1142.
28. Asahara H, Himeno H, Tamura K, Nameki N, Hasegawa T, Shimizu M. *Escherichia coli* seryl-tRNA synthetase recognizes $tRNA^{ser}$ by its characteristic tertiary structure. J Molec Biol 1994;236:738–748.
29. Borel F, Vincent C, Leberman R, Härtlein M. Seryl-tRNA synthetase from *Escherichia coli*: implication of its N-terminal domain in aminoacylation activity and specificity. Nucl Acid Res 1994;22:2963–2969.
30. Rould MA, Perona JJ, Steitz TA. Structural basis of anticodon loop recognition by glutaminyl-tRNA synthetase. Nature 1991;352:213–218.
31. Cavarelli J, Moras D. Recognition of tRNAs by aminoacyl-tRNA synthetases. FASEB J 1993;7:79–86.
32. Benner SA, Cohen MA, Gonnet GH, Berkowitz DB, Johnsson KP. Reading the Palimpsest: contemporary biochemical data and the RNA world. In: Gesteland RF, Atkins JF (eds) "The RNA World". Cold Spring Harbor: Cold Spring Harbour Laboratory Press, 1993;27–70.
33. Landès C, Perona JJ, Brunie S, Rould MA, Zelwer C, Steitz TA, Risler J-L. A structure based multiple sequence alignment of all class I aminoacyl-tRNA synthetases. Biochimie 1995;77:194–203.
34. Murzin AG. Oligonucleotide/oligosaccharide binding fold: common structural and functional solution for nonhomologous sequences. EMBO J 1993;12:861–867.
35. Delarue M, Poterszman A, Nikonov S, Garber M, Moras D, Thierry J-C. Crystal structure of a prokaryotic aspartyl-tRNA synthetase. EMBO J 1994;13:3219–3229.
36. Artymiuk PJ, Rice DW, Poirrette AR, Willett P. A tale of two synthetases. Nature Struc Biol 1994;1:758–760.

37. Wilson KP, Shewchuk LM, Brennan RG, Otsuka AJ, Matthews BW. *E. coli* biotin holoenzyme synthetase/bio repressor crystal structure delineates the biotin- and DNA- binding domains. Proc Natl Acad Sci USA 1992;89:9257–9261.
38. Schimmel P, Giegé R, Moras D, Yokoyama S. An operational RNA code for amino acids and possible relationship to genetic code. Proc Natl Acad Sci USA 1993;90:8763–8768.
39. Hopfield JJ. Origin of the genetic code: a testable hypothesis based on tRNA structure, sequence, and kinetic proofreading. Proc Natl Acad Sci USA 1978;75:4334–4338.
40. Gilbert W. The RNA World. Nature 1986;319:618.
41. Joyce GF, Orgel LE. Prospects for understanding the origin of the RNA world. In: Gesteland RF, Atkins JF (eds) "The RNA World". Cold Spring Harbour: Cold Spring Harbour Laboratory Press, 1993;1–25.
42. Noller HF, Hoffarth V, Zimniak L. Unusual resistance of peptidyl transferase to protein extraction procedures. Science 1992;256:1416–1419.
43. Piccirilli JA, McConnell TS, Zaug AJ, Noller HF, Cech TR. Aminoacyl esterase activity of the *Tetrahymena* ribozyme. Science 1992;256:1420–1424.
44. Sassanfar M, Szostak JW. An RNA motif that binds ATP. Nature 1993;364:550–552.
45. Yarus M. Specificity of arginine binding by the Tetrahymena intron. Biochemistry 1989;28:980–988.
46. Mayerfeld I, Yarus M. An RNA pocket for an aliphatic hydrophobe. Nature Struc Biol 1994;1:287–292.
47. Illangesekare M, Sanchez G, Nickles T, Yarus M. Aminoacyl-RNA synthesis catalyzed by an RNA. Science 1995;267:643–647.
48. Orgel LE. The origin of polynucleotide-directed protein synthesis. J Molec Evol 1989;29:465–474.
49. Gibson TJ, Lamond AI. Metabolic complexity in the RNA world and implications for the origin of protein biosynthesis. J Molec Evol 1992;30:7–15.
50. Higgins DG, Bleasby AJ, Fuchs R. Clustal V: improved software for multiple sequence alignment. Comput Appl Biochem Sci 1992;8:189–195.

Tracing Biological Evolution in Protein and Gene Structures.
M. Gō and P. Schimmel, editors.

Module structure and function of glutamyl-tRNA synthetase

Masaru Tateno[1,2], Masateru Mizutani[1], Kei Yura[1], Osamu Nureki[3], Shigeyuki Yokoyama[2] and Mitiko Gō[1]

[1]Department of Biology, Faculty of Science, Nagoya University, Nagoya; [2]Department of Biophysics and Biochemistry, School of Science, University of Tokyo, Tokyo; and [3]Crystallography Laboratory, The Institute of Physical and Chemical Research (RIKEN), Saitama, Japan

Abstract. *Background.* A protein module is a contiguous polypeptide segment with a compact conformation within globular domains. We identified the module structure of glutamyl-tRNA synthetase (GluRS) from *Thermus thermophilus* (*T. thermophilus*) and investigated the correlation between the function and the module organization of the synthetase.

Methods. Using computer modeling techniques, a model of $tRNA^{Glu}$ and then a preliminary model of the GluRS and $tRNA^{Glu}$ complex were constructed identifying modules which contribute to the tRNA recognition in the GluRS N-terminal half (domain I, II). To assign function of modules in the GluRS C-terminal half (domain III, IV) molecular dynamics (MD) simulation was performed, since domain IV was too distant from $tRNA^{Glu}$ to bind together in the preliminary docking model.

Results. The MD simulation revealed that the GluRS C-terminal domain (domain IV) could approach and bind to the anticodon, and then form hydrogen-bond networks with it. Most of the amino acid residues interacting with the anticodon were concentrated on modules 31 and 32, thus we termed them "anticodon recognition modules".

Conclusions. Module 31 corresponds to the helix-turn-helix (HTH) motif, and therefore it is suggested that the motif could bind to RNA as well as to DNA. This finding further implies that the HTH module could behave as an active functional segment interacting with RNA in the early stages of the evolutionary processes of life.

Key words: class I aminoacyl-tRNA synthetase, exon shuffling, HTH module, molecular evolution, molecular dynamics.

Introduction

Globular proteins and domains can decompose into several modules. A module is a contiguous polypeptide segment which is a structural unit with a compact conformation [1]. Correspondence of module boundaries with intron positions of the genes encoding the proteins has been detected in hemoglobin [1], lysozyme [2], cytochrome c [3], ovomucoid third domain [3] and triose phosphate isomerase [4,5]. This correspondence suggests that a module was encoded by a single exon. Since exon shuffling [6] changes the combination of modules proteins could have evolved by module shuffling and acquired their own functions.

Address for correspondence: Mitiko Gō, Department of Biology, Faculty of Science, Nagoya University, Chikusa-ku, Furo-cho, Nagoya 464-01, Japan. Tel.: +81-52-789-2976. Fax: +81-52-789-2977.

Aminoacyl-tRNA synthetases (aaRSs) are significant enzymes to investigate the role of modules in the function and evolution of proteins. Aminoacyl-tRNA synthetases strictly recognize and ligate their specific tRNA(s) and amino acid, and thus play the key role in the fidelity of genetic information. The tRNA identity (the specificity of a given tRNA toward an amino acid) depends only on tRNA recognition by their cognate aaRS [7]. Therefore, the evolution of aaRSs must have been of significance in the origin of life through generation of the protein biosynthesis system. From these points of view, it is interesting to examine the correlation between the module structure, which was presumably the structural unit in the protein evolution, and the molecular mechanism of the substrate recognition of aaRSs. One can trace the process through which such strict recognition mechanisms of aaRSs evolved. For this purpose, correlation of the module structure with function of the synthetases was studied and the function-structure relationship was compared among aaRSs.

We identified the module structure of several aaRSs classified into class I whose crystal structures have been solved (M. Mizutani et al., unpublished data). In addition, we identified the module structure of glutamyl-tRNA synthetase (GluRS) (M. Tateno et al., manuscript in preparation), since the crystal structure of GluRS from *Thermus thermophilus* (*T. thermophilus*) has recently been determined by Nureki et al. [8]. In this article, we discuss the correlation between the module organization and the mechanism related to the recognition of the specific tRNA by GluRS.

Since the three-dimensional (3-D) structure of the GluRS N-terminal half (domain I, II) has significant similarity with that of glutaminyl-tRNA synthetase (GlnRS) from *Escherichia coli* (*E. coli*), the crystal structure of which was determined in the form of a complex with the cognate tRNA [9], the molecular mechanism of the tRNA recognition by GluRS was predicted by comparing both structures, and the prediction was confirmed in site-directed mutagenesis experiments [8]. In contrast, the 3-D structure of the GluRS C-terminal half (domain III, IV), which was indicated to be involved in the anticodon recognition, differs from that of GlnRS [8]. For this reason, little has been known in detail about the molecular mechanism of recognition of the anticodon of the cognate tRNA. Accordingly, we investigated the interaction between the GluRS C-terminal half and the anticodon of the cognate tRNA, using molecular modeling techniques coupled to molecular dynamics (MD) simulation (M. Tateno et al., manuscript in preparation). In this article we focus on the anticodon recognition by GluRS and discuss the correlation between the molecular mechanism of the recognition and the module structure.

Identification of the module structure of GluRS from *T. thermophilus*

A module is defined as a compact polypeptide segment contiguous in a globular domain [1]. In other words, a module is a segment in which each amino acid residue is not distant from other residues in the segment. Therefore, a boundary between two successive modules is not distant from the amino acid residues in both of the two modules. Thus, a module boundary is located in the center of a local segment of a globular domain. To detect such local centers, we use centripetal profiles *Fi* [10], which are calculated as the mean square distance between the C_α atom of the *i*th residue and that of the other C_α atoms within the range of a window of $\pm w$ residues (window range) from the *i*th residue. The local minima of *Fi* against *i* are candidates for module boundaries.

The position of a local minimum of the centripetal profile sometimes depends on the window range $\pm w$. An unstable minimum merges into a close stable minimum. A minimum close to the N- and C-terminals of a sequence is neglected even if it is stable, since one cannot always make use of *w* residues around the terminals when doing calculations. The lower limit of module length was set at five. Thus, we assign stable local minima of centripetal profiles to module boundaries. The process is run through automatically on a computer (M. Gō et al., manuscript in preparation).

Using this method, the module structure of GluRS from *T. thermophilus* was identified and the synthetase was found to be composed of 33 modules (N- and C-terminal half is composed of 22 and 11 modules, respectively) (Fig. 1).

Comparison of the module organization of the N-terminal halves between GluRS and GlnRS

Aminoacyl-tRNA synthetases recognize characteristic nucleotides of the cognate tRNA(s) (known as identity determinants), resulting in discrimination of the cognate tRNA(s) [11–15]. We tried to assign the modules in the GluRS N-terminal half (domain I, II) which contribute to recognition of determinants of $tRNA^{Glu}$.

Though the overall 3-D structures of the N-terminal halves of GluRS and GlnRS are remarkably similar, there are slight differences in their 3-D structures and in their mode of discriminating their specific substrates [8]. To determine how differences in their 3-D structures correlate with those of their module organizations, we superimposed the 3-D structures by calculating root mean square deviation (RMSD) of the best fit between them. As a result, about three-fourths of the modules of the N-terminal half (domain I, II) of GluRS are shared with GlnRS. On the other hand, most of the unshared and thus unique modules contribute to the tRNA recognition and discriminate the specific tRNA. Such unique modules were presumably inserted/deleted or recombined in the evolutionary processes of the synthetases corresponding to the position of the identity determinants of their specific tRNA(s), so as to strictly recognize them.

Through a comparison of the module organization between the closely related aaRSs one can understand not only molecular mechanisms of tRNA recognition but also trace plausible evolutionary processes leading to acquisition of recognition mechanisms. However, one cannot apply a similar method to anticodon recognition by the GluRS C-terminal half (domain III, IV), since its 3-D structure differs from that of GlnRS. Accordingly, using molecular modeling techniques coupled with molecular dynamics (MD) simulation, we investigated interactions between the GluRS C-terminal half and the anticodon of the cognate tRNA (M. Tateno et al., manuscript

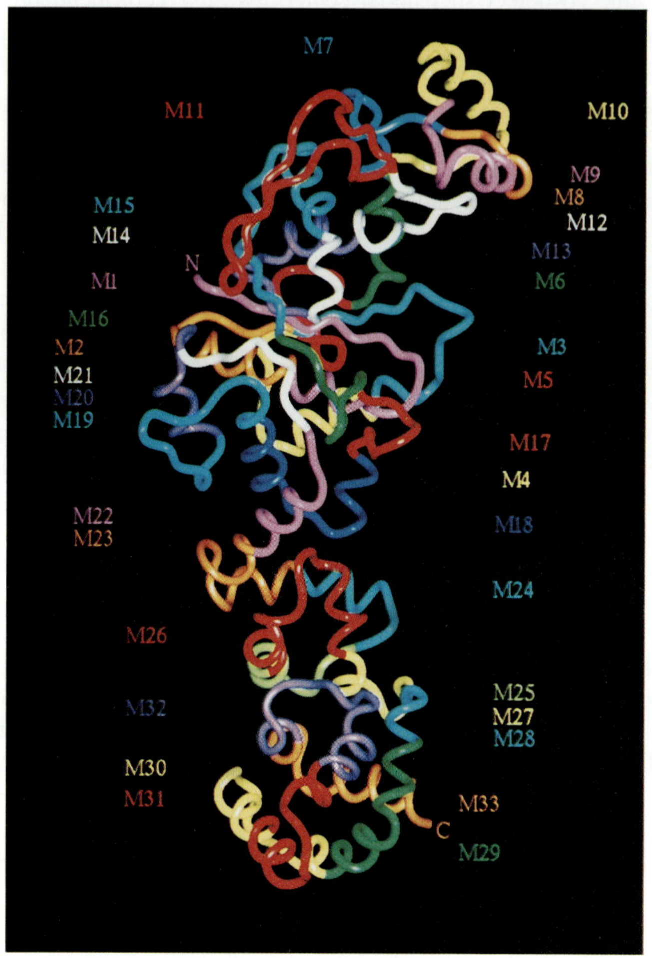

Fig. 1. Module structure of GluRS. Each module is painted in different colors.

in preparation). We constructed the starting structure of the GluRS and $tRNA^{Glu}$ complex for analysis.

Preliminary docking model of GluRS and $tRNA^{Glu}$

Since the 3-D structure of the GluRS N-terminal half (domain I, II) is similar to that of GlnRS, we superimposed the C_α atoms of the N-terminal halves between GluRS and GlnRS by the method of Rossmann and Argos [16].

A model of $tRNA^{Glu}$ was initially constructed using the crystal structure of yeast $tRNA^{Phe}$ as its template and was partly modified by molecular modeling coupled with distance-constrained MD calculations. The phosphates of the modeled $tRNA^{Glu}$ were then superimposed on those of GlnRS-bound $tRNA^{Gln}$.

Thus, we obtained the starting structure of the GluRS and $tRNA^{Glu}$ complex (Fig. 2A). In this stage there is no serious steric overlap between atoms of the two molecules, such as crossing over of two chemical bonds. After the structure was carefully optimized by several steps of energy minimization, there were no amino acids and nucleotides which locate at local minima with high energy. In fact, the values of all energy terms are low, especially the van der Waals energy and the electrostatic energy. The total energy of the structure is also remarkably low. In addition, RMSD for atoms of each residue between the starting and the minimized structures of the modeled complex are significantly small (0.12 Å). This means that overlaps between atoms of GluRS and $tRNA^{Glu}$ could be reduced with little conformational change in each residue.

According to site-directed mutagenesis experiments [8], identity determinants of the $tRNA^{Glu}$ molecule are present in the acceptor stem, D-arm and anticodon. In protection experiments using *N*-nitroso-*N*-ethylurea, the minor groove side of the D-stem helix is strongly protected by GluRS (S. Sekine et al., submitted to J Molec Biol), and it binds tightly to the synthetase. The contact region of the D-stem of $tRNA^{Glu}$ with GluRS in the modeled structure is in good agreement with the experimental result. In addition, amino acid residues of GluRS which were indicated to interact with $tRNA^{Glu}$, e.g., Ser276, Glu282, Ser299, Lys309 and Trp312 [8], make contact with the identity determinants of $tRNA^{Glu}$. In this way it was shown that the

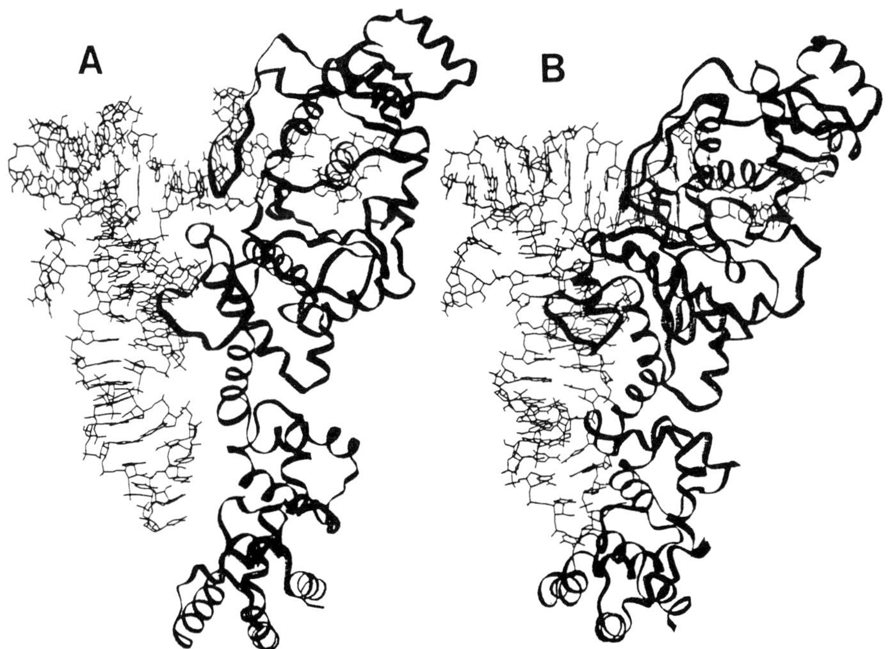

Fig. 2. A: The starting structure of the docking model of GluRS-$tRNA^{Glu}$. B: The structure obtained by MD calculation on the GluRS C-terminal half and the anticodon of $tRNA^{Glu}$.

model of the GluRS and $tRNA^{Glu}$ complex is consistent with the experimental results with no contradiction.

By analyzing this model one can investigate mechanisms of recognition of other identity determinants in the $tRNA^{Glu}$ molecule at the atomic resolution. However, the acceptor stem, which is included in the GluRS recognition sites, did not take an active form in the model, and the -CCA extremity of $tRNA^{Glu}$ did not reach the catalytic site of GluRS. Accordingly, we carried out another computer modeling, in which we investigated the interaction between GluRS and identity determinants in the acceptor stem of $tRNA^{Glu}$, using the model mentioned above and the coordinates of GlnRS-bound $tRNA^{Gln}$ (M. Tateno et al., manuscript in preparation). According to results of this modeling, we can explain differences in mechanisms of the tRNA recognition between GluRS and GlnRS. As for the other site of identity determinants in the $tRNA^{Glu}$ molecule, anticodon, one cannot elucidate recognition mechanisms of it by GluRS using the model mentioned above since the anticodon was too distant from GluRS to bind.

According to another site-directed mutagenesis experiment, the GluRS variant whose Lys426 is replaced with Gln (K426Q) had a 50 times higher Michaelis constant (Km) for $tRNA^{Glu}$ than did the wild-type of GluRS. Lys426 is in the C-terminal domain (domain IV) of GluRS and it confronts the anticodon, though they locate at the distance of more than 10 Å, in the model. Therefore Lys426 is presumed to contribute to the anticodon recognition by the conformational change of GluRS and/or $tRNA^{Glu}$. Interestingly, GluRS has a helix-turn-helix (HTH) motif in the C-terminal domain (domain IV) [8] which corresponds well to module 31 (i.e., HTH module) and Lys426 belongs to this module. The 3-D structure of the HTH module of GluRS is remarkably similar to that of the 434Cro protein. We calculated the RMSD between the two HTH modules superimposing the C_α atoms. The value is 0.43 Å, therefore they show a striking resemblance (M. Tateno et al., manuscript in preparation).

How does Lys426 in the GluRS C-terminal domain interact with the anticodon of $tRNA^{Glu}$? Does the HTH module (module 31) contribute to the anticodon recognition? If so, how is it related to evolution of GluRS and the HTH module? We made use of molecular dynamics (MD) simulation to investigate such problems.

Molecular dynamics simulation of the GluRS C-terminal half and the anticodon of $tRNA^{Glu}$

To prepare an interaction model between GluRS and the anticodon of $tRNA^{Glu}$, we searched the conformational space of the GluRS and $tRNA^{Glu}$ complex by performing MD simulation of 50 ps in a vacuum at 300 K (M. Tateno et al., manuscript in preparation). Atoms of the GluRS C-terminal half (domain III, IV; 306-468) and the anticodon (pU34U35C36p) of $tRNA^{Glu}$ were allowed to move but the others were fixed. No other constraints, such as distance constraints, were included in this simulation. All the calculations of energy minimization and MD simulation were carried out using AMBER 4.0 [17].

During the simulation, the GluRS C-terminal domain (domain IV) did rotate around the hinge which links it to the upper domain (domain III) (Fig. 2B), and also a conformational change of the anticodon nucleotides was induced with the C-terminal domain (domain IV) (Fig. 3). Since GluRS has a basic patch in the domain, this domain motion is presumed to be caused by electrostatic interactions between the basic residues of the domain and the phosphates of $tRNA^{Glu}$.

The anticodon nucleotides were taken up into pockets of GluRS (Fig. 4) and formed hydrogen bonds with the amino acid residues (Fig. 5). Interestingly, the HTH module of GluRS surrounds and binds to U34 (the first letter of the anticodon) (Fig. 6). Lys426, which located in the HTH module and was considered to contribute to the anticodon recognition, formed a hydrogen bond with the phosphate of U34 (Fig. 3).

Since RMSD among structures of 30 to 50 ps in the interval of 100 fs were small, the complex was confirmed to have reached a dynamically stable state. We performed energy minimization of the 50 ps MD structure and searched a minimized structure around it. Since the bond energy of the complex was slightly higher than that of the starting structure, we analyzed the Ramachandran map of the modeled GluRS and found no amino acid residues with high-energy conformations. Though conformational change of the anticodon nucleotides slightly increased the energy, this also holds for the case of GlnRS-bound $tRNA^{Gln}$.

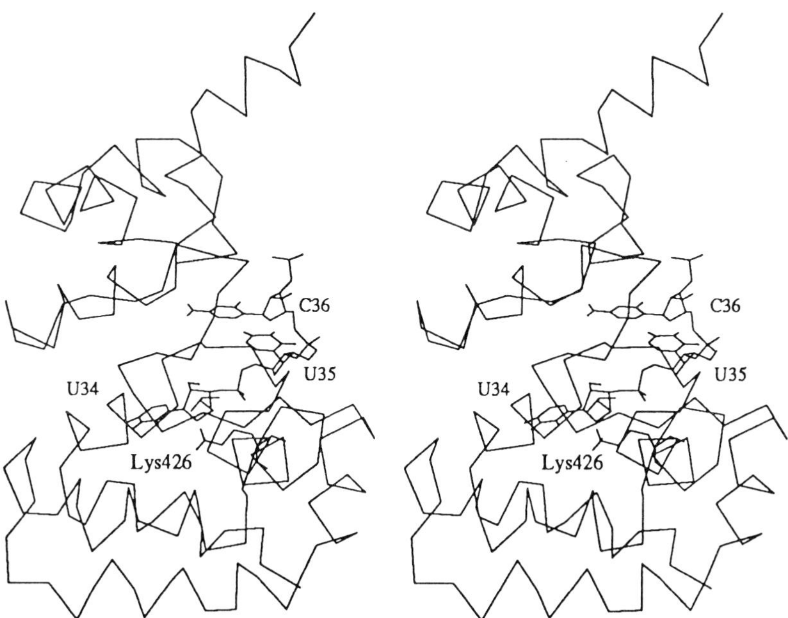

Fig. 3. Predicted interaction between the GluRS C-terminal half and the anticodon nucleotides of $tRNA^{Glu}$. Lys426 is predicted to interact with the anticodon, based on experimental results. It binds to the phosphate of U34.

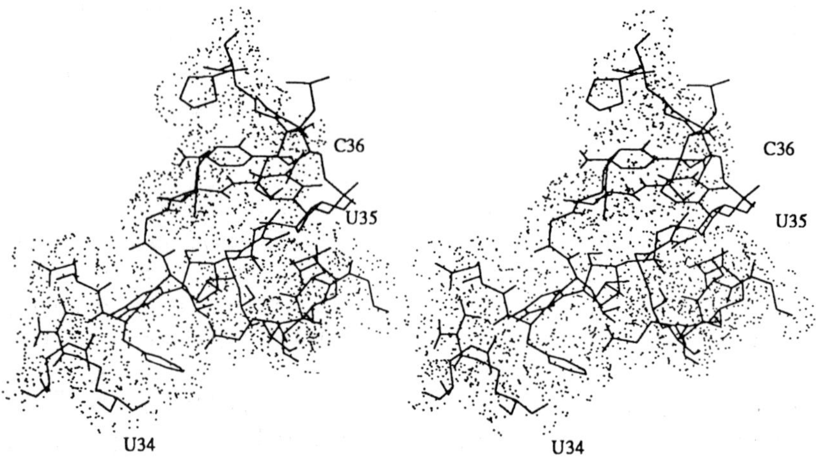

Fig. 4. Anticodon nucleotides bind into pockets of GluRS and form a hydrogen-bond network with amino acid residues.

Interaction between GluRS and the anticodon of tRNA^{Glu} proposed by the MD simulation

In the modeled structure of the complex anticodon nucleotides are bound to the pocket of GluRS and form a hydrogen-bond network with the amino acid residues. Some of the hydrogen bonds are specific to the bases and seem to stabilize the complex of GluRS and the cognate tRNA. Through this scheme we realized how GluRS recognizes and discriminates the anticodon of the specific tRNA from others.

We also investigated the relationship between the module organization and the amino acid residues that interact with the anticodon nucleotides. Such residues are located mainly on modules M31 and M32, though a few are present on other modules. Therefore, we termed the two modules "anticodon-recognition modules". Since module M31 belongs to the HTH module, it is suggested that the HTH module could bind to RNA as well as to DNA. As far as we know, this is the first time to propose not only the possibility but also a bound state between RNA and the HTH module at the atomic resolution.

Evolutionary implication of the anticodon recognition module of GluRS

What does it mean that the HTH module of an "old" protein such as GluRS could bind to RNA? Aminoacyl-tRNA synthetase is an enzyme essential for the protein biosynthesis, and therefore, for life. Standing on the ground, aaRSs could have already existed and functioned in early stages of the evolutionary processes of life, i.e., the RNA world, when genes were composed of RNA. The HTH module of an old protein such as GluRS could be its origin. Actually, through computer modeling

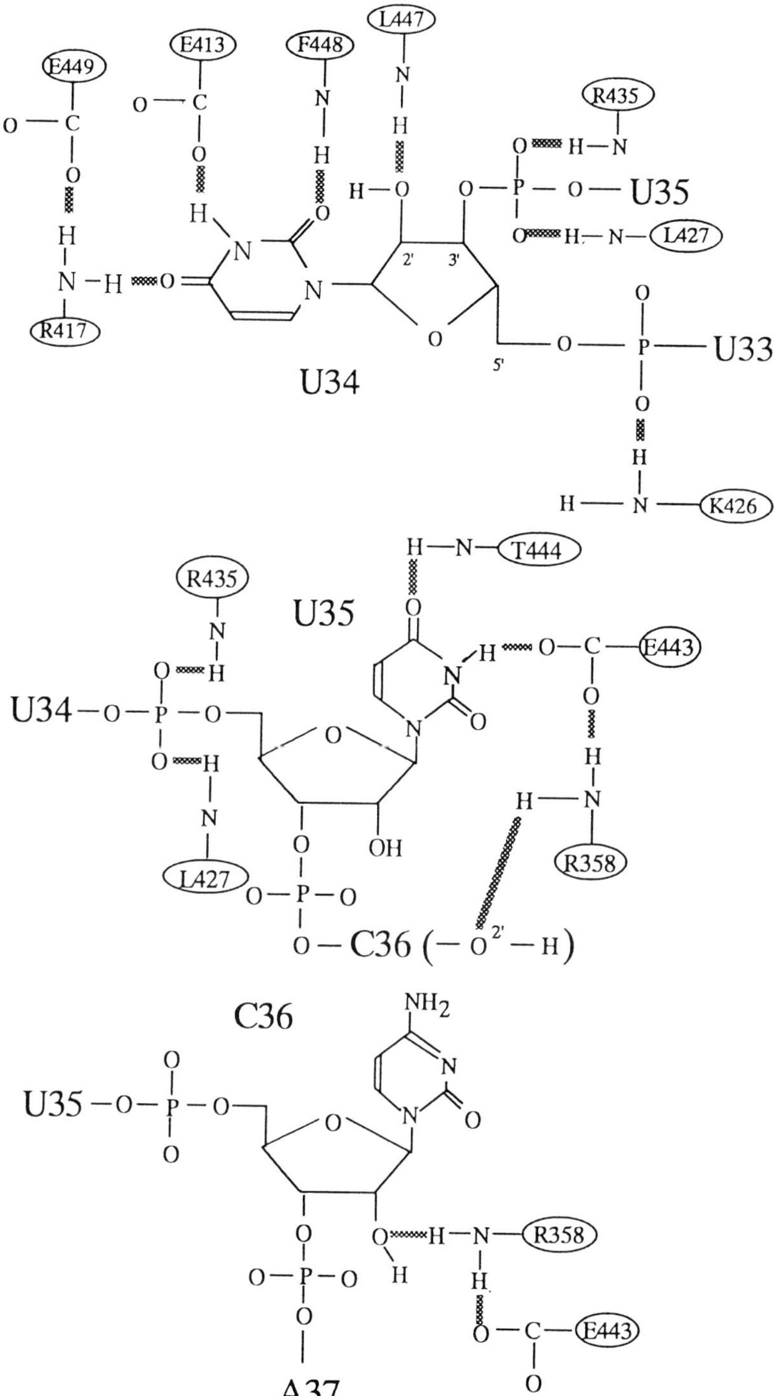

Fig. 5. Schematic drawing of the hydrogen-bond networks around each anticodon nucleotide.

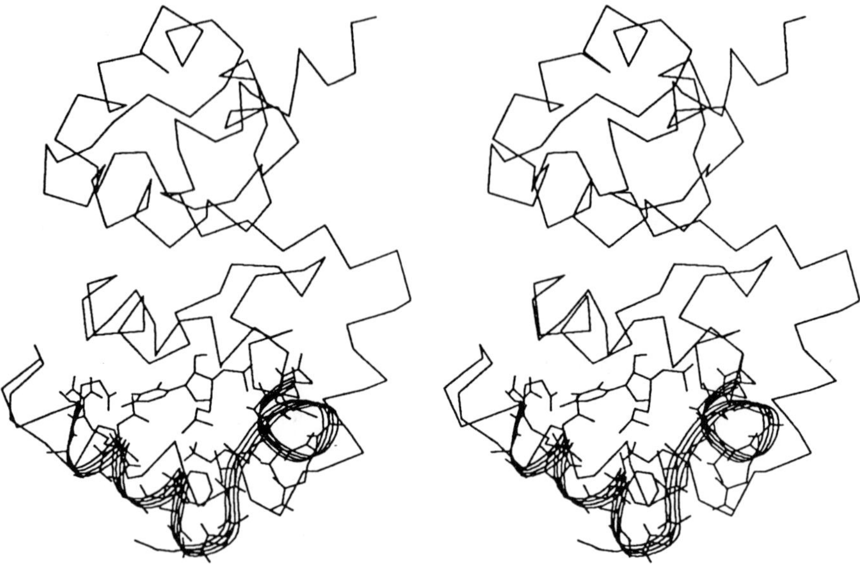

Fig. 6. The helix-turn-helix motif (helix-turn-helix module (M31)) of GluRS binds to U34 (the first letter of the anticodon).

of the interaction between GluRS and the anticodon of $tRNA^{Glu}$, we found that the module of GluRS could bind to the anticodon of $tRNA^{Glu}$ and have an important role for the anticodon recognition by the synthetase. The view that the HTH module could bind to RNA and behave as an active functional segment in the RNA world is thus given support.

Investigation of the molecular mechanism of the tRNA recognition by aaRSs can also contribute to studies on their molecular evolution. Recently, Mirande et al. proposed a hypothesis of horizontal gene transfer, namely that the genes of eukaryotic GlnRSs have a prokaryotic origin acquired by genetic exchange between the primary kingdoms [18]. By elucidating the recognition mechanism of $tRNA^{Glu}$ by GluRS and by comparing it with the case of GlnRS, it is expected to shed light on such questions of molecular evolution.

Acknowledgements

We thank M. Ohara for critical comments on the manuscript. Part of this work was supported by Grants-in-Aid from the Ministry of Education, Science and Culture of Japan to K.Y. and M.G.

References

1. Gō M. Correlation of DNA exonic regions with protein structural units in haemoglobin. Nature

1981;291:90–92.
2. Gō M. Modular structural units, exons, and function in chicken lysozyme. Proc Natl Acad Sci USA 1983;80:1964–1968.
3. Gō M. Protein structures and split genes. Adv Biophys 1985;19:91–131.
4. Gilbert W, Marchionni M, McKnight G. On the antiquity of introns. Cell 1986;46:151–154.
5. Gō M, Nosaka M. Protein architecture and the origin of introns. Cold Spring Harbor Symp Q Biol 1987;52:915–924.
6. Gilbert W. Why genes in pieces? Nature 1978;271:501.
7. Normanly J, Ogden RC, Horvath SJ, Abelson J. Changing the identity of a transfer RNA. Nature 1986;321:213–219.
8. Nureki O, Vassylyev DG, Katayanagi K, Shimizu T, Sekine S, Kigawa T, Miyazawa T, Yokoyama S, Morikawa K. Architecture of class-defining and specific domains of glutamyl-tRNA synthetase. Science 1995;267:1958–1965.
9. Rould MA, Perona JJ, Soll D, Steitz TA. Structure of *E. coli* glutamyl-tRNA synthetase complexed with $tRNA^{Gln}$ and ATP at 2.8 Å resolution. Science 1989;246:1135–1142.
10. Noguti T, Sakakibara H, Gō M. Localization of hydrogen-bonds within modules in barnase. Proteins 1993;16;357–363.
11. Yarus M. tRNA identity: a hair of the dogma that bit us. Cell 1988;55:739–741.
12. Normanly J, Abelson J. tRNA identity. Ann Rev Biochem 1989;58:1029–1049.
13. Schimmel P. Parameters for the molecular recognition of transfer RNAs. Biochemistry 1989;28: 2747–2759.
14. Schulman LH. Recognition of tRNAs by aminoacyl-tRNA synthetases. Prog Nucl Acid Res Molec Biol 1991;41:23–87.
15. Giegé R, Puglisi JD, Florentz C. Recognition of tRNAs by aminoacyl-tRNA synthetases. Prog Nucl Acid Res Molec Biol 1993;45:129–206.
16. Rossman MG, Argos P. Exploring structural homology of proteins. J Molec Biol 1975;250:7525–7532.
17. Pearlman DA, Case DA, Caldwell JC, Seibel GL, Singh UC, Weiner P, Kollman PA. AMBER 4.0, University of California, San Francisco.
18. Lamour V, Quevillon S, Diriong S, N'Guyen VC, Lipinski M, Miranda M. Evolution of the Glx-tRNA synthetase family: the glutaminyl enzyme as a case of horizontal gene transfer. Proc Natl Acad Sci USA 1994;91:8670.

Tracing Biological Evolution in Protein and Gene Structures.
M. Gō and P. Schimmel, editors.

Ribosomal RNA in pieces: a modern paradigm of the primordial ribosome

Michael W. Gray, Spencer J. Greenwood, Donald S. Smallman, David F. Spencer and Murray N. Schnare
Program in Evolutionary Biology, Canadian Institute for Advanced Research, Department of Biochemistry, Dalhousie University, Halifax, Nova Scotia, Canada

Abstract. We argue here that small RNA components, associating via intermolecular base-pairing interactions, must have preceded the structurally and functionally complex large subunit (LSU) and small subunit (SSU) rRNA molecules that exist in contemporary ribosomes. We describe a stepwise model for the generation of present-day rRNA molecules from their evolutionary antecedents. The stages we envisage in the evolutionary transformation of many small rRNAs to a few large rRNAs are:
Stage I. Physically and transcriptionally separate rRNA coding modules giving rise to individual pre-rRNA transcripts that are independently processed.
Stage II. Physically and transcriptionally linked rRNA coding modules arrayed in a nonstandard transcriptional order, with extensive processing of the resulting long co-transcripts.
Stage III. Physically and transcriptionally linked rRNA coding modules arranged in a standard order, again with extensive processing of long co-transcripts.
Stage IV. Physically and transcriptionally linked rRNA coding modules arrayed in a standard pattern, with limited posttranscriptional processing.
We have studied in detail the naturally fragmented SSU and/or LSU RNAs in a number of unicellular eukaryotes (protists), and we suggest that these discontinuous rRNAs provide a useful modern paradigm of rRNA evolution. Each of the cases we present typifies one of the above-postulated stages in the evolutionary transformation of "split" to "unsplit" rRNAs. Distinctive and novel features of rRNA gene arrangement and expression in each of these cases are described. We propose that these modern examples of split rRNAs recapitulate various stages in the evolution of large rRNAs from smaller components, thereby supporting the overall feasibility of the proposed model.

Key words: discontinuous rRNA, evolution, fragmented rRNA, RNA world, split rRNA.

Introduction

At the heart of translation is the ribosome, the multicomponent ribonucleoprotein complex that is the cellular site of protein biosynthesis. As evidence has mounted that many of the key interactions (and perhaps even reactions) of translation may be mediated by the RNA components of the ribosome [1–4], there has been an upsurge

Address for correspondence: Michael W. Gray, Department of Biochemistry, Sir Charles Tupper Medical Building, Dalhousie University, Halifax, Nova Scotia B3H 4H7, Canada. Tel.: +1-902-494-2521. Fax: +1-902-494-1355. E-mail: mgray@ac.dal.ca

of interest in the structure, function and evolution of the large subunit (LSU) and small subunit (SSU) rRNAs. Comparative analysis of primary sequence and potential secondary structure has identified several distinctive features of LSU and SSU rRNA molecules that need to be accommodated in any scheme describing their origin and evolution. These key features include the following:

1. *Large size.* Ribosomal RNA molecules are long, covalently continuous polynucleotide chains, e.g., *Escherichia coli* (*E. coli*) 16S and 23S rRNAs, which are often regarded as archetype SSU and LSU rRNAs, are 1,542 and 2,904 nucleotides in length, respectively. Most eukaryotic nucleocytoplasmic rRNAs are even longer than their prokaryotic counterparts; certain mitochondrial homologs are shorter, but still many hundreds of nucleotides long [5].
2. *Interspersed pattern of conserved and variable regions.* Regions that are conserved in both primary sequence and secondary structure are flanked by regions that are extremely variable in size and structure [1,6–8].
3. *Highly conserved long-range base-pairing interactions.* These pairings bring together noncontiguous regions that are distant in the primary sequence [1,6,9], thereby forming a universal core of higher order structure [10,11].
4. *Structural and functional complexity.* The long-range base-pairing interactions define a characteristic domain structure [1,9], with particular functional interactions being associated with particular structural domains of the rRNA [2,3].

In considering how the translation apparatus must have evolved, we assume that it did not arrive full-blown upon the evolutionary scene, but that it must have started small and simple, with incremental increases in the structural and functional complexity that underlie the current accuracy and efficiency of the system. In the same vein, we consider that the primordial ribosome did not possess the structural and functional complexity of the contemporary version, but was a substantially simpler entity. In keeping with the view of an ancient RNA world (see [12]), and the recognition of the importance of rRNA in ribosome function [1–4], it has been suggested that the original ribosome may have consisted largely or solely of RNA [13–16], with ribosomal proteins being added later in evolution. In this context, the issue of origin and evolution of the ribosome is reduced to a question of origin and evolution of rRNA.

Modular construction of ribosomal RNA molecules: split rRNAs

As noted above, SSU and LSU rRNA molecules display a characteristic interspersed pattern of conserved and variable regions [1,6–8], with the conserved regions being brought together by intramolecular base-pairing interactions to form a universal secondary structure core. Within this core occur virtually all of the functional interactions between the ribosome and other translation components [1–3]. Herein, too, lies the peptidyl transferase center (PTC), the region of LSU rRNA implicated in peptide bond synthesis [2]. Virtually all of the posttransciptionally altered nucleotides in both SSU and LSU rRNA (pseudouridine, base-methylated and sugar-methylated residues) are also found within the universal core of these rRNAs [1,17], suggesting an important (if still obscure) role for these modified components in rRNA and ribosome function.

Although prokaryotic (eubacterial and archaebacterial) SSU and LSU rRNAs are typically single polynucleotide chains, with *E. coli* 16S and 23S rRNAs serving as exemplars in this regard, there are a number of instances in which SSU and LSU rRNAs are discontinuous (split). Perhaps the most widely recognized case of a split LSU rRNA is the 5.8S:28S complex in the cytoplasmic ribosome of virtually all eukaryotes. The 5.8S rRNA component is homologous with the first ~160 residues of *E. coli* 23S rRNA, with the 28S rRNA species being equivalent to the remainder of its *E. coli* 23S rRNA counterpart [18–21]. In eukaryotes, an additional internal transcribed spacer (ITS), absent in the *E. coli* 23S rRNA precursor [22,23], separates the 5.8S and 28S coding regions in the pre-rRNA, and is excised during posttranscriptional processing of the latter [24]. This extra processing step results in the physical separation of the 5.8S module from the rest of the LSU rRNA structure (the 28S rRNA). Thus, intramolecular base-pairing interactions between the 5.8S and 28S rRNA regions in the pre-rRNA are converted by ITS excision to intermolecular base-pairing interactions that precisely mimic the intramolecular interactions between the corresponding regions in *E. coli* 23S rRNA [25]. A number of other considerably more extreme cases of split rRNAs have now been documented in the literature, and some of these are considered below.

When the naturally occurring breaks in SSU and LSU rRNAs are mapped to their respective secondary structures, a striking picture emerges, in that discontinuities in rRNA primary sequence are localized to those regions that are designated as variable on the basis of primary and secondary structure comparisons [26]. These variable regions, which may or may not be covalently continuous, effectively define individual modules of rRNA structure. As we shall see, examples in which rRNA genes are fragmented and rearranged at the level of the genome reinforce this concept of a modular pattern of rRNA structure.

Observations gleaned from naturally fragmented rRNAs are important for two reasons: 1) they indicate that an rRNA molecule need not be covalently continuous in order to be functional; and 2) they demonstrate that fragments of rRNA are able to form a noncovalent network that preserves the conserved core of a covalently continuous homolog. It should be emphasized, however, that the locations at which phosphodiester breaks occur within an rRNA molecule are of critical importance. Discontinuities within variable regions of SSU and LSU rRNAs appear to be tolerated with few or no functional consequences. In marked contrast, even single breaks within the conserved core may be sufficient to inactivate ribosome function [27].

A stepwise model of ribosomal RNA evolution

It seems highly unlikely that long polyribonucleotide chains could have existed at the earliest stages of biochemical evolution, yet it is equally improbable that a process such as translation would have made its appearance only after long, covalently continuous rRNA molecules had evolved. The advantage of a modular rRNA structure is that a functioning ribosome could have been assembled at an early stage in evolution through the noncovalent interaction of the various separate building blocks that would eventually become incorporated into contemporary rRNA molecules.

Over the past decade, we have identified and characterized a number of naturally fragmented SSU and LSU rRNAs and have investigated the organization and expression of their coding units and the processing of the resulting transcripts. As a result of these studies, some of which are summarized below, we have developed a stepwise model for the evolution of SSU and LSU rRNA molecules [26,28–30]. The essential basis of this model is the assumption that the primordial ribosome was composed of a number of separate small RNAs that had the capacity to associate, through complementary base-pairing interactions, to form a noncovalent RNA network. Through such interactions, a functional core of secondary and tertiary structure, equivalent to that in a contemporary ribosome, could be established. The model outlines a series of stages by which many small rRNAs could have been converted, over the course of evolution, to a few large rRNAs, i.e., the model describes how split rRNAs, characterized by intermolecular interactions, could have been transformed into unsplit rRNAs, characterized by intramolecular interactions.

The various stages in the model are shown in Fig. 1 and their salient characteristics are summarized in Table 1.

Stage I: physically and transcriptionally separate rRNA coding modules, individual pre-rRNA transcripts, limited posttranscriptional processing

We postulate that the earliest functioning ribosome would have comprised or contained a number of small RNA components whose intermolecular interactions (complementary base-pairing) served to establish a functional core within which peptide bond synthesis could be accomplished. In effect, each of these RNA components would comprise a separate module of rRNA structure and would be specified by and transcribed from an independent coding module at the level of the genome. Each transcript would be independently processed by removal of extraneous sequences at its 5′- and/or 3′-ends.

Stage II: physically and transcriptionally linked rRNA coding modules, nonstandard transcriptional order, extensive processing of long co-transcripts

At a later stage in rRNA evolution we envisage the gathering together at the genome level of the various rRNA coding modules, under the control of one or a few promoters. The evolutionary pressure driving the system towards such physical linkage could conceivably have been a requirement to produce interacting small RNAs in stoichiometric amounts, thereby minimizing or avoiding the loss of rate-limiting components into nonproductive complexes. In contrast to stage I, rRNA coding modules are assumed to have been physically and transcriptionally linked at

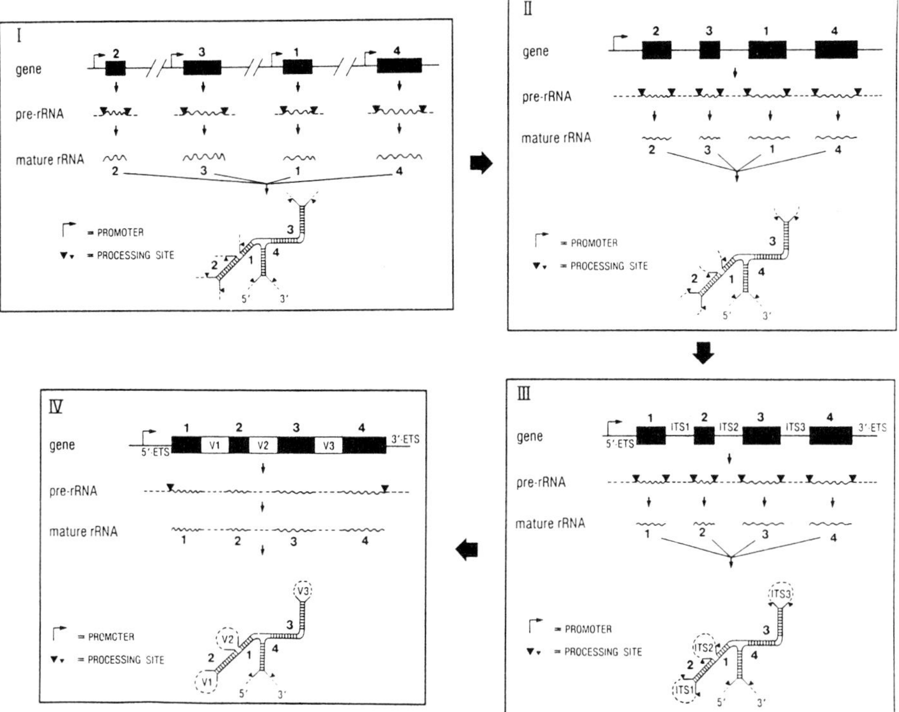

Fig. 1. Proposed model for the evolutionary conversion of split rRNA genes to continuous ones (see [30]). The figure shows a hypothetical rRNA composed of four initially separate modules (solid rectangles), numbered in the transcriptional order (5′ to 3′) in which these modules later became organized. Solid lines denote core regions of the mature rRNAs and their precursors, whereas dashed lines denote pre-rRNA sequences that are either eliminated during processing (I–III) or that become incorporated into the mature rRNA as variable regions in the course of evolution (IV). A detailed description of the various stages is given in the text and elsewhere [29,30]. Abbreviations: V = variable region; ETS = external transcribed spacer; ITS = internal transcribed spacer.

stage II; however, they need not have been in the same transcriptional orientation as in contemporary, conventional rRNA genes. At stage II, transcription would have been accompanied or followed by extensive processing of the resulting pre-rRNAs; processing sites are assumed to be those originally associated with the transcripts of the individual modules in stage I. Thus, the end result of expression and posttranscriptional processing is basically the same in stages I and II: a collection of small rRNA modules that interacted noncovalently.

Stage III: physically and transcriptionally linked rRNA coding modules, standard transcriptional order, extensive processing of long co-transcripts

The transition from stage II to stage III is characterized by rearrangement of the coding modules in the genome, such that the modules are now arrayed in the standard

Table 1. Characteristics of postulated stages in rRNA evolution.

Stage	Physical linkage?	Transcriptional linkage?	Standard order?	Processing
I	no	no	–	limited
II	yes	yes	no	extensive
III	yes	yes	yes	extensive
IV	yes	yes	yes	limited

5′ to 3′ order. The driving force for such a rearrangement could have been the development of a close coupling between transcription, processing and ribosome assembly, such that ordered transcription of modules became essential. Extensive posttranscriptional processing would still have occurred at this stage to generate a collection of small rRNAs similar to that seen at stages I and II. Thus, although the pathways by which the rRNA products are assumed to have been generated are substantially different at these three stages, the form of the resulting mature rRNAs is basically the same.

Stage IV: physically and transcriptionally linked rRNA coding modules, standard transcriptional order, limited processing of long co-transcripts

It is at this stage that separate coding modules of rRNA structure are melded together at the genome level, as a result of loss of processing sites and/or processing enzymes involved in the posttranscriptional removal of ITSs at stage III. Inability to eliminate a particular ITS would leave that sequence as part of the mature rRNA, converting it to a nonexcised variable region. Such a variable region would effectively constitute a covalent link between two originally separate rRNA components (modules). Sporadic loss of processing sites/enzymes would lead to the progressive conversion of ITSs to variable regions, and the transition from a collection of many small rRNA components (and intermolecular interactions) to a few larger rRNA species (and intramolecular interactions).

Contemporary examples of split rRNAs and rearranged rDNA coding modules

The above model, while largely speculative, provides a plausible explanation for how contemporary rRNA molecules could have evolved. In developing the model, we have drawn upon present-day examples of unusual rRNA structures and rDNA arrangements that exemplify the various stages outlined above. To further illustrate the model, examples of stage III, II and I rRNA systems will now be described briefly, in that order (i.e., from more to less conventional).

Stage III

Two examples of a stage III arrangement are provided by the nuclear DNA-encoded cytoplasmic LSU rRNAs of the trypanosomatid protist, *Crithidia fasciculata* (*C. fasciculata*) [31], and the euglenoid protist, *Euglena gracilis* (*E. gracilis*) [32,33]. In these cases, additional ITSs are present in the rDNA, compared to other eukaryotes, and these are localized in the LSU rRNA coding region. In *C. fasciculata* five additional ITSs give rise to a seven-component LSU rRNA, one component being the 5.8S rRNA homolog and the remaining six being equivalent to 28S rRNA [31]. In *E. gracilis* 12 additional ITSs result in a 14-component LSU rRNA, the most extensively fragmented LSU rRNA characterized so far [32,33].

When mapped within their respective LSU rRNA coding regions, the individual mature rRNA fragments are found to be in the same relative transcriptional order as their counterparts in a conventional LSU rRNA gene. In other words, at the genome level, the *C. fasciculata* and *E. gracilis* nuclear rDNA units appear fundamentally similar in organization to those of other eukaryotes. Detailed examination of rRNA processing in *E. gracilis* (S.J. Greenwood, unpublished) has identified a high-molecular-weight precursor that contains the coding regions of all of the mature SSU and LSU rRNA components, as well as of 5′ and 3′ external transcribed spacers (ETSs). In this respect, rDNA transcription is similar to other eukaryotic nuclear rDNA systems, in that a long precursor is initially synthesized; only at the level of posttranscriptional processing, with the removal of many novel ITSs, do the differences between the *E. gracilis* and *C. fasciculata* systems, on the one hand, and other eukaryotic systems become manifest.

Secondary structures for the *C. fasciculata* [31] and *E. gracilis* [32] LSU rRNA complexes have been proposed; in these, individual rRNA fragments are seen to be associated through the conserved long-range interactions characteristic of the folding of a covalently continuous LSU rRNA. By subjecting isolated *E. gracilis* rRNA samples to controlled denaturation, it has been possible to adduce evidence consistent with the proposed pairing scheme. It has also been possible to reconstitute, from mixtures of individual rRNA species, partial complexes that correspond to discrete LSU rRNA structural domains. Such experiments (D.S. Smallman, unpublished) demonstrate that rRNA fragments have the potential, through intermolecular interactions, to form higher order structures that are normally established through intramolecular pairings between the same sequences.

Stage II

This stage is characterized by a nonstandard ordering of modules of rRNA structure at the genome level. The first described example of a stage II arrangement is the mitochondrial LSU rRNA gene in the ciliate protist, *Tetrahymena pyriformis* (*T. pyriformis*). In *T. pyriformis* mitochondria both the SSU and LSU rRNAs are bipartite, consisting of a small (α) fragment and a much larger (β) fragment [34,35]. In both cases the α fragment corresponds to the first 200 (SSU) to 280 (LSU) nucleotides of the respective rRNA, and in both cases the α and β fragments have the capacity to base pair with one another, with the discontinuity between the two rRNA pieces being localized to a variable region in the rRNA secondary structure.

In *T. pyriformis* mtDNA, there is a single copy of the SSU rRNA gene but two copies of the LSU rRNA gene, located in subterminal inverted repeats near the ends of the linear molecule (see [35]). The organization of the SSU rDNA is conventional in the sense that the SSUα- and SSUβ-coding modules are transcriptionally oriented as expected (i.e., α is upstream of β) [34]. The rDNA therefore contains a novel ITS whose processing at the level of the pre-rRNA generates the mature SSUα and SSUβ rRNAs. In contrast, both copies of the LSU rDNA have an unusual organization, in that the LSUα module is located downstream of the LSUβ one in the direction of transcription, and is separated from it by a $tRNA^{Leu}$ gene [35]. Northern hybridization experiments suggest that the entire LSU rDNA region is transcribed into a high-molecular-weight precursor containing β- and α-coding regions, with subsequent processing to yield the individual rRNAs [35]. We have suggested a processing scheme in which the LSUα portion folds back upon and base pairs with the LSUβ portion, with subsequent excision of the intervening $tRNA^{Leu}$ and flanking spacer regions [36]. In this way pairing between LSUα (which is equivalent to, but longer than, 5.8S rRNA) and LSUβ could occur intramolecularly to generate the expected contacts; however, such pairing would take place late in the transcription process, whereas the analogous interactions normally occur early in transcription.

An even more dramatic example of a stage II arrangement is seen in the 15.8 kbp linear mtDNA of the green alga, *Chlamydomonas reinhardtii* (*C. reinhardtii*). Here, both SSU and LSU rRNA genes are fragmented, with subgenic modules of each being rearranged and interspersed with one another and with protein coding and tRNA genes over a 6-kbp region of the mtDNA [37–39]. Because all of the rDNA modules and other genes are located on the same DNA strand [38], the entire rDNA region has the potential to be transcribed from a single promoter. In fact, RNA mapping experiments have identified precursors and abutting transcripts that cover this region, indicating that long co-transcripts are indeed synthesized, with mature products being generated by precise endonucleolytic cleavages [37–39]. Northern hybridization analyses have revealed abundant small transcripts whose sizes approximate those of the rDNA coding modules evident in the genome; i.e., no splicing of mature rRNA pieces occurs [38,39]. As in other cases of discontinuous rRNAs, it is possible to model networked secondary structures for the SSU and LSU rRNA pieces; these contain all of the expected long-range base-pairing interactions and universal core regions, with breaks between the various interacting species falling in regions that are normally variable [38]. A similar situation has been described for the mitochondrial rRNAs and their genes in another *Chlamydomonas* species, *C. eugametos* [40]. In this case, however, although all SSU and LSU rDNA modules are also encoded by the same strand, their order and pattern of interspersion are quite different than what is seen in *C. reinhardtii* mtDNA.

It is not yet known how the mitochondrial ribosome in *C. reinhardtii* and *C. eugametos* is assembled. It is theoretically possible that components of the SSU and LSU rRNAs begin to interact while still being part of long co-transcripts, before these are processed to the individual mature RNAs. However, because SSU and LSU rRNA coding sequences are scrambled and are not only interspersed with each other but with protein coding and tRNA sequences as well, it is difficult to devise a folding scheme that will allow the required SSU and LSU rRNA interactions to occur intramolecularly, and, at the same time, position cleavage sites suitably for processing. The alternative possibility is that ribosome assembly per se is dissociated temporally from processing, and that the mature rRNA pieces assemble after processing via the appropriate intermolecular interactions.

Stage I

Fragmented and dispersed rRNA coding modules have also been described in the exceedingly small (6–7 kbp) mtDNAs of three apicomplexan species of protozoa: *Plasmodium yoelii* [41], *Plasmodium falciparum* (the human malaria parasite) [42] and *Theileria parva* [43]. In these cases, in contrast to the situation in *C. reinhardtii* and *C. eugametos* mtDNAs, both strands of the mtDNA encode rRNA modules. Transcript mapping experiments in *P. falciparum* mitochondria have provided evidence that these modules are transcribed and that the resulting RNAs are processed [42]. In this case, because rRNA transcripts emanate from both strands, it is necessary to postulate that at least some rRNA pieces are transcribed from separate promoters and that they associate intermolecularly in the course of ribosome assembly.

Recent studies of *E. gracilis* mitochondrial rRNAs (D.F. Spencer, unpublished) have now yielded direct evidence of independent transcription of rRNA species that represent portions of conventional SSU and LSU rRNAs. As in *T. pyriformis* mitochondria, SSU and LSU rRNAs are both bipartite in *E. gracilis* mitochondria, although the two pieces (designated "5′" and "3′") are more equal in size than are the *Tetrahymena* α and β components. Capping experiments have demonstrated that all four *E. gracilis* mitochondrial rRNA species (SSU-5′, SSU-3′, LSU-5′, LSU-3′) accept label from [α-^{32}P]GTP in the presence of guanylyltransferase, indicating they contain unprocessed (i.e., pp... or ppp...) 5′-termini. Cloned *Xba*I restriction fragments containing rRNA-hybridizing sequences have been isolated and characterized; one such clone contains the coding region for the SSU-5′ rRNA, whereas the other encodes the SSU-3′ rRNA. Consideration of the physical state of *E. gracilis* mtDNA (which appears to exist in the form of a highly heterodisperse collection of gene- or subgene-sized linear pieces) makes it unlikely that these two SSU rRNA-encoding *Xba*I clones are physically linked in the native *E. gracilis* mitochondrial genome. Whether they are or not, the capping experiments clearly show that they must represent separate SSU rDNA transcriptional units.

Conclusions

By demonstrating the feasibility of each of the postulated stages in our model of rRNA evolution, the novel systems described above together lend credence to the overall proposed pathway. It is important to note, however, that we do not consider any of these examples of split rRNAs, or split and rearranged rRNA genes, to represent primitive traits retained from a primordial ancestor. Rather, because such discontinuous rRNAs/rRNA genes are highly restricted phylogenetically, it is more reasonable to view them as derived traits. On the other hand, it is also reasonable to suggest that they may represent a reversion to a primitive condition that existed during the evolution of rRNA and its genes. If, as we suggest, variable regions in contemporary rRNAs are remnants of ITSs that have been incorporated into mature rRNAs in the course of evolution, and if for that reason they are largely dispensible for core rRNA functions, then they may well be able to sustain newly evolved cleavages or even be lost altogether (as in some mitochondrial rRNAs), without detriment to overall ribosome function. Modules of rRNA coding sequence, defined by flanking variable regions, have evidently been able to be rearranged at the genome level in some systems, without precluding assembly of functional ribosomes. Such observations strongly support the concept that rRNA molecules have a fundamentally modular structure.

The evolutionary scheme outlined here is able to account for the distinctive features of LSU and SSU rRNA molecules enumerated at the beginning of this article: large size; interspersion of conserved and variable regions; existence of highly conserved long-range base-pairing interactions; and structural and functional complexity. Also, it illustrates how a primordial ribosome could have been assembled from and functioned as a collection of small RNAs, such as are likely to have existed early in the evolutionary history of the translation system. Finally, it suggests a mechanism for an incremental increase in ribosome complexity, through the progressive addition of new modules (i.e., new small RNAs) of structure and function.

Acknowledgements

Work in the authors' laboratory on "Ribosomal RNA in Pieces: Biosynthesis and Evolution" is currently supported by an operating grant (MT-11212) from the Medical Research Council of Canada to M.W.G., who is a Fellow in the Program in Evolutionary Biology of the Canadian Institute for Advanced Research.

References

1. Noller HF. Structure of ribosomal RNA. Ann Rev Biochem 1984;53:119–162.
2. Noller HF. Ribosomal RNA and translation. Ann Rev Biochem 1991;60:191–227.
3. Dahlberg AE. The functional role of ribosomal RNA in protein synthesis. Cell 1989;57:525–529.
4. Noller HF, Hoffarth V, Zimniak L. Unusual resistance of peptidyl transferase to protein extraction procedures. Science 1992;256:1416–1419.
5. Gray MW. Organelle origins and ribosomal RNA. Biochem Cell Biol 1988;66:325–348.
6. Stiegler P, Carbon P, Ebel J-P, Ehresmann C. A general secondary-structure model for procaryotic and eucaryotic RNAs of the small ribosomal subunits. Eur J Biochem 1981;120:487–495.
7. Gerbi SA. The evolution of eukaryotic ribosomal DNA. BioSystems 1986;19:247–258.
8. Raué HA, Klootwijk J, Musters W. Evolutionary conservation of structure and function of high

molecular weight ribosomal RNA. Prog Biophys Molec Biol 1988;51:77–129.

9. Gutell RR, Weiser B, Woese CR, Noller HF. Comparative anatomy of 16-S-like ribosomal RNA. Prog Nucl Acid Res Molec Biol 1985;32:155–216.
10. Gray MW, Sankoff D, Cedergren RJ. On the evolutionary descent of organisms and organelles: a global phylogeny based on a highly conserved structural core in small subunit ribosomal RNA. Nucl Acid Res 1984;12:5837–5851.
11. Cedergren R, Gray MW, Abel Y, Sankoff D. The evolutionary relationships among known life forms. J Molec Evol 1988;28:98–112.
12. Darnell JE, Doolittle WF. Speculations on the early course of evolution. Proc Natl Acad Sci USA 1986;83:1271–1275.
13. Santer M. Ribosomal RNA on the surface of ribosomes. Science 1963;141:1049–1050.
14. Crick FHC. The origin of the genetic code. J Molec Biol 1968;38:367–379.
15. Woese CR. The emergence of genetic organization. In: Ponnamperuma C (ed) Exobiology. Amsterdam: North-Holland, 1972;301–341.
16. Woese CR. Just So Stories and Rube Goldberg machines: speculations on the origin of the protein synthetic machinery. In: Chambliss G, Craven GR, Davies J, Davis K, Kahan L, Nomura M (eds) Ribosomes: Structure, Function, and Genetics. Baltimore: University Park Press, 1980;357–373.
17. Maden BEH. The numerous modified nucleotides in eukaryotic ribosomal RNA. Prog Nucl Acid Res Molec Biol 1990;39:241–303.
18. Nazar RN. A 5.8 S rRNA-like sequence in prokaryotic 23 S rRNA. FEBS Lett 1980;119:212–214.
19. Cox RA, Kelly JM. Mature 23 S rRNA of prokaryotes appears homologous with the precursor of 25-28 rRNA of eukaryotes. Comments on the evolution of 23-28 rRNA. FEBS Lett 1981;130:1–6.
20. Jacq B. Sequence homologies between eukaryotic 5.8S rRNA and the 5′ end of prokaryotic 23S rRNA: evidence for a common evolutionary origin. Nucl Acid Res 1981;9:2913–2932.
21. Walker WF. Proposed sequence homology between the 5′-end regions of prokaryotic 23 S rRNA and eukaryotic 28 S rRNA. Relevance to the hypothesis that 5.8 S rRNA is homologous to the 5′-end region of 23 S rRNA. FEBS Lett 1981;126:150–151.
22. Nazar RN. Evolutionary relationship between eukaryotic 29-32 S nucleolar rRNA precursors and the prokaryotic 23 S rRNA. FEBS Lett 1982;143:161–162.
23. Clark CG, Gerbi SA. Ribosomal RNA evolution by fragmentation of the 23S progenitor: maturation pathway parallels evolutionary emergence. J Molec Evol 1982;18:329–336.
24. Perry RP. Processing of RNA. Ann Rev Biochem 1976;45:605–629.
25. Gutell RR, Gray MW, Schnare MN. A compilation of large subunit (23S and 23S-like) ribosomal RNA structures: 1993. Nucl Acid Res 1993;21:3055–3074.
26. Gray MW, Schnare MN. Evolution of the modular structure of rRNA. In: Hill WE, Dahlberg A, Garrett R, Moore PB, Schlessinger D, Warner JR (eds) The Ribosome: Structure, Function, & Evolution, Chapter 52. Washington: American Society for Microbiology, 1990;589–597.
27. Wool IG. The mechanism of action of the cytotoxic nuclease α-sarcin and its use to analyse ribosome structure. Trends Biochem Sci 1984;9:14–17.
28. Gray MW, Boer PH, Collings JC, Heinonen TYK, Spencer DF, Schnare MN. Ribosomal RNA genes in pieces. In: Kotyk A, Škoda J, Pačes P, Kostka V (eds) Highlights of Modern Biochemistry. Zeist: VSP International, 1989;521–530.
29. Gray MW. Split rRNAs and modified nucleosides in ribosome evolution. In: Hartman H, Matsuno K (eds) The Origin and Evolution of the Cell. Singapore: World Scientific, 1992;333–358.
30. Gray MW, Schnare MN. Evolution of rRNA gene organization. In: Zimmermann RA, Dahlberg AE (eds) Ribosomal RNA: Structure, Evolution, Gene Expression and Function in Protein Synthesis. Boca Raton: CRC Press, 1995;49–69.
31. Spencer DF, Collings JC, Schnare MN, Gray MW. Multiple spacer sequences in the nuclear large subunit ribosomal RNA gene of *Crithidia fasciculata*. EMBO J 1987;6:1063–1071.
32. Schnare MN, Gray MW. Sixteen discrete RNA components in the cytoplasmic ribosome of *Euglena gracilis*. J Molec Biol 1990;215:73–83.
33. Schnare MN, Cook JR, Gray MW. Fourteen internal transcribed spacers in the circular ribosomal

DNA of *Euglena gracilis*. J Molec Biol 1990;215:85–91.
34. Schnare MN, Heinonen TYK, Young PG, Gray MW. A discontinuous small subunit ribosomal RNA in *Tetrahymena pyriformis* mitochondria. J Biol Chem 1986;261:5187–5193.
35. Heinonen TYK, Schnare MN, Young PG, Gray MW. Rearranged coding segments, separated by a transfer RNA gene, specify the two parts of a discontinuous large subunit ribosomal RNA gene in *Tetrahymena pyriformis*. J Biol Chem 1987;262:2879–2887.
36. Heinonen TYK, Schnare MN, Gray MW. Sequence heterogeneity in the duplicate large subunit ribosomal RNA genes of *Tetrahymena pyriformis* mitochondrial DNA. J Biol Chem 1990;265: 22336–22341.
37. Gray MW, Boer PH. Organization and expression of algal (*Chlamydomonas reinhardtii*) mitochondrial DNA. Phil Trans R Soc Lond B 1988;319:135–147.
38. Boer PH, Gray MW. Scrambled ribosomal RNA gene pieces in *Chlamydomonas reinhardtii* mitochondrial DNA. Cell 1988;55:399–411.
39. Boer PH, Gray MW. Genes encoding a subunit of respiratory NADH dehydrogenase (ND1) and a reverse transcriptase-like protein (RTL) are linked to ribosomal RNA gene pieces in *Chlamydomonas reinhardtii* mitochondrial DNA. EMBO J 1988;7:3501–3508.
40. Denovan-Wright EM, Lee RW. Comparative structure and genomic organization of the discontinuous mitochondrial ribosomal RNA genes of *Chlamydomonas eugametos* and *Chlamydomonas reinhardtii*. J Molec Biol 1994;241:298–311.
41. Vaidya AB, Akella R, Suplick K. Sequences similar to genes for two mitochondrial proteins and portions of ribosomal RNA in tandemly arrayed 6-kilobase-pair DNA of a malarial parasite. Molec Biochem Parasitol 1989;35:97–108.
42. Feagin JE, Werner E, Gardner MJ, Williamson DH, Wilson RJM. Homologies between the contiguous and fragmented rRNAs of the two *Plasmodium falciparum* extrachromosomal DNAs are limited to core sequences. Nucl Acid Res 1992;20:879–887.
43. Kairo A, Fairlamb AH, Gobright E, Nene V. A 7.1 kb linear DNA molecule of *Theileria parva* has scrambled rDNA sequences and open reading frames for mitochondrially encoded proteins. EMBO J 1994;13:898–905.

Tracing Biological Evolution in Protein and Gene Structures.
M. Gō and P. Schimmel, editors.

Structure and function of RNA-binding proteins in chloroplasts and cyanobacteria

Mamoru Sugita, Masaru Ohta and Masahiro Sugiura
Center for Gene Research, Nagoya University, Nagoya, Japan

Abstract. Chloroplasts contain their own genetic system which has both prokaryotic and eukaryotic features. We have identified five nuclear-encoded chloroplast RNA-binding proteins or ribonucleoproteins (RNPs) from tobacco that may be involved in the posttranscriptional processes of gene expression in chloroplasts. These RNPs are found complexed to multiple chloroplast RNA species. In addition, the size distribution of the RNP complexes in the stroma indicates that these proteins bind to (pre-)RNAs prior to the formation of polysomes. Cyanobacteria also have eukaryotic-type RNA-binding proteins which contain a conserved RNA-binding domain resembling that of its chloroplast counterpart. This indicates a close evolutionary relationship between cyanobacterial and chloroplast RNA-binding proteins. Thus, these proteins may have an important role in RNA metabolism that is unique to chloroplasts and cyanobacteria.

Key words: consensus sequence-type RNA-binding domain, endosymbiosis, ribonucleoproteins, RNA processing and splicing, *Synechococcus* PCC 6301, tobacco.

Introduction

Chloroplasts are intracellular organelles that are specialized for photosynthesis and are unique to plants and algae. Chloroplasts have their own genome and a transcription-translation machinery which is distinct from that of the nucleocytoplasm [1,2]. The entire nucleotide sequence of the chloroplast genome was first determined for tobacco, then liverwort and subsequently for rice [2]. Unlike nuclear genes, most chloroplast genes are co-transcribed as polycistronic pre-RNAs which are then extensively processed into shorter RNA species [2]. Chloroplast RNA processing steps include RNA cutting, 3′-trimming, *cis/trans* splicing and RNA editing [2]. However, little is known about the molecular mechanisms involved in these RNA processes in chloroplasts.

In order to begin to elucidate the molecular mechanisms of RNA processing in chloroplasts, we have isolated and characterized five nuclear-encoded chloroplast RNA-binding proteins with molecular weights between 28 and 33 kDa from tobacco [3,4]. Each of the isolated chloroplast proteins contains two RNP consensus sequence RNA-binding domains (CS-RBD) which have been identified in proteins involved in

Address for correspondence: Mamoru Sugita, Center for Gene Research, Nagoya University, Furo-cho, Chikusa-ku, Nagoya 464-01, Japan. Fax: +81-52-789-3081.

various steps of RNA metabolism such as capping, processing, polyadenylation and splicing of pre-RNAs [5,6]. However, at present, the precise function(s) of the chloroplast RNA-binding proteins remains unknown.

Chloroplasts are thought to have originated from a cyanobacterium-like progenitor. Recently, two 12 kDa RNA-binding proteins have been identified in *Synechococcus* sp. PCC 6301, each of which contains a single CS-RBD [7,8]. In this review, possible functions of the isolated chloroplast and cyanobacterial RNA-binding proteins will be considered.

Chloroplast genes and their expression

Chloroplast DNAs of higher plants are circular molecules between 120 to 160 kbp [2]. To date, the entire nucleotide sequences of chloroplast DNAs from six land plants (rice, maize, tobacco, beech drop, black pine and liverwort) and two algae (*Euglena gracilis* and *Porphyra purpurea*) have been accomplished [9]. These studies indicate that chloroplast DNAs of higher plants contain all the chloroplast rRNA genes (four genes), 30–32 tRNA genes and about 100 genes encoding polypeptides. The chloroplast genetic system has a number of prokaryotic-like features. For example, chloroplast ribosomes are 70S in size and rRNAs and tRNAs are very similar to their *Escherichia coli* (*E. coli*) counterparts. Chloroplast mRNAs also contain triphosphates at their 5′ ends and lack poly(A) tails. In addition, the upstream regions of many transcription initiation sites contain DNA sequences similar to the prokaryotic "-10" and "-35" consensus sequences. However, one example of the eukaryotic features is that no chloroplast tRNA gene has been found which codes for a 3′-CCA end. Interestingly, some tRNA genes do not require their 5′ upstream regions for transcription at least in vitro [10].

Chloroplast genes are generally co-transcribed and so far nearly 60 such transcription units have been identified in tobacco [2]. One example, the *psbB* operon, consists of the genes *psbB, psbH*, *petB* and *petD* encoding the photosystem II 47 kDa protein, the 10 kDa phosphoprotein, the cytochrome b_6/f complex b_6 subunit and IV subunit, respectively, and is transcribed from a bacterial-type promoter as a tetracistronic precursor of 5.6 kb. In addition, two splicing events and multiple processing steps are necessary to form the functional shorter RNA species [2].

Introns of chloroplast genes

Several chloroplast genes are interrupted by introns. In the tobacco chloroplast genome, 18 genes are known to contain introns (503–2526 bp), six for tRNA genes and 12 for protein-encoding genes. Two of the latter genes, *clpP* which encodes the proteolytic subunit of an ATP-dependent protease and IRF168 (intron-containing unidentified reading frame of 168 codons), contain two introns each, indicating the presence of a total of 20 *cis*-introns [2]. Most introns, except for the intron of the tRNA^{Leu} gene, can be folded into secondary structures similar to that of the group II intron class of fungal mitochondrial genes. Chloroplast group II introns have conserved boundary sequences GTGYGRY··· at the 5′ ends and RYCNAYY(Y)YNAY at their 3′ ends, which are similar in part to those of nuclear pre-mRNA introns [11]. The 5′ ends of group II introns of the chloroplast tRNA^{Ile} (GAU) and tRNA^{Ala} (UGC) genes, however, exhibit the unique sequence TTG(C/G)GTC···. The intron of the tRNA^{Leu} gene can be classified into the group I intron class of fungal mitochondrial genes which includes the *Tetrahymena* LSU rRNA gene [12]. The intron of the 23S rDNA from *Chlamydomonas reinhardtii* (*C. reinhardtii*) also belongs to this group and encodes a double-stranded DNA endonuclease [13]. A new class of short introns, designated group III intron, appears to be unique to *Euglena* chloroplasts [14].

The most striking feature among the intron-containing chloroplast genes is found in the *rps12* gene encoding the ribosomal protein CS12 in land plants [2] and the *psaA* gene coding for the photosystem I P700 apoprotein of *C. reinhardtii* [15]. The genes are each divided into three separate and distantly located parts and require a *trans*-splicing process to form the functional mRNA.

Small RNAs involved in RNA splicing and processing in chloroplasts

Efficient self-splicing of pre-RNAs from split chloroplast genes has been demonstrated for the 23S pre-rRNA and *psbA* pre-mRNA introns in *Chlamydomonas* [16,17], but has not yet been demonstrated for candidate intron-containing genes of higher plant chloroplasts. It is well known that splicing of nuclear pre-mRNAs and processing of nuclear pre-rRNAs are catalyzed by protein-RNA complexes [18,19]. Therefore, it is possible that protein-RNA complexes are also involved in splicing and processing of pre-RNAs in chloroplasts. Recently, two examples have been found. The *tscA* RNA of *C. reinhardtii* (430 nt) is encoded in the chloroplast genome and is thought to be required for *trans* splicing of exon 1 and exon 2 of *psaA* pre-mRNAs [20]. In tobacco, a small RNA (spRNA, 218 nt) is encoded by *sprA*, which is located 69 bp downstream from the termination codon of the ribosomal protein CL32 gene, *rpl32* [21]. Part of the spRNA sequence exhibits the potential for base pairing with the leader sequence of pre-16S rRNA, suggesting a role for spRNA in chloroplast ribosome biogenesis, i.e., 16S rRNA maturation [21]. Additional small RNA species might be encoded in the chloroplast genome. However, no in vitro RNA splicing systems from chloroplasts are currently available. Thus, this makes it difficult to analyze individual steps in RNA splicing in chloroplasts and to detect the various factors involved.

Chloroplast ribonucleoproteins

It is interesting to know whether there are hnRNP-like proteins in chloroplasts, and if this is the case, whether these proteins are involved in splicing and/or processing of chloroplast RNAs. Five tobacco chloroplast proteins were found to bind strongly to ssDNA columns: three proteins, cp28, cp31 and cp33, eluted at 2 M NaCl, whereas cp29A and cp29B eluted at 0.6 M NaCl [3,4]. These proteins are encoded in the nuclear genome of *Nicotiana sylvestris* (*N. sylvestris*) and are synthesized as transit peptide-containing precursor polypeptides in the cytoplasm, and then imported into chloroplasts. This has been verified experimentally [3,4]. Each of the five precursors is composed of four domains: a transit peptide, an acidic amino terminal domain (AD) and two consensus sequence-type RNA-binding domains (CS-RBD I and II). Therefore, we named these proteins chloroplast ribonucleoproteins (RNPs) or chloroplast RNA-binding proteins. The domain structure of cp31 is shown in Fig. 1.

Cp28 and cp31 are highly conserved, having 79% amino acid identity over their entire sequences while cp33 has 37% amino acid identity to both cp28 and cp31 proteins. Cp29A and cp29B are also highly conserved (85%). Cp29A shows amino acid identities of 57, 57 and 46% with cp28, cp31 and cp33, respectively. In addition to the two RNA-binding domains, each of the proteins contains an acidic N-terminal domain which is unique to RNPs of chloroplasts. Forty-two percent of the first 64 amino acids of the cp31 protein are acidic and this region lacks any basic amino acids. A chloroplast RNA-binding protein gene family similar to that of tobacco has also been found in *Arabidopsis thaliana* (*A. thaliana*) [22]. Similar chloroplast proteins have been found in spinach (28RNP) [23], maize (NBP or nucleic acid binding protein) [24] and *N. plumbaginifolia* (CP-RNP30 and 31) [25]. The chloroplast proteins can be categorized into three groups: I (cp28, cp31, 28RNP and NBP), II (cp29A and B, cp29, CP-RNP30 and 31) and III (cp33). The conservation of the three types of chloroplast RNPs in a variety of plants suggests that these proteins are required for chloroplast functions, at least in dicotyledonous plants [22].

In vitro nucleic acid binding experiments showed that the five tobacco chloroplast RNPs have higher affinities for RNA homopolymers, poly(G) and poly(U) than those for ssDNA and dsDNA, suggesting that they bind preferentially to U or G-rich stretches in RNA molecules in vivo [26,27]. The nucleic acid-binding properties of these proteins provide additional evidence for the similarity of chloroplast RNA-

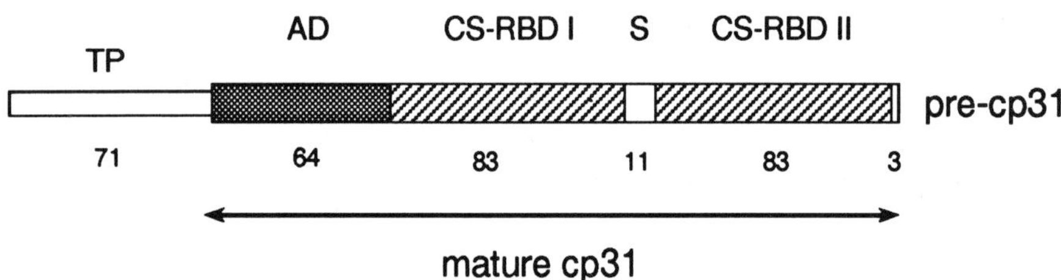

Fig. 1. The schematic representation of chloroplast RNP protein cp31. TP = transit peptide; AD = acidic domain; CS-RBD I and II = RNA-binding domains I and II; S = spacer. Numbers of amino acids of the AD, RBDs, spacer and C-terminal parts are shown below. The mature cp31 is produced by proteolytic cleavage of pre-cp31.

binding proteins to HeLa hnRNP proteins. The deletion mutant cp29A and B proteins containing only one RNA-binding domain (either domain I or II) lose their affinity for ssDNA or dsDNA and remain bound only slightly to poly(G) and poly(U), indicating that both RNA-binding domains I/II and the spacer in cp29A and cp29B (possibly also in other chloroplast RNPs) are necessary for RNA-binding at least in vitro [27].

The chloroplast RNP genes

Each of the five chloroplast RNPs are encoded by single-copy nuclear genes in *N. sylvestris*. The genes for cp31, cp33 and cp29B consist of four exons and three introns which are inserted at the same sites in all three genes [4,28,29]. The first and second introns are located in the first CS-RBD while the third intron is in the second CS-RBD. There is no intron in the region coding for the transit peptide and the acidic amino terminal domain. The second and third introns interrupt the two CS-RBDs at the same position, suggesting that the occurrence of an intron within an ancestral gene encoding a CS-RBD preceded the duplication of this domain. In contrast to the human hnRNP A1 [30] and mouse nucleolin genes [31], there is no intron between the first and second CS-RBDs. The position of the first intron is precisely the same as the second intron of human hnRNP A1 genes [30], the third intron of *A. thaliana* poly(A)-binding protein [32], and the introns of tobacco RGP-1a, b, c [33]. This supports the hypothesis that introns are as ancient as the genes themselves and that the present day prokaryotes have lost most of them [34,35].

No alternatively spliced transcripts have been detected in the tobacco RNP genes [28], a situation which is unlike that of human hnRNP proteins, whose diversity is mainly generated by alternative splicing [30]. Diversity of chloroplast RNA-binding proteins, on the one hand, may perhaps be attributed to multiple genes generated by the duplication of a protogene.

Possible function of the chloroplast RNPs

The transcripts for cp28, cp31 and cp33 accumulate both in leaves and in roots, but the levels in leaves are 6 to 10-fold higher than that in roots [3]. The same transcripts are also detected in nonphotosynthetic cultured cells (*N. tabacum* BY-2 cell line) [28,29]. Leaves, roots and the cultured cells contain chloroplasts, amyloplasts and proplastids, respectively. The detection of their gene expression in these multiple cell types suggests that chloroplast RNA-binding proteins are essential for plastid functions.

Our preliminary studies indicate that the chloroplast RNPs are a major component of the soluble proteins found in chloroplasts and are distributed in the nonpolysomal fraction. In addition, the tobacco RNP proteins were immunoprecipitated with a variety of chloroplast mRNA species and nonribosomal rRNAs (Ohta et al.

unpublished). These observations strongly suggest that, in analogy to hnRNP particles, the chloroplast RNPs are associated with pre-RNAs as stable RNA-protein complexes before being replaced by ribosomes for translation or being assembled with ribosomal proteins for ribosome formation. In spinach, it has been shown that the 28RNP protein is required for plastid mRNA 3′ end processing and/or stability [23]. The 28RNP binds specifically to a stem-loop structure in the 3′ untranslated regions of *psbA, rbcL* or *petD* mRNAs [23]. However, similar results have not been observed for tobacco chloroplast RNPs (unpublished results).

Eukaryotic CS-RBD-type RNA-binding proteins are present in cyanobacteria

Chloroplasts are thought to have originated from bacteria-like endosymbionts similar to cyanobacteria [36,37]. The fact that the chloroplast RNPs are encoded in the nuclear genome raises the question whether the chloroplast proteins are of a cyanobacterial or of primordial nuclear origin. To investigate the possibility of a cyanobacterial origin of chloroplast CS-RBD proteins, we analysed proteins isolated using an ssDNA column from the unicellular cyanobacterium *Synechococcus* sp. PCC 6301 [7,8]. Of several ssDNA-binding proteins eluting between 0.3 and 0.6 M NaCl, abundant proteins of 12 kDa were identified as CS-RBD proteins. We designated the identified proteins as 12RNP1 and 12RNP2, respectively.

12RNP1 is 110 amino acids long while 12RNP2 consists of 99 amino acids. The two 12RNP proteins each contain a single CS-RBD of 82 amino acids which show a 64% identity with each other. 12RNP1 has a short glycine-rich region of 18 amino acids (78% Gly) that is appended to the CS-RBD, while 12RNP2 does not have such a glycine-rich sequence. As shown in Fig. 2, 12RNP1 contains two highly conserved motifs, IYVGNL and RGFGFVEM, which resemble the chloroplast consensus sequence motifs (I/L)(Y/F)VGNL and RGFGFVTM, respectively. The RNP-CS (RGFAFIEM) of 12RNP2 is less similar to those of chloroplast RNPs. Overall, the CS-RBD of 12RNP1 shows 40–51% amino acid identities, 78–94% similarities to those of tobacco cp28, cp29A and cp31 and maize chloroplast NBP. Likewise, identities to CS-RBDs of the RNA-binding, glycine-rich proteins [33,38] are 36–40%, while their identity to those of the potato U2-snRNP B" is only 22% [39]. The cyanobacterial protein shows 18–34% identity with those of poly(A)-binding proteins, nuclear hnRNP proteins and snRNP proteins from nonplant sources (Fig. 2). The 12RNP2 protein is the smallest of the CS-RBD-containing proteins identified so far. Recently, genes for the proteins RbpA and RbpB which are homologous to 12RNP1 have also been found in heterocyst-forming filamentous cyanobacteria, namely *Anabaena* sp. PCC 7120, *Chlorogloeopsis* sp. PCC 6912 [40], and *Anabaena variabilis* M3 [41].

The high similarity found among the cyanobacteria and chloroplast CS-RBDs shows that chloroplast RNPs are closely related to their homologs from cyanobacteria, and thus provides additional evidence for the endosymbiotic theory [36]. From these observations, it appears that the cyanobacteria-like endosymbionts contained CS-

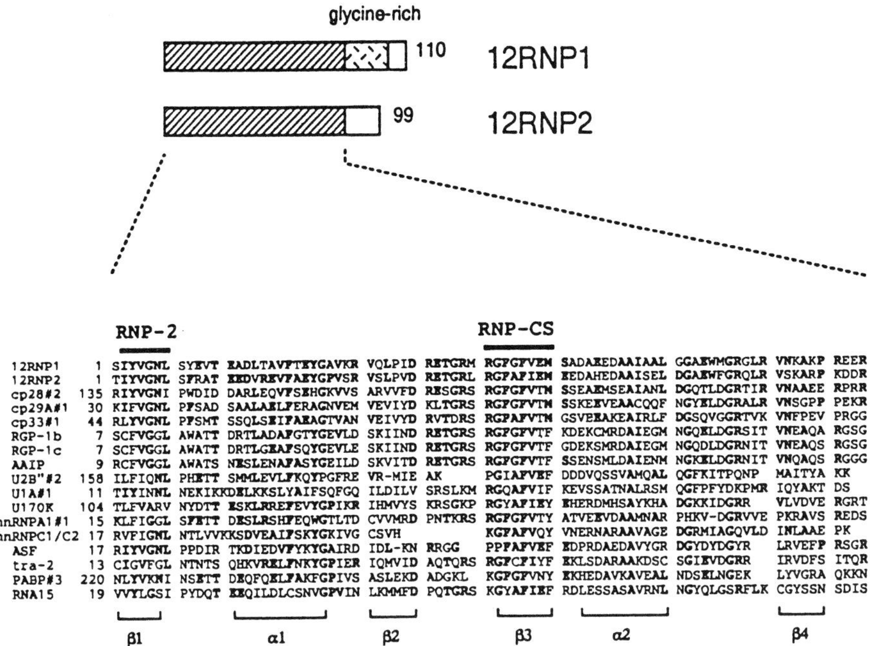

Fig. 2. The schematic diagram of 12RNP1 and 12RNP2 (upper part) and sequence alignment of CS-RBD-containing proteins. The regions of two conserved motifs, RNP-2 and RNP-CS, are shown in bars. # means one of more than two CS-RBDs. The position in the primary sequence of the first amino acid on every line is shown. Organism sources are *Synechococcus* (12RNP1 and 2), tobacco (cp28, cp29A, RGP-1b and -1c), maize (AAIP), potato (snRNP U2B"), human (snRNPs U1A, U1 70K, hnRNPs A1, C1/C2), fruit fly (tra-2) and yeast (PABP, RNA15). The regions of β-sheets and α-helices are referred to in a review by Kenan et al. [5].

RBD-type protein gene(s) in their own genomes, and that this gene(s) was subsequently transferred to the nucleus. Consequently, multiple genes encoding either one or two RNA-binding domains could possibly have been produced by duplication of this gene or by fusion to genes already present in the nuclear DNA. The two domain-type protein genes may have evolved to chloroplast RNP genes after acquiring a DNA region encoding a transit peptide.

Possible function of RNA-binding proteins in cyanobacteria

The *Synechococcus* 12RNP genes are transcribed as monocistronic mRNAs, the mRNA level of 12RNP1 being at least 20-fold higher than that of 12RNP2 [8]. The results of a gene-disruption experiment indicate that deletion of the 12RNP1 gene is lethal while 12RNP2 is not essential under normal growth conditions at 30°C (Sugita, unpublished). Interestingly, upon exposure of *Synechococcus* PCC 6301 cells to lower temperatures, the 12RNP1 mRNA level increases from 2-fold (20°C for 3 h) [8] to 12-fold (10°C for 2 h; unpublished), while the 12RNP2 mRNA level does not change.

A similar increase has also been observed for the gene-encoding RbpA of the filamentous cyanobacterium *A. variabilis* M3 [41]. In addition, the 12RNP1 transcript accumulates in the stationary phase of cell growth (unpublished). The 12RNP1 and 12RNP2 proteins show high binding specificities to poly(G) and poly(U) [8], suggesting that 12RNP proteins may play an important role in cell viability or in RNA metabolism, especially under conditions of environmental stresses. Since snRNP-like components are suggested to exist in cyanobacteria [42], the 12RNP proteins may also be components in snRNP-like particles which could be involved in group I and/or group II intron splicing in cyanobacteria [43–45].

Identification of RNA sequences recognized by the chloroplast and cyanobacterial proteins may provide clues to their function. Further studies are necessary to understand the role of these RNA-binding proteins in the posttranscriptional steps of gene expression.

Acknowledgements

We thank Drs T. Wakasugi and J.Y. Suzuki for useful comments and critical reading of the manuscript. A part of this work was supported by a Grant-in-Aid from the Ministry of Education, Science and Culture (Japan) and by the Special Coordination Funds of the Science and Technology Agency (Japan).

References

1. Gruissem W. Chloroplast gene expression: how plants turn their plastids on. Cell 1989;56:161–170.
2. Sugiura M. The chloroplast genome. Plant Molec Biol 1992;19:149–168.
3. Li Y, Sugiura M. Three distinct ribonucleoproteins from tobacco chloroplasts: each contains a unique amino terminal acidic domain and two ribonucleoprotein consensus motifs. EMBO J 1990;9:3059–3066.
4. Ye L, Li Y, Fukami-Kobayashi K, Gō M, Konishi T, Watanabe A, Sugiura M. Diversity of a ribonucleoprotein family in tobacco chloroplasts: two new chloroplast ribonucleoproteins and a phylogenetic tree of ten chloroplast RNA-binding domains. Nucl Acid Res 1991;19:6485–6490.
5. Kenan DJ, Query CC, Keene JD. RNA recognition: towards identifying determinats of specificity. Trends Biochem Sci 1991;16:214–220.
6. Dreyfuss G, Matunis MJ, Pinol-Roma S, Burd CG. hnRNP proteins and the biogenesis of mRNA. Ann Rev Biochem 1993;62:289–321.
7. Sugita M, Sugiura M. Nucleic acid-binding proteins containing a consensus sequence-type RNA-binding domain of the cyanobacterium *Anacystis nidulans*. In: Murata N (ed) Research in Photosynthesis, Vol. III. Dordrecht: Kluwer Academic Publishers, 1992;275–278.
8. Sugita M, Sugiura M. The existence of eukaryotic ribonucleoprotein consensus sequence-type RNA-binding proteins in a prokaryote, *Synechococcus* 6301. Nucl Acid Res 1994;22:25–31.
9. Sugiura M. The chloroplast genome. In: Tipton KF (ed) Essays in Biochemistry. London: Portland Press, 1995;(in press).
10. Gruissem W, Elsner-Menzel C, Latshaw S, Narita JO, Schaffer MA, Zurawski G. A subpopulation of spinach chloroplast tRNA genes does not require upstream promoter elements for transcription. Nucl Acid Res 1986;14:7541–7556.
11. Jacquier A. Self-splicing group II and nuclear pre-mRNA introns:how similar are they? Trends

Biochem Sci 1990;15:351–354.
12. Cech TR. Conserved sequences and structures of group I introns: building an active site for RNA catalysis – a review. Gene 1988;73:259–271.
13. Dürrenberger F, Rochaix JD. Chloroplast ribosomal intron of *Chlamydomonas reinhardtii*: in vitro self-splicing, DNA endonuclease activity and in vivo mobility. EMBO J 1991;10:3495–3501.
14. Copertino DW, Hallick RB. Group II and group III introns of twintrons: potential relationships with nuclear pre-mRNA introns. Trends Biochem Sci 1993;18:467–471.
15. Kück U, Choquet Y, Schneider M, Dron M, Bennoun P. Structural and transcription analysis of two homologous genes for the P700 chlorophyll a-apoproteins in *Chlamydomonas reinhardtii*: evidence for in vivo *trans* splicing. EMBO J 1987;6:2185–2195.
16. Herrin DL, Chen YF, Schmidt GW. RNA splicing in *Chlamydomonas* chloroplasts: self-splicing of 23S preRNA. J Biol Chem 1990;265:21134–21140.
17. Herrin DL, Bao Y, Thompson AJ, Chen YF. Self-splicing of the *Chlamydomonas* chloroplast *psbA* introns. Plant Cell 1991;3:1095–1107.
18. Sharp PA. Split genes and RNA splicing. Cell 1994;77:805–815.
19. Fournier MJ, Maxwell ES. The nucleolar snRNAs: catching up with the spliceosomal snRNAs. Trends Biochem Sci 1993;18:131–135.
20. Goldschmidt-Clermont M, Choquet Y, Girard-Bascou J, Michel F, Schirmer-Rahire M, Rochaix JD. A small chloroplast RNA may be required for *trans*-splicing in *Chlamydomonas reinhardtii.* Cell 1991;65:135–143.
21. Vera A, Sugiura M. A novel RNA gene in the tobacco plastid genome: its possible role in the maturation of 16S rRNA. EMBO J 1994;13:2211–2217.
22. Ohta M, Sugita M, Sugiura M. Three types of nuclear genes encoding chloroplast RNA-binding proteins (cp29, cp31 and cp33) are present in *Arabidopsis thaliana*: presence of cp31 in chloroplasts and its homologue in nuclei/cytoplasms. Plant Molec Biol 1995;27:529–539.
23. Schuster G, Gruissem W. Chloroplast mRNA 3′ end processing requires a nuclear-encoded RNA-binding protein. EMBO J 1991;10:1493–1502.
24. Cook WB, Walker JC. Identification of a maize nucleic acid-binding protein (NBP) belonging to a family of nuclear-encoded chloroplast proteins. Nucl Acid Res 1992;20:359–364.
25. Mieszak M, Klahre U, Levy JH, Goodall GJ, Filipowicz W. Multiple plant RNA binding proteins identified by PCR: expression of cDNAs encoding RNA binding proteins targeted to chloroplasts in *Nicotiana plumbaginifolia.* Molec Gen Genet 1992;234:390–400.
26. Li Y, Sugiura M. Nucleic acid-binding specificities of tobacco chloroplast ribonucleoproteins. Nucl Acid Res 1991;19:2893–2896.
27. Ye L, Sugiura M. Domains required for nucleic acid binding activities in chloroplast ribonucleoproteins. Nucl Acid Res 1992;20:6275–6279.
28. Li Y, Ye L, Sugita M, Sugiura M. Tobacco nuclear gene for the 31 kDa chloroplast ribonucleoprotein: genomic organization, sequence analysis and expression. Nucl Acid Res 1991;19:2987–2991.
29. Li Y, Nagayoshi S, Sugita M, Sugiura M. Structure and expression of the tobacco nuclear gene encoding the 33 kDa chloroplast ribonucleoprotein. Molec Gen Genet 1993;239:304–309.
30. Biamonti G, Buvoli M, Bassi MT, Morandi C, Cobianchi F, Riva S. Isolation of an active gene encoding human hnRNP protein A1: evidence for alternative splicing. J Molec Biol 1989;207:491–503.
31. Bourbon HM, Lapeyre B, Amalric F. Structure of the mouse nucleolin gene: the complete sequence reveals that each RNA binding domain is encoded by two independent exons. J Molec Biol 1988;200:627–638.
32. Hilson P, Carroll KL, Masson PH. Molecular characterization of PAB2, a member of the multigene family coding for poly(A)-binding proteins in *Arabidopsis thaliana.* Plant Physiol 1993;103:525–533.
33. Hirose T, Sugita M, Sugiura M. cDNA structure, expression and nucleic acid-binding properties of three RNA-binding proteins in tobacco: occurence of tissue-specific alternative splicing. Nucl Acid Res 1993;21:3981–3987.
34. Darnell JE, Doolittle WF. Speculations on the early course of evolution. Proc Natl Acad Sci USA

1986;83:1271–1275.
35. Gilbert W, Marchionni M, McKnight G. On the antiquity of introns. Cell 1986;46:151–154.
36. Margulis L. Symbiosis in Cell Evolution. San Francisco: W.H. Freeman and Co., 1981.
37. Gray MW. The evolutionary origins of organelles. Trends Genet 1989;5:294–299.
38. Mortenson E, Dreyfuss G. RNP in maize protein. Nature 1989;337:312.
39. Simpson GG, Vaux P, Clark G, Waugh R, Beggs JD, Brown JWS. Evolutionary conservation of the spliceosomal protein, U2B". Nucl Acid Res 1991;19:5213–5217.
40. Mulligan ME, Jackman DM, Murphy ST. Heterocyst-forming filamentous cyanobacteria encode proteins that resemble eukaryotic RNA-binding proteins of the RNP family. J Molec Biol 1994;235:1162–1170.
41. Sato N. A cold-regulated cyanobacterial gene cluster encodes RNA-binding protein and ribosomal protein S21. Plant Molec Biol 1994;24:819–823.
42. Kovacs SA, O'Neil J, Watcharapijarn J, Moe-Kirvan C, Vijay S, Silva V. Eubacterial components similar to small nuclear ribonucleoproteins: identification of immunoprecipitable proteins and capped RNAs in a cyanobacterium and a gram-positive eubacterium. J Bacteriol 1993;175:1871–1878.
43. Xu MQ, Kathe SD, Goodrich-Blair H, Nierzwicki-Bauer SA, Shub DA. Bacterial origin of a chloroplast intron: conserved self-splicing group I introns in cyanobacteria. Science 1990;250:1566–1570.
44. Kuhsel MG, Strickland R, Palmer JD. An ancient group I intron shared by eubacteria and chloroplasts. Science 1990;250:1570–1573.
45. Ferat JL, Michel F. Group II self-splicing introns in bacteria. Nature 1993;364:358–361.

Tracing Biological Evolution in Protein and Gene Structures.
M. Gō and P. Schimmel, editors.

Mechanism of alternative RNA splicing

Kunio Inoue and Yoshiro Shimura
Department of Biophysics, Faculty of Science, Kyoto University, Kyoto, Japan

Abstract. Expression of the *Drosophila* somatic sex-determination genes, *Sex-lethal (Sxl), transformer (tra)* and *doublesex (dsx)*, is regulated by sex-specific splicing. *Sxl* and *tra* produce their functional protein products only in females, while *dsx* expresses distinct functional proteins in both sexes. Focusing on the mechanisms of the sex-specific splicing, we have shown the following:

1. Non-sex-specific splicing of *tra* pre-mRNA occurs in a default manner in both sexes. In contrast, female-specific splicing is induced by *Sxl* protein; *Sxl* protein binds to the U-rich sequence at the non-sex-specific splice site, resulting in the repression of non-sex-specific splicing.
2. In females, *Sxl* protein also induces female-specific splicing of its own pre-mRNA, maintaining the production of functional *Sxl* protein.
3. Male-specific splicing of *dsx* pre-mRNA occurs in a default manner, whereas female-specific splicing is promoted by *Tra* and *Tra*-2 proteins. These regulator proteins bind to the repeated sequences in the female-specific exon, activating the suboptimal female-specific splice site.

We will discuss the current idea that the similar mechanism of splice site selection is involved in both constitutive and alternative splicing.

Key words: *Drosophila*, sex determination, splice site, *Sxl*, *tra*, *dsx*.

Introduction

In higher eukaryotes, alternative splicing of messenger RNA precursors (pre-mRNAs) plays an important role in regulating gene expression; it can lead to the production of multiple mRNA species from a single gene during development and differentiation. Hence, one can easily imagine that studies on alternative splicing are indispensable for the understanding of developmental phenomena. Moreover, how certain combinations of 5′ and 3′ splice sites are correctly selected is a fundamental question for elucidation of the mechanism of constitutive as well as alternative pre-mRNA splicing.

Expression of the *Drosophila* somatic sex-determination genes (Fig. 1), *Sex-lethal (Sxl), transformer (tra)*, and *doublesex (dsx)*, is regulated by sex-specific splicing (Fig. 2) [1–3]. It has been suggested that *Sxl* and *tra* produce the functional proteins only in females, while *dsx* generates the distinct proteins in both sexes. Genetical studies have shown that *Sxl* controls expression of *tra*, and *tra* subsequently controls expression of *dsx* in collaboration with *tra-2* (Fig. 1) [4]. *Sxl* and *tra-2* have been

Address for correspondence: Dr Kunio Inoue, Department of Biophysics, Faculty of Science, Kyoto University, Kyoto 606-01, Japan.

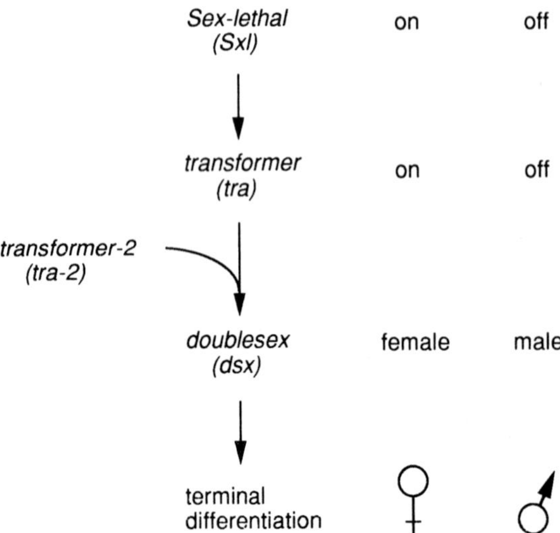

Fig. 1. The *Drosophila* somatic sex-determination pathway.

revealed to encode putative RNA-binding proteins, suggesting that they function as regulators of sex-specific splicing [1,5,6]. Focusing on these *Drosophila* sex-determination genes, we have studied the mechanisms of how alternative splicing is regulated.

Sex-specific splicing of *tra* pre-mRNA

Sex-specific splicing of *tra* pre-mRNA is achieved by the alternative usage of two 3′ splice sites (Fig. 2B) [2]. Choice of the upstream 3′ splice site leads to the inclusion of the second exon that contains a translational stop codon, generating nonfunctional mRNA in both female and male flies. The second exon is excluded only in females by the use of the downstream 3′ splice site, thereby producing mRNA that encodes functional *Tra* protein [2].

To clarify whether sex-specific splicing of *tra* pre-mRNA is controlled by *Sxl* gene products, we performed co-transfection experiments in which *Sxl* cDNA and the *tra* gene are expressed in *Drosophila* Kc cells [7]. When a plasmid expressing the *tra* gene under the control of copia-element LTR (copia-*tra* plasmid) was transfected into Kc cells, the non-sex-specific *tra* mRNA was exclusively generated. Only non-sex-specific splicing was observed when the copia-*tra* DNA was co-transfected with the male-specific *Sxl* cDNA located downstream of the *Drosophila* heat shock gene 70 promoter (hsp-*Sxl* M1). In contrast, female-specific splicing of *tra* pre-mRNA was observed on co-expression of the copia-*tra* plasmid with the female-specific *Sxl* cDNA (hsp-*Sxl* F1) or with a mixture of hsp-*Sxl* M1 and hsp-*Sxl* F1. These results led us to the conclusion that non-sex-specific splicing of *tra* occurs in a default manner, while

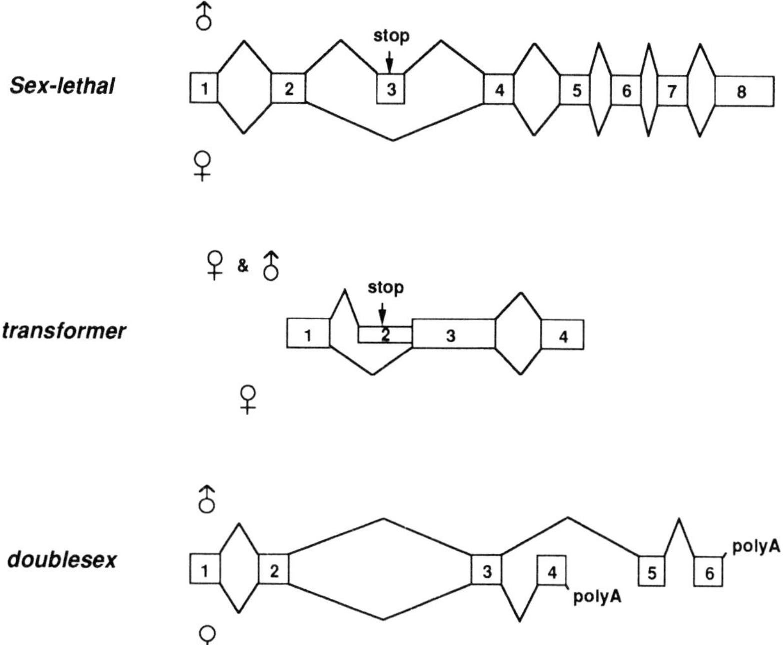

Fig. 2. Sex-specific splicing of *Sxl*, *tra*, and *dsx* pre-mRNAs. Boxes represent the exon sequences; stop: translational stop codon; polyA: polyadenylation site.

female-specific splicing is induced in the presence of female-specific (functional) *Sxl* protein. Moreover, it was confirmed that the male-specific *Sxl* mRNA does not encode a functional protein.

As shown in Fig. 3, we can think of two mechanisms how female-specific splicing of *tra* pre-mRNA is induced: blockage of the non-sex-specific 3′ splice site and activation of the female-specific 3′ splice site. To examine which is the case, we constructed two plasmids containing a mutant *tra* gene in which either one of the two 3′ splice sites had been deleted [7]. When the mutant plasmid copia-*tra*ΔN containing a deletion of the non-sex-specific 3′ splice site was transfected into Kc cells, splicing occurred efficiently at the female-specific 3′ splice site irrespective of the presence of *Sxl* protein. By contrast, with the mutant copia-*tra*ΔF, in which the female-specific splice site had been deleted, splicing occurred at the non-sex-specific 3′ splice site only in the absence of the functional *Sxl* protein. These results indicated that non-sex-specific splicing is inhibited in the presence of *Sxl* protein.

Since *Sxl* encodes a putative RNA-binding protein, we would expect that *Sxl* protein binds specifically to the *tra* pre-mRNA at or near the non-sex-specific 3′ splice site. We tested the binding activity of the *Sxl* protein by Northwestern blotting analysis [7]. The *Sxl* protein was expressed in *Escherichia coli* (*E. coli*) using the T7 phage promoter system. All of the proteins from *E. coli* overproducing the *Sxl* protein were subjected to SDS-PAGE and then blotted onto nitrocellulose membrane. The

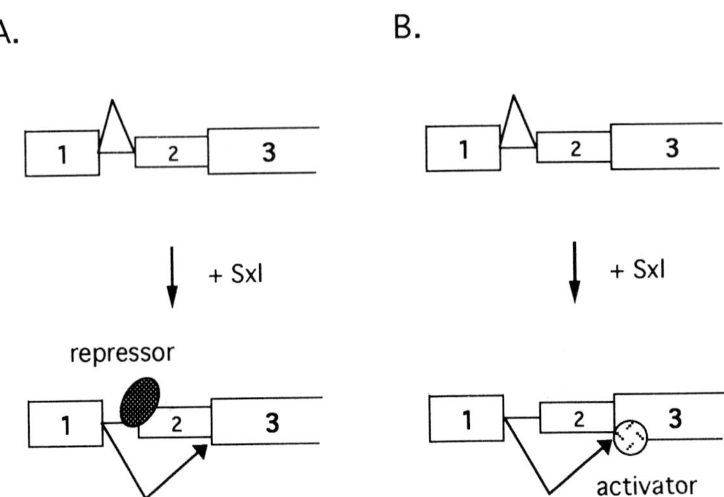

Fig. 3. Two possible mechanisms for female-specific splicing of *tra* pre-mRNA. Boxes and lines between boxes represent the exon and intron sequences, respectively. Upper line shows non-sex-specific splicing of *tra* pre-mRNA, while lower line shows female-specific splicing in the presence of the female-specific *Sxl* protein.

blotted membrane was probed with various *tra*-derived RNAs synthesized in vitro (Fig. 4). The results obtained were very consistent with our idea; the bacterially produced *Sxl* protein binds specifically to the U-rich sequence which lies in the pyrimidine cluster at the non-sex-specific 3′ splice site of *tra* pre-mRNA. It is known that the pyrimidine cluster is indispensable for the splicing reaction and recognized by some splicing factors such as U2AF (for review, see [8]). When a mutant *tra* gene in which U-to-C substitutions had been introduced into the U-rich

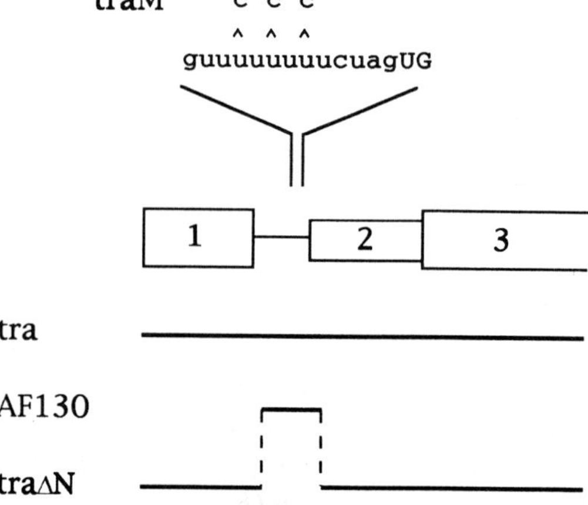

Fig. 4. Schematic representation of probe RNAs used in Northwestern blotting experiments. The bacterially produced *Sxl* protein bound to the *tra* and AF130 probes that contain the U-rich (U-octamer) sequence. *tra*M has U-to-C base substitutions in the U-rich sequence.

sequences was expressed in flies [9] and Kc cells [7], non-sex-specific splicing occurred irrespective of the presence of *Sxl* protein (Fig. 4). Therefore, *Sxl* protein functions as a negative regulator to compete out the splicing factors at the non-sex-specific 3′ splice site, resulting in the use of the female-specific splice site of *tra* pre-mRNA in female flies.

Autoregulation of sex-specific splicing of *Sxl* pre-mRNA

It has been suggested that expression of *Sxl* gene is maintained by autoregulation of the sex-specific splicing (Fig. 2) [10]. Male-specific *Sxl* mRNAs include the third exon which contains the translational stop codon, generating nonfunctional proteins, while female-specific mRNAs exclude this exon and encode the functional proteins [1]. Using transfection experiments, our group has revealed that *Sxl* protein induces female-specific splicing of *Sxl* pre-mRNA by inhibiting male-specific splicing [11]. Moreover, multiple *cis*-elements which include U-rich sequences are required for the regulation of sex-specific splicing [11].

Sex-specific splicing of *dsx* pre-mRNA

Pre-mRNA *dsx* undergoes sex-specific processing (splicing and polyadenylation reactions), which leads to the production of two distinct sex-specific polypeptides (Fig. 2) [3]. In females, the third exon is spliced to the female-specific fourth exon and the cleavage/polyadenylation reaction occurs immediately downstream of the fourth exon. In contrast, in males, the fourth exon is skipped, and the third exon is fused to the male-specific fifth exon followed by the sixth exon. Using a co-transfection system, we have shown that both the functional *Tra* and *Tra*-2 proteins are required for production of female-specific *dsx* mRNA, whereas male-specific splicing occurs in a default manner [12]. As shown in Fig. 5, there are several possibilities to explain how female-specific splicing is induced: 1) activation of the female-specific 3′ splice site; 2) activation of the cleavage/polyadenylation reaction using the female-specific polyadenylation signal; and 3) inhibition of the male-specific 3′ splice site. To determine which of the above mechanisms is operative, we constructed two chimeric plasmids (Fig. 6): copia-*dtz* contains the region of *dsx* that extends from the third exon to the female-specific fourth exon, followed by part of the intron and the second exon of *fushitarazu* (*ftz*); copia-*dsx*35m contains the region of *dsx* that extends from the third exon to the female-specific fifth exon, but lacks the entire female-specific exon and portions of its adjacent introns [12]. When copia-*dtz* was transfected into Kc cells, we observed almost exclusively the product of splicing between the *dsx* third exon and the *ftz* exon. In contrast, when copia-*dtz* was co-expressed with the female-specific *tra* cDNA and the *tra-2* cDNA, the female-specific splicing product was generated. Because copia-*dtz* does not contain either the female-specific polyadenylation signal or the male-specific 3′ splice site, we could exclude

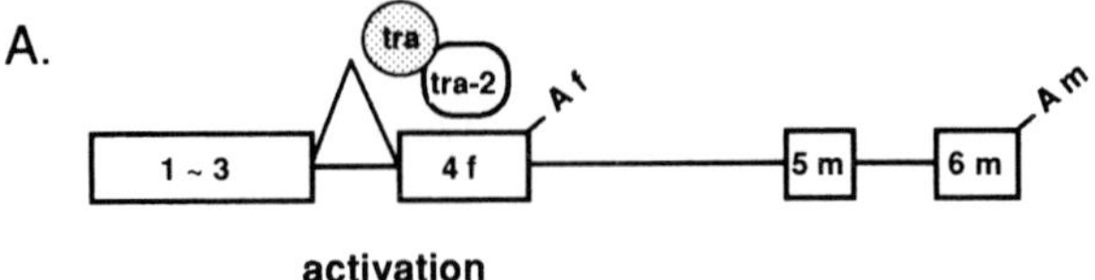

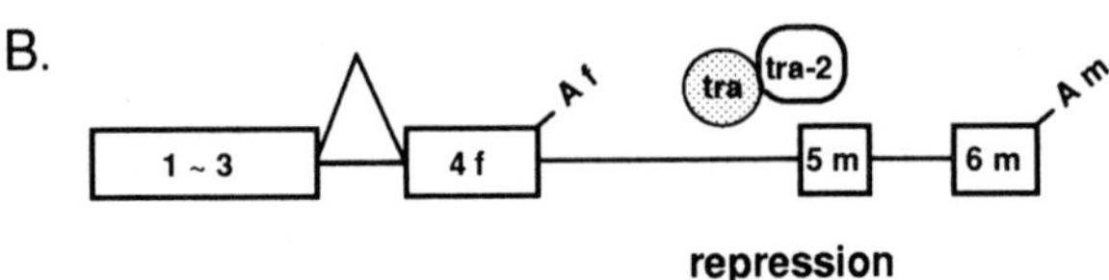

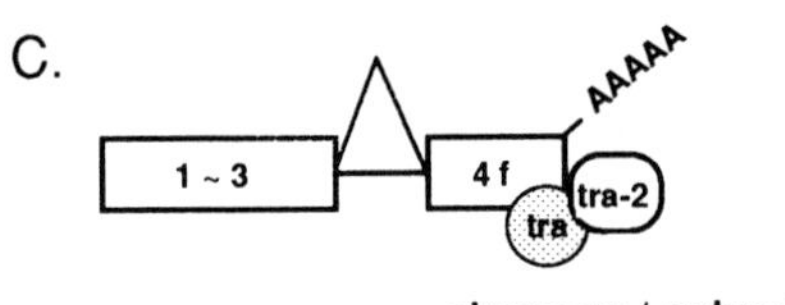

Fig. 5. Possible mechanisms for female-specific splicing of *dsx* pre-mRNA. Af, Am: the female- and male-specific polyadenylation signals.

possibilities 2 and 3. The results obtained with copia-*dsx*35m also support this notion; splicing between the third and the fifth exons occurs irrespective of the presence of *Tra* and *Tra*-2. Thus, we concluded that the female-specific processing results from activation of the female-specific 3′ splice site by the *Tra* and *Tra*-2 proteins.

It has been pointed out that the pyrimidine cluster of the female-specific 3′ splice site is interrupted by purine residues [13]. We introduced Pu-to-Py substitutions into the female-specific splice site of *dsx* minigene, which increased the length of the pyrimidine cluster (Fig. 7). When the resultant minigene was transfected, female-specific splicing occurred efficiently even in the absence of the *Tra* and *Tra*-2 proteins [12]. The results obtained suggest that the female-specific splice site has a

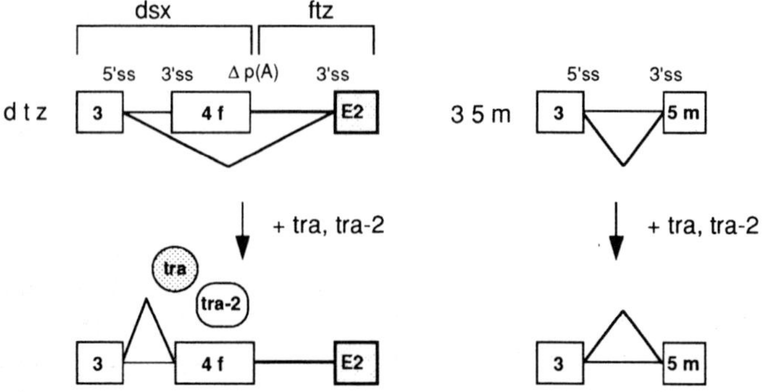

Fig. 6. Splicing of *dtz* and *dsx*35m pre-mRNAs. For *dtz* construct, the female-specific polyadenylation signal was deleted (Δp(A)). 5′ss, 3′ss: 5′ and 3′ splice sites.

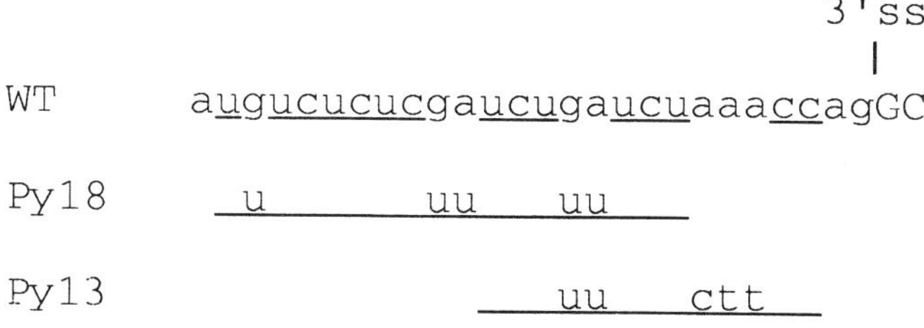

Fig. 7. Sequence of the pyrimidine cluster at the female-specific 3′ splice site of *dsx* pre-mRNA. Nucleotide sequences of wild (WT) or mutant (Py18, Py13) *dsx* minigenes were shown. Underlines represent pyrimidine nucleotides. 3′ss: the female-specific 3′ splice site.

suboptimal pyrimidine cluster, which could explain why the female-specific splice site is not used in male flies.

The female-specific fourth exon contains six tandemly interspersed repeats of 13-nucleotide (nt) sequence. We have shown that deletion or base-substitutions of the repeat sequences results in the failure of female-specific splicing [12,14]. These results led to the conclusion that the repeat sequences function as *cis*-elements for the female-specific splicing of *dsx* pre-mRNA.

To determine whether either or both of the *Tra* and *Tra*-2 proteins directly associate with *dsx* pre-mRNA, we performed UV-cross-linking experiments [14]. The bacterially produced *Tra* and *Tra*-2 proteins were incubated with probe RNAs synthesized in vitro. The reaction mixtures were irradiated with UV-light and then electrophoresed in SDS-PAGE. When *dsx* pre-mRNA was used as a probe, both *Tra* and *Tra*-2 bound to the probe RNA. Moreover, we found that these proteins bound specifically to the 13-nt sequences. These results strongly suggest that the female-specific splicing of *dsx* pre-mRNA is activated by binding of the female-specific *Tra* protein and *Tra*-2 protein to the 13-nt sequences.

Discussion

Studies on the *Drosophila* sex-determination pathway led us to the finding of two regulatory mechanisms of alternative splicing. In the negative control (Fig. 8A), the *Sxl* protein binds to its own and *tra* pre-mRNAs, blocking usage of the default splice sites [7,11]. On the other hand, in the positive regulatory system (Fig. 8B), *Tra* and *Tra*-2 bind to the exon sequence of *dsx* pre-mRNA, activating the suboptimal splice site [12,14]. Recently, in vitro studies have shown that *Tra* and *Tra*-2 function by recruiting general splicing factors, including members of the SR protein family, that are essential for constitutive splicing [15,16]. Members of the SR protein family have similar structural features, containing one or two RBDs and an RS domain (for review, see [17,18]). It is most likely that the RS domains in *Tra*, *Tra*-2, and the SR proteins are involved in protein-protein interactions.

It has been shown that exon sequences are involved in general splice site selection.

A. *tra* pre-mRNA

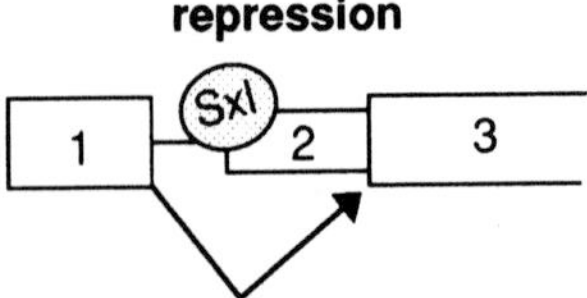

B. *dsx* pre-mRNA

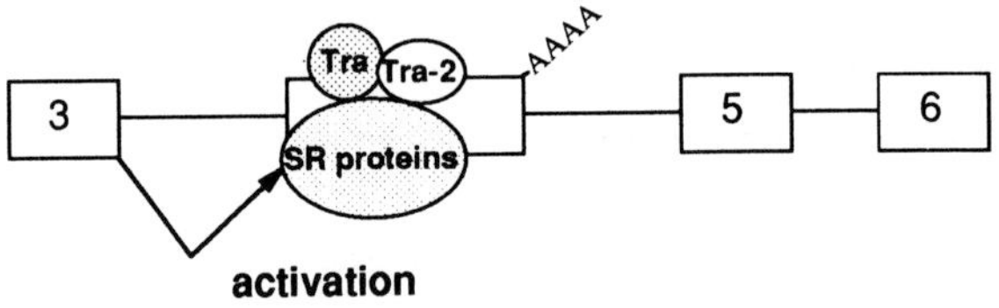

Fig. 8. Negative and positive controls of splicing.

For example, the purine-rich sequence located within the last exon of the mouse immunogloblin μ gene promotes splicing of distant upstream introns regardless of cell types [19]. Furthermore, if the 13-nt sequences of *dsx* pre-mRNA is replaced by the purine-rich exon sequence, it can activate the usage of the female-specific 3′ splice site even in the absence of *Tra* and *Tra*-2 [20]. Such exonic enhancer sequences are thought to be most important when the splice sites to be selected are suboptimal or weak [20,21]. Interestingly, it was demonstrated that one of the SR protein family, SF2/ASF, binds to the Pu-rich sequence in the last exon of bovine growth hormone pre-mRNA, thereby stimulating splicing of the upstream intron [22]. The stimulation is counteracted by the addition of hnRNP A1. Another SR protein, SC35, neither binds to the sequence nor stimulates splicing of the intron [22]. The SR proteins are differentially expressed in a variety of tissues [17,18]. Therefore, it is possible that alternative splicing in many genes is controlled directly by combinations of the SR proteins and other RNA-binding proteins such as hnRNP A1; no specific factors are involved (for review see [23]). In contrast, specific regulatory proteins may be required for positive controls of alternative splicing in some cases [23]. It is likely that such specific proteins associate with pre-mRNAs and recruit general splicing factors, such as the SR proteins, to activate specific splice sites.

Acknowledgements

This work was supported in part by Grants-in-Aid from the Ministry of Education, Science, and Culture of Japan and from the Mitsubishi Foundation. K.I. was also supported by a grant from Senri Life Science Foundation.

References

1. Bell LR, Maine EM, Schedle P, Cline TW. Sex-lethal, a *Drosophila* sex determination switch gene, exhibits sex-specific RNA binding and sequence similarity to RNA binding proteins. Cell 1988;55:1037–1046.
2. Boggs RT, Gregor P, Idriss S, Belote JM, McKeown M. Regulation of sexual differentiation in *D. melanogaster* via alternative splicing of RNA from the transformer gene. Cell 1987;50:739–747.
3. Burtis KC, Baker BS. *Drosophila* doublesex gene controls somatic sexual differentiation by producing alternatively spliced mRNAs encoding related sex-specific polypeptide. Cell 1989;56:997–1010.
4. Nagoshi RN, McKeown M, Burtis KK, Belote JM, Baker BS. The control of alternative splicing at genes regulating sexual differentiation in *D. melanogaster*. Cell 1988;53:229–236.
5. Amrein H, Gorman M, Nöthiger R. The sex-determining gene tra-2 of *Drosophila* encodes a putative RNA binding protein. Cell 1988;55:1025–1035.
6. Goralski TJ, Edstrom J-E, Baker BS. The sex determination locus transformer-2 of *Drosophila* encodes a polypeptide with similarity to RNA binding proteins. Cell 1989;56:1011–1018.
7. Inoue K, Hoshijima K, Sakamoto H, Shimura Y. Binding of the *Drosophila* sex-lethal gene product to the alternative splice site of transformer primary transcript. Nature 1990;344:461–463.
8. Green MR. Biochemical mechanisms of constitutive and regulated pre-mRNA splicing. Ann Rev Cell Biol 1991;7:559–599.
9. Sosnowski BA, Belote JM, McKeown M. Sex-specific alternative splicing of RNA from the transformer gene results from sequence-dependent splice site blockage. Cell 1989;58:449–459.
10. Bell LR, Horabin JI, Schedle P, Cline TW. Positive autoregulation of sex-lethal by alternative splicing maintains the female determined state in *Drosophila*. Cell 1991;65:229–239.
11. Sakamoto H, Inoue K, Higuchi I, Ono Y, Shimura Y. Control of *Drosophila* sex-lethal pre-mRNA splicing by its own female-specific product. Nucl Acid Res 1992;20:5533–5540.
12. Hoshijima K, Inoue K, Higuchi I, Sakamoto H, Shimura Y. Control of doublesex alternative splicing by transformer and transformer-2 in *Drosophila*. Science 1991;252:833–836.
13. Nagoshi RN, Baker BS. Regulation of sex-specific splicing at the *Drosophila* doublesex gene: *cis*-acting mutation in exon sequences alter sex-specific RNA splicing patterns. Genes Dev 1990;4:89–97.
14. Inoue K, Hoshijima K, Higuchi I, Sakamoto H, Shimura Y. Binding of the *Drosophila* transformer and transformer-2 proteins to the regulatory elements of doublesex primary transcript for sex-specific RNA processing. Proc Natl Acad Sci USA 1992;89:8092–8096.
15. Tian M, Maniatis T. Positive control of pre-mRNA splicing in vitro. Science 1992;256:237–240.
16. Tian M, Maniatis T. A splicing enhancer complex controls alternative splicing of doublesex pre-mRNA. Cell 1993;74:105–114.
17. Birney E, Kumar S, Krainer AR. Analysis of the RNA-recognition motif and RS and RGG domains: conservation in metazoan pre-mRNA splicing factors. Nucl Acid Res 1993;21:5803–5816.
18. Moore MJ, Query CC, Sharp PA. In: Gesteland RF, Atkins JF (eds) The RNA World. New York: Cold Spring Harbor Laboratory Press, 1993;303–357.
19. Watakabe A, Sakamoto H, Shimura Y. Repositioning of an alternative exon sequence of mouse IgM pre-mRNA activates splicing of the preceding intron. Gene Expression 1991;1:175–184.
20. Watakabe A, Tanaka K, Shimura Y. The role of exon sequences in splice site selection. Genes Devel 1993;7:407–418.
21. Watakabe A, Shimura Y. The role of exon sequences in splice site selection. Molec Cell Biol 1994;14:1347–1354.
22. Sun Q, Mayeda A, Hampson RK, Krainer AR, Rottman FM. General splicing factor SF2/ASF promotes alternative splicing by binding to an exonic splicing enhancer. Genes Devel 1993;7:2598–2608.
23. Inoue K, Ohno M, Shimura Y. Aspects of splice site selection in constitutive and alternative pre-mRNA splicing. Gene Expression 1995;4:177–182.

Tracing Biological Evolution in Protein and Gene Structures.
M. Gō and P. Schimmel, editors.

Spliceosomal introns in the spliceosomal small nuclear RNA genes

Tokio Tani, Yuzi Takahashi, Seiichi Urushiyama and Yasumi Ohshima

Department of Biology, Faculty of Science, Kyushu University, Hakozaki, Higashi-Ku, Fukuoka, Japan

Abstract. Small nuclear RNAs (snRNAs) U1, U2, U4, U5 and U6 are a group of metabolically stable RNA molecules involved in pre-mRNA splicing. We have cloned the U6 snRNA gene in the fission yeast *Schizosaccharomyces pombe* (*S. pombe*) and found an intron with a structure characteristic for a nuclear pre-mRNA intron (a spliceosomal intron). To investigate the origin of the U6 intron, we analyzed the U6 snRNA genes of 52 organisms using the polymerase chain reaction, and found that the yeast species *Rhodotorula hasegawae* (*R. hasegawae*) (*Erythrobasidium hasegawianum*) and *Rhodosporidium dacryoidum* (*R. dacryoidum*) also have spliceosomal introns at different positions in their U6 snRNA genes. No intron was found within the region of the U6 snRNA genes examined in all the other organisms tested. All the U6 introns are located within the conserved region, which is strikingly similar in structure to the catalytic center of the negative strand of the satellite RNA of tobacco ring-spot virus ((-)sTRSV) and to the conserved core region of the group I self-splicing intron. We also found a spliceosomsal intron in the U2 snRNA gene of *R. hasegawae.* The intron of the U2 snRNA gene is present in the highly conserved region immediately downstream of the branch site recognition domain, which can form a novel base pairing with U6 snRNA. These findings support the idea that the U2/U6 complex plays a catalytic role in pre-mRNA splicing, and that the spliceosomal introns in the snRNA genes occurred through reverse splicing of an intron excised from an mRNA precursor into a catalytic region of the snRNA complex.

Key words: intron, pre-mRNA splicing, U2 snRNA, U6 snRNA.

Introduction

The nuclei of all types of eukaryotic cells contain a class of metabolically stable small RNAs known as small nuclear RNAs (snRNAs) [1]. Six major snRNAs, U1–U6 snRNA, were identified in mammalian cells. These RNAs form small nuclear ribonucleoprotein particles (snRNPs) with proteins. Of these, U4 and U6 snRNAs are base paired via two extensive intermolecular RNA helices and are present in the same snRNP. The U1, U2, U5 and U4/U6 snRNPs are involved in pre-mRNA splicing (for reviews, see [2–4]) and U3 snRNP is required for the processing of ribosomal RNAs [5].

As shown in Fig. 1, splicing of pre-mRNA is performed by a two-step transesterification mechanism [2–4]. In the first step the 5′ splice site is cleaved to generate a 5′ exon and an intron-second exon in a lariat configuration. The lariat structure is the result of a 2′-5′ phosphodiester bond between the guanosine at the 5′ splice

---

Address for correspondence: Dr Tokio Tani, Department of Biology, Faculty of Science, Kyushu University, Hakozaki, Higashi-Ku, Fukuoka 812-81, Japan. Fax: +81-092-632-2741.
E-mail: ttaniscb@mbox.nc.kyushu-u.ac.jp

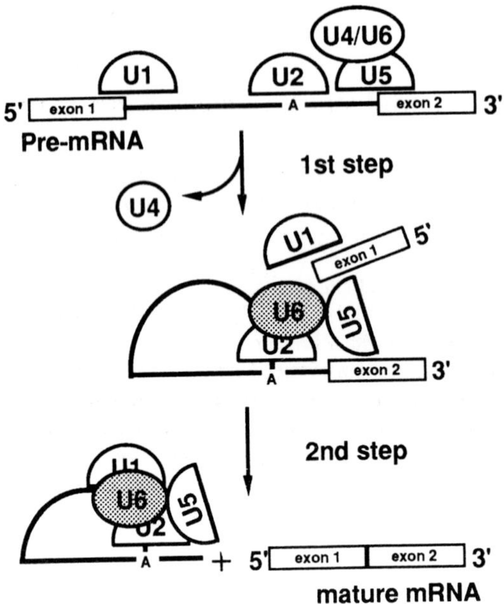

Fig. 1. Schematic representation of the pre-mRNA splicing pathway and the involvement of snRNPs in the reaction. The exons are shown as boxes and the intron as a thick line.

site and the 2′-hydroxyl of an adenosine near the 3′ splice site of the intron (the branch site). In the second step the 3′ splice site is cleaved to generate ligated exons and an excised lariat intron. These reactions take place in a large protein-RNA complex, termed the spliceosome, which contains U1, U2, U5 and U4/ U6 snRNPs.

Assembly of spliceosome on pre-mRNAs in vitro involves sequential binding of the U1 snRNP, the U2 snRNP, and U4/ U6 and U5 snRNPs as a preformed U4/ U5/ U6 tri-snRNP. U1 snRNA recognizes the 5′ splice site through the base pairing with the consensus sequence at the 5′ splice site. U2 snRNA forms base pairings with the branch site sequence in an intron of a pre-mRNA. U5 snRNA interacts with exon nucleotides adjacent to the 5′ and 3′ splice sites, and is thought to recognize the 5′ splice site in cooperation with U1 snRNA and orient correctly the 5′ and 3′ exons before the ligation reaction [4]. Prior to the 5′ splice site cleavage, the base pairing between U4 and U6 snRNAs is unwound, reducing the affinity of U4 snRNP for the spliceosome [6,7]. UV cross-linking experiments revealed that the conserved sequence of U6 snRNA interacts with the sequence downstream of the 5′ splice site [8,9].

U6 snRNA has several unique properties. First, the U6 snRNA is the most highly conserved spliceosomal snRNA. Second, a cap structure in U6 snRNA is γ-monomethyl phosphate [10], which is not present in other snRNAs. Third, U6 snRNA is the only spliceosomal snRNA that lacks an Sm binding site. Forth, U6 snRNA is transcribed by RNA polymerase III [11], whereas other snRNAs are transcribed by RNA polymerase II.

The apparent constraints on both the size and the sequence of U6 snRNA among species argue for a central role of U6 snRNA in the splicing process. To elucidate the function of U6 snRNA, we isolated the U6 snRNA gene of the fission yeast, *Schizosaccharomyces pombe* (*S. pombe*). *S. pombe* is as convenient for genetic analysis and gene manipulation as the budding yeast, *Saccharomyces cerevisiae* (*S. cerevisiae*), and the splicing machinery of *S. pombe* is thought to be closer than that of *S. cerevisiae* to the machinery of higher eukaryotes. The analysis of the gene structure revealed the presence of a spliceosomal intron in the *S. pombe* U6 snRNA gene. Systematic analysis of the U6 snRNA genes in other organisms by PCR (polymerase chain reaction) and the analysis of the gene for *Rhodotorula hasegawae* (*R. hasegawae*) U2 snRNA, which forms the complex with U6 snRNA in the spliceosome, revealed the presence of spliceosomal introns in the regions important for the functions of these snRNAs. The possible implications of these findings for the mechanisms of intron acquisition and for the role of U2/U6 snRNA complex in the pre-mRNA splicing reaction will be discussed.

Materials and Methods

Preparation of total DNAs and RNAs

Total DNAs from various organisms were prepared as described [12] or provided by other researchers. Total RNAs from yeasts were prepared using glass beads as described [13].

Northern blot, Southern blot and RNA sequencing

Northern and Southern blot analyses were performed as described [14]. Hybridization was done at 42°C in a solution containing 6 X SSC, 50 mM sodium phosphate, 5 X Denhardt's solution, 0.1% SDS, 2 mM EDTA, and 20 µg/ml of tRNA. After hybridization, a filter was washed for 20 min in a solution containing 6 X SSC and 0.1% SDS at room temperature, and then 42 and 50°C for 20 and 10 min, respectively.

RNA sequencing was done using the dideoxy method with a reverse transcriptase. The oligonucleotide primer complementary to the presumed exon sequence was kinase labeled and annealed with 10 µg of total RNA. The sample was split into four aliquots and used to direct cDNA synthesis in the presence of ddATP, ddGTP, ddCTP, or ddTTP. The products were analyzed on an 8% polyacrylamide/8.3 M urea gel.

PCR reactions

PCR reactions were done in a solution containing 50 mM KCl, 10 mM Tris-HCl (pH 8.5), 2.5 mM $MgCl_2$, 0.1% Triton X-100, 200 mM dNTP, 0.1 µM of each primer, 2.5 units of Taq polymerase (Promega) and 1 µg of total DNA. Amplification was carried out by 30 cycles of denaturation for 1 min at 90°C, annealing for 2 min at 50°C and extension for 2 min at 70°C. The products were run on a 10% polyacrylamide gel.

For inverse PCR, the total DNA was digested by Sau3A or HhaI, circularized with T4 DNA ligase and subjected to PCR using the reverse primers corresponding to the intron region as described [14].

Results

A spliceosomal intron in the U6 snRNA gene of S. pombe

The nucleotide sequence of U6 snRNA is highly conserved among species. We have screened the *S. pombe* genomic gene library using the 30-mer oligonucleotide complementary to the most conserved region of U6 snRNA, and isolated the 2.0 kb DNA fragment containing the gene coding for U6 snRNA [15]. Figure 2 shows the nucleotide sequence of the *S. pombe* U6 snRNA gene. The 5′ end of the coding region was determined by primer extension and RNA sequencing using the dideoxy method [15]. The 3′ end of the gene was tentatively assigned to be at +149 by comparison with rat U6 snRNA sequence.

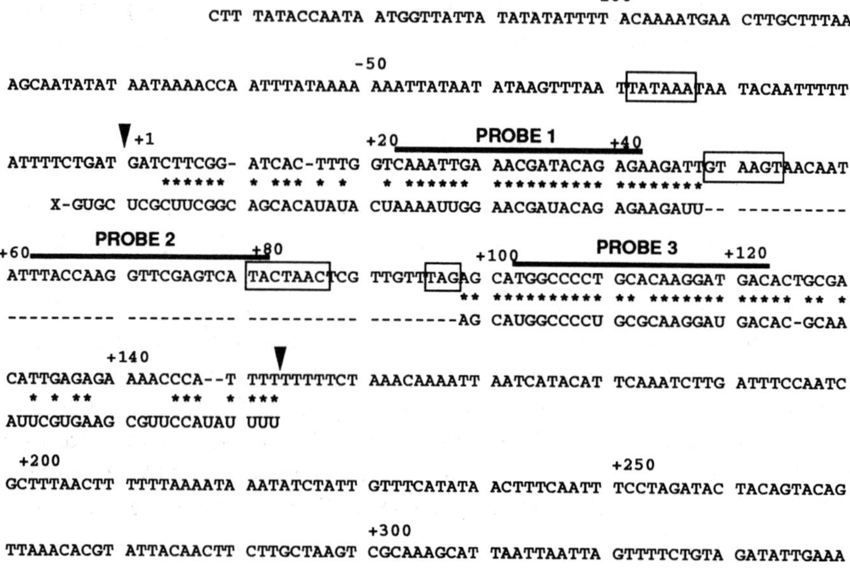

Fig. 2. Nucleotide sequence of the *S. pombe* U6 snRNA gene [15]. The upper line shows the sequence of the noncoding DNA strand. The lower line shows the sequence of the rat U6 snRNA. The identical nucleotides are indicated by asterisks. The arrowheads show the 5′ and 3′ ends of the gene. A putative TATA box for the transcription and the sequences similar to the consensus sequences of donor, acceptor and branch sites are boxed. Lines over the sequences represent the regions complementary to the probes used in Northern blot analysis.

Comparison of the nucleotide sequence of the coding region with that of rat U6 snRNA revealed an unexpected structure of the *S. pombe* gene. Sequence identity with rat U6 snRNA is abruptly interrupted at +47, and is resumed at +97. Interestingly, the 5′ and 3′ ends of the interruption are GTAAGT and TAG, respectively. These are identical to the consensus sequences of the 5′ and 3′ splice sites of eukaryotic protein-coding genes. In addition, the sequence 5′-TACTAAC-3′, which is located 12 base pairs upstream of the 3′ boundary, matches the potential branch site sequence 5′-CTRAY-3′ present 6–18 residues upstream of the 3′ splice sites of *S. pombe* genes [16]. Moreover, this sequence is identical to the TACTAAC box, the strictly conserved branch site sequence found in *S. cerevisiae*. The distance between GT and AG is similar to the length of many *S. pombe* introns. All these structural features suggest that region +47 to +96 is a spliceosomal intron.

To demonstrate that the intron-like sequence is actually an intron, we performed several experiments described below. First, Northern blot analysis was performed using the oligonucleotides complementary to three parts of the gene as probes. As shown in Fig. 3A, probes 1 and 3, which are complementary to the presumed exons, hybridized to RNA of about the same size as 100 nucleotides. In contrast, no hybridization band was detected with the probe 2, which is complementary to the intron-like region. These results suggest that the middle part of the transcript had been removed. Second, the partial nucleotide sequence of the *S. pombe* U6 snRNA was determined by the dideoxy method to identify the precise splice site (Fig. 3B).

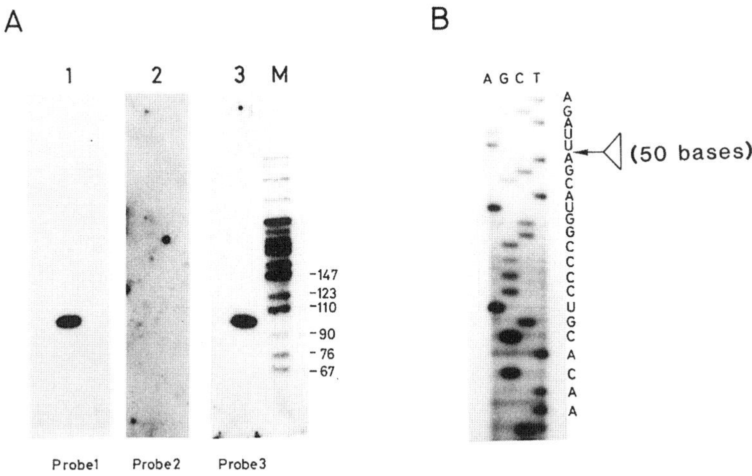

Fig. 3. A: Northern blot analysis using *S. pombe* total RNA [15]. Blots were probed separately with the probe 1 (lane 1), probe 2 (lane 2) and probe 3 (lane 3) shown in Fig. 2. Markers were ^{32}P-labeled Hpa II fragments of pBR322. B: Nucleotide sequence of the mature U6 snRNA near the splice junction. RNA sequencing was done by the dideoxy method.

The nucleotide sequence of the mature U6 snRNA indicates that the region +47 to +96 is indeed spliced out from a primary transcript. Third, Southern blot analysis using the *S. pombe* genomic DNA showed the single hybridizing band, demonstrating that the single U6 snRNA gene is present in *S. pombe* (Fig. 4). This result excludes the possibility that another intron-less U6 snRNA gene is present in the *S. pombe* genome. Fourth, the accumulation of a putative U6 snRNA precursor of about 150 nucleotides was observed when *S. pombe* cells had been heat shocked at 43°C for 30 min to inhibit the splicing reaction in vivo (Fig. 5). In addition, splicing of the U6 snRNA precursor was shown to be impaired in *S. pombe* ts$^-$ mutants that are defective in pre-mRNA splicing [17], suggesting that splicing of the pre-U6 snRNA

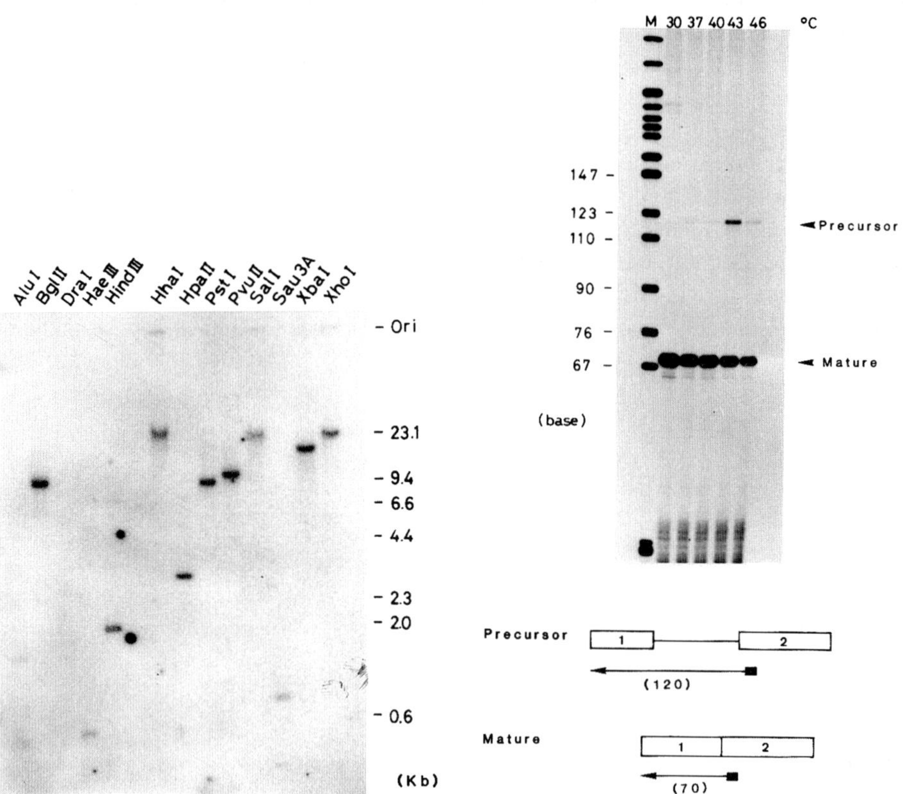

Fig. 4. Southern blot analysis of the total genomic DNA from *S. pombe*. The total *S. pombe* DNA was digested with each restriction enzyme indicated, blotted and probed with the probe 1 complementary to the exon 1. The single band was observed in all cases, suggesting that the single U6 snRNA gene is present in the *S. pombe* genome.

Fig. 5. Detection of the precursor of U6 snRNA. Primer extension was performed using ^{32}P-labeled probe 3 shown in Fig. 2. Total RNA prepared from the *S. pombe* cells exposed to the indicated temperature for 30 min was hybridized with the probe 3, reverse transcribed using AMV reverse transcriptase and electrophoresed on an 8% polyacrylamide/8.3 M urea gel. Schematic structures of the predicted products from the precursor and mature U6 snRNAs are shown in a lower panel.

is performed by a mechanism common to that for pre-mRNA splicing. Taken together, we concluded that the intron-like insertion found in the *S. pombe* U6 snRNA gene is a spliceosomal intron.

This was the first example of an snRNA gene containing an intron. It was surprising that the spliceosomal intron is found in the U6 snRNA gene, because it is not a protein-coding gene and U6 snRNA itself is involved in removal of an intron. The intron is present in the highly conserved region of U6 snRNA. Based on the discovery of the spliceosomal intron in a highly conserved region of the *S. pombe* U6 snRNA gene, Brow and Guthrie [18] proposed the hypothesis that U6 snRNA functions as a catalytic element during the pre-mRNA splicing reaction and that the U6 intron was generated by erroneous insertion of an intron excised from a pre-mRNA into U6 snRNA. In that scenario, the site of intron insertion reflects the catalytic center of U6 snRNA [18].

Analysis of the U6 snRNA genes in various organisms by PCR

In relation to the hypothesis described above, we examined whether U6 genes in organisms other than *S. pombe* have an intron. DNAs from phylogenetically diverse organisms were subjected to the polymerase chain reaction (PCR) to examine structures of the U6 genes [14]. We used as PCR primers 20 mer oligonucleotides corresponding to the highly conserved regions flanking the site of the *S. pombe* U6 intron. The predicted size of the amplified products from the intron-less U6 gene is approximately 50 bp. We first analyzed 24 kinds of organisms from animals, plants, and fungi. As shown in Fig. 6A, organisms other than *S. pombe* produced amplified products of approximately 50 bp, indicating that there are no introns in the U6 snRNA genes of these organisms. The longer products from *Drosophila*, horseshoe crab, and *Tetrahymena* DNAs are generated because of the tandem repetition of the U6 snRNA genes in these organisms [14,19].

Then, we analyzed U6 snRNA genes from various kinds of yeasts. DNAs from 18 yeast genera were analyzed as shown in Fig. 6B. Interestingly, DNAs from *Rhodosporidium dacryoidum* (*R. dacryoidum*) and *R. hasegawae* (*Erythrobasidium hasegawianum*) produced the amplified products of 130 and 110 bp, respectively, suggesting that the U6 snRNA genes in these yeasts have introns.

Structure of the R. dacryoidum U6 gene

Sequence analysis of the amplified products from *R. dacryoidum* DNA revealed the presence of an intron-like insertion of 76 bp in the U6 snRNA gene. The partial nucleotide sequence of the *R. dacryoidum* U6 snRNA gene obtained by the inverse PCR method is presented in Fig. 7. The insertion contains GT and AG at the 5′ and 3′ ends, suggesting that it is a spliceosomal intron. Northern blot analysis and RNA sequencing demonstrated that the intron-like sequence is indeed an intron [14].

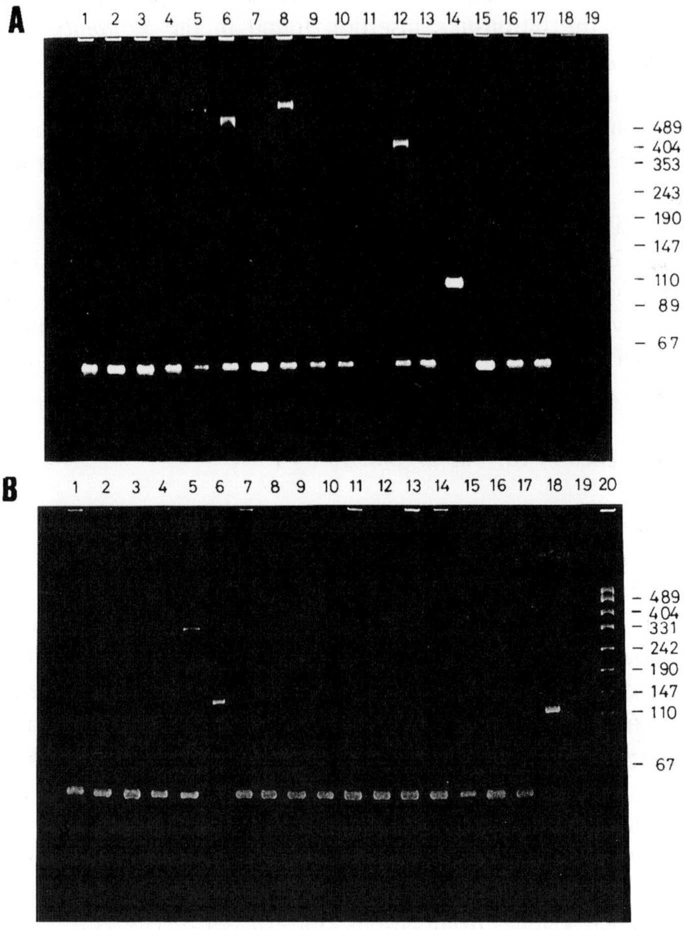

Fig. 6. PCR analysis of U6 genes in various organisms [14]. A: Analysis of U6 genes of phylogenetically diverse organisms. A part of the U6 snRNA gene was amplified with an upstream primer (corresponding to +21 – +40 of the *S. pombe* U6 snRNA gene) and a downstream primer (+57 – +71). Genomic DNAs examined were as follows: Human (lane 1), rat (lane 2), pig (lane 3), calf (lane 4), salmon (lane 5), *Drosophila melanogaster* (lane 6), *Caenorhabditis elegans* (lane 7), horseshoe crab (lane 8), squid (lane 9), sea anemone (lane 10), sponge (lane 11), *Tetrahymena thermophila* (lane 12), *Physarum polycepharum* (lane 13), *S. pombe* (lane 14), *Dictyostelium discoideum* (lane 15), carrot (lane 16), rice (lane 17), *Escherichia coli* (lane 18), mock reaction without DNA (lane 19). *Planaria, Neurospora crassa, Aspergillus oryzae,* seaweed, *Ascaris suum* and *Lentinus* DNAs produced only 50 bp bands (data not shown). B: Analysis of the U6 snRNA genes in various kinds of yeasts. DNAs examined were as follows: *Hanseniaspora guilliermondii* (lane 1), *Saccharomycopsis lipolytica* (lane 2), *Hansenula anomala* (lane 3), *Nadsonia commutata* (lane 4), *Cryptococcus laurentii* (lane 5), *R. dacryoidum* (lane 6), *Selenozyma peltata* (lane 7), *Schizoblastosporion kobayasii* (lane 8), *Filobasidium capsuligenum* (lane 9), *Ambrosiozyma philentoma* (lane 10), *Stephanoascus ciferrii* (lane 11), *Lipomyces tetrasporus* (lane 12), *Kluyveromyces lactis* (lane 13), *Leucosporidium scottii* (lane 14), *Arthroascus javanensis* (lane 15), *Oosporidium margaritiferum* (lane 16), *Leucosporidium antarcticum* (lane 17), *R. hasegawae* (lane 18), mock reaction without DNA (lane 19), pUC19/Hpa II (lane 20).

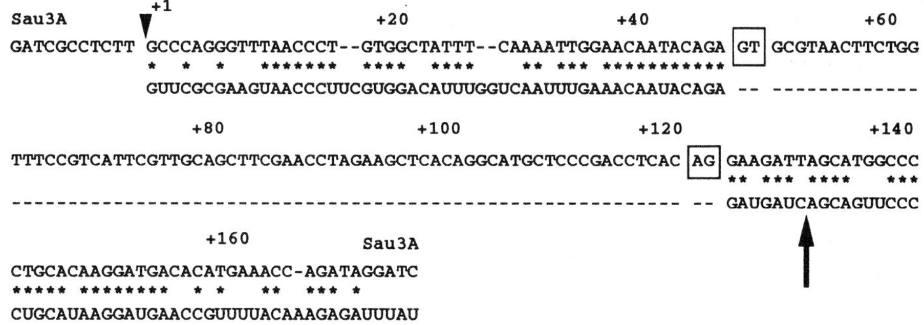

Fig. 7. Nucleotide sequence of the *R. dacryoidum* U6 snRNA gene [14]. The upper line shows the noncoding strand of the *R. dacryoidum* U6 snRNA gene, and the lower line shows the sequence of *S. cerevisiae* U6 snRNA. The 5′ end of the gene indicated by an arrowhead was determined with RNA sequencing. The GT and AG sequences found at the splice sites are boxed. The vertical arrow denotes the position of the intron in the *S. pombe* U6 snRNA gene.

Four spliceosomal introns are present in the R. hasegawae U6 snRNA gene

Sequence analysis of the amplified products from *R. hasegawae* DNA revealed a 63 bp intron-like insertion in the U6 snRNA gene. Figure 8 shows the complete nucleotide sequence of the *R. hasegawae* U6 snRNA gene obtained by the inverse PCR. Surprisingly, comparison of the gene sequence with the *S. pombe* U6 snRNA sequence revealed the presence of four spliceosomal introns, which have GT and AG at the 5′ and 3′ splice sites, in the *R. hasegawae* U6 snRNA gene.

The RNA sequencing of the mature *R. hasegawae* U6 snRNA (Fig. 9A) and Northern blot analysis using oligonucleotide probes complementary to each presumed exon and intron (Fig. 9B) demonstrated that the intron-like insertions are removed in the mature U6 snRNA in *R. hasegawae*. Based on these results, we concluded that *R. hasegawae* U6 snRNA gene has four spliceosomal introns. Introns were not found in the U6 snRNA genes in other species of *Rhodosporidium* and *Rhodotorula* [14].

Structure of the gene for U2 snRNA in R. hasegawae

Our next question was whether the presence of a spliceosomal intron is limited to a U6 snRNA gene or not. To address this question we analyzed structures of the genes for the spliceosomal snRNAs other than U6 snRNA in *R. hasegawae*. *R. hasegawae* has four spliceosomal introns in the U6 snRNA gene, even though the mature U6 snRNA is only 111 bases in length. This raises the possibility that a mechanism that generates and maintains many introns might be present in this organism.

We cloned the gene coding for U2 snRNA in *R. hasegawae* using an oligonucleotide complementary to the phylogenetically conserved U2 region which contains a domain forming base pairs with the branch site sequence [20]. Figure 10 represents the nucleotide sequence of the *R. hasegawae* U2 snRNA gene. Interestingly, the *R. hasegawae* U2 snRNA gene also has a spliceosomal intron of 60 bp. The U2 intron

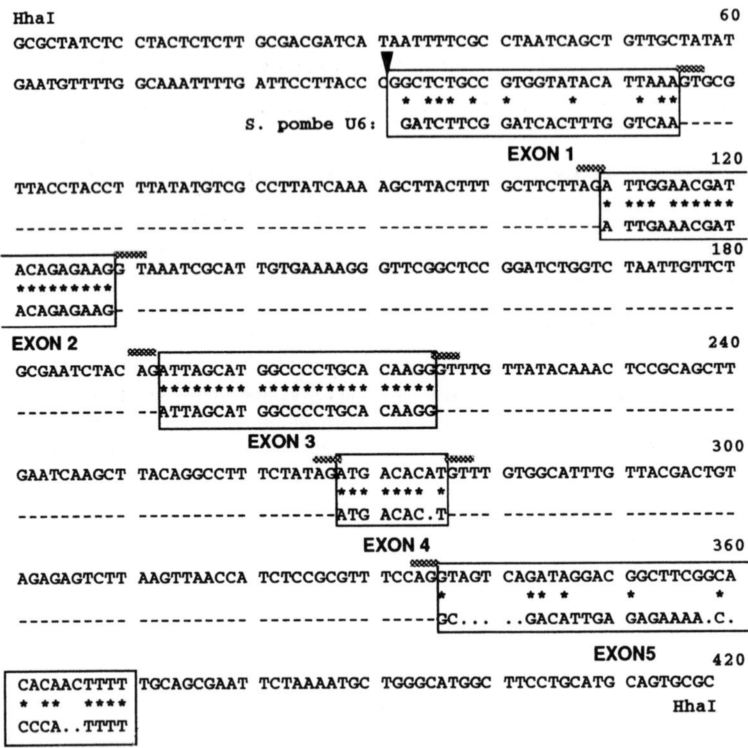

Fig. 8. The nucleotide sequence of the *R. hasegawae* U6 snRNA gene [14]. The upper and lower lines represent the noncoding strand sequence of the *R. hasegawae* U6 snRNA gene and the sequence of *S. pombe* U6 snRNA, respectively. The presumed exon sequences are boxed. Thick bars over nucleotides indicate the GT and AG sequences at the ends of the possible introns. The 5′ end was determined by RNA sequencing and shown by an arrowhead.

was found in the highly conserved region immediately downstream of the branch site recognition domain.

Discussion

Structural similarity between U6 snRNA and the catalytic domain of the ribozymes

We demonstrated that the U6 snRNA genes in three kinds of yeasts have spliceosomal introns, although they are not protein-coding genes. All of the spliceosomal introns in the U6 snRNA genes are found within the highly conserved region in the presumed secondary structure of U6 snRNA shown in Fig. 11A. Two hairpin structures can be formed in common among U6 snRNAs from human to *S. cerevisiae*. One possible hairpin structure at the 5′ end of U6 snRNA is required for the capping of U6 snRNA with a γ-methyltriphosphate cap [21]. The other possible hairpin

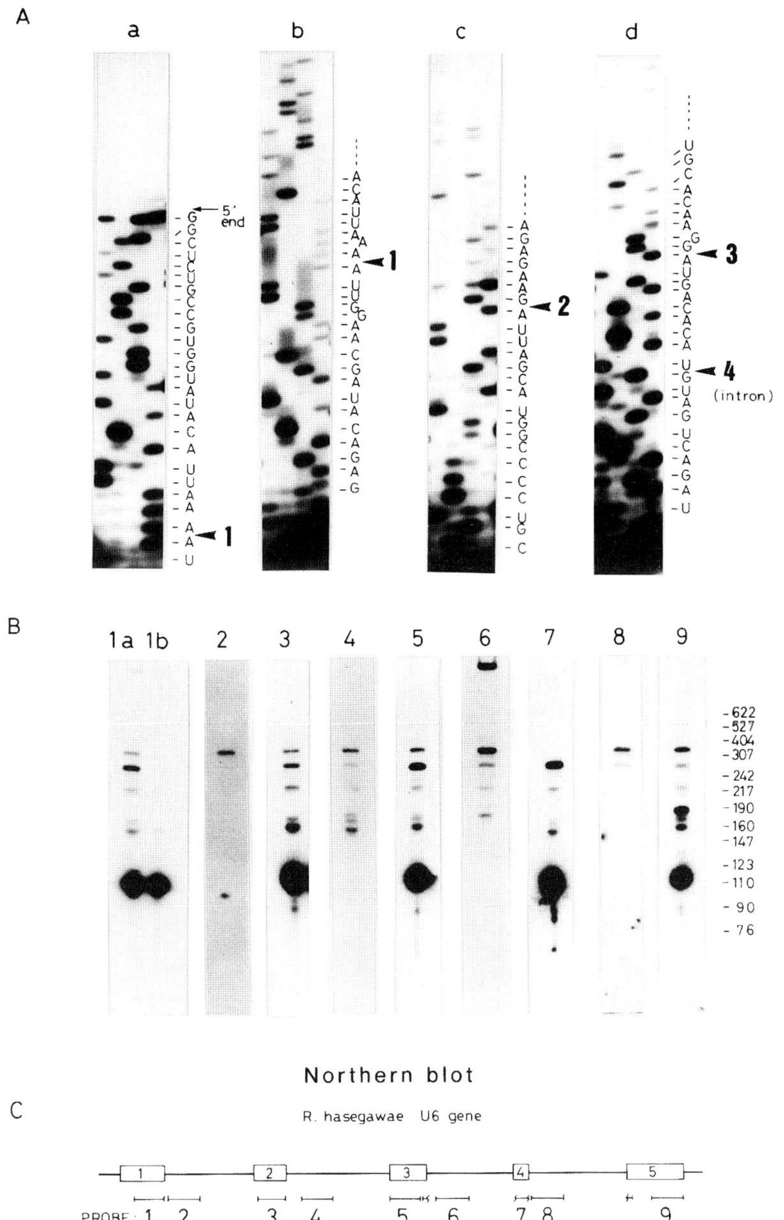

Fig. 9. A: Nucleotide sequence of the mature U6 snRNA of *R. hasegawae* determined by RNA sequencing [14]. Arrowheads represent the position of the intron sequences. B: Northern blot analysis of total RNA from heat-shocked (lanes 1a and 2 to 9) or normal (lane 1b) *R. hasegawae* cells. The probes used are shown in C. The lane numbers correspond to the probe numbers used for hybridization. The mature U6 snRNA is about 115 bases in length. The upper bands represent the U6 precursor or intermediate products during the splicing reaction. C: Schematic representation of the exon/intron organization of the *R. hasegawae* U6 snRNA gene.

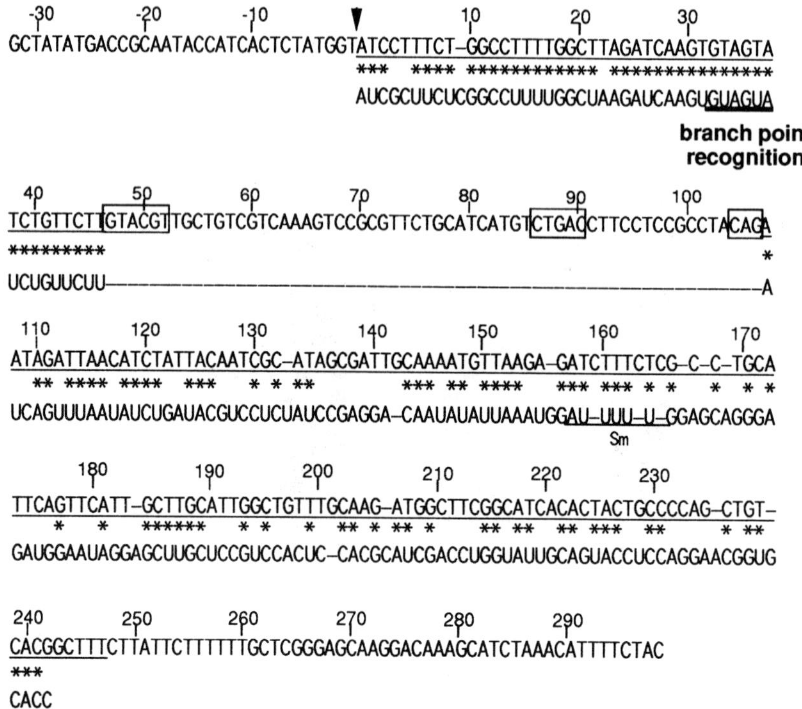

Fig. 10. The nucleotide sequence of the *R. hasegawae* U2 snRNA gene [20]. The sequence of human U2 snRNA is shown in the lower line. Asterisks denote the identical nucleotides. The 5′ end of the gene determined by RNA sequencing is indicated by an arrowhead. The donor, acceptor and the possible branch site sequences are boxed.

structure is formed downstream of the invariant sequence ACAGAGA in U6 snRNA [22,23]. Most of the U6 introns are located within the region from the ACAGAGA sequence to the second hairpin structure.

Interestingly, we found that the conserved region including the second hairpin structure of U6 snRNA has a remarkable structural similarity with the catalytic domain of the negative strand of the satellite RNA of tobacco ring-spot virus ((-)sTRSV) and the elements P and Q in the conserved core region of the group I self-splicing intron [14,24] (Fig. 11B,C). The (-)sTRSV is 359 bases in length and undergoes self-catalyzed cleavage at the 5′ side of a GU sequence, which is identical to the 5′ splice site consensus of pre-mRNA, during replication [25]. Ligation of cleaved segments of (-)sTRSV has also been observed in vitro. The minimum catalytic domain of (-)sTRSV consists of a 50-base RNA known as a hairpin ribozyme [26] (Fig. 11B). The catalytic domain of (-)sTRSV has a sequence ACAGAGAAG close to the cleavage site. The identical ACAGAGAAG sequence and a similar possible hairpin structure are also present in the U6 snRNA.

The conserved core region of the group I intron is involved in the catalysis of self-splicing. The catalytic core region of the group I intron also has CAGAGA sequence,

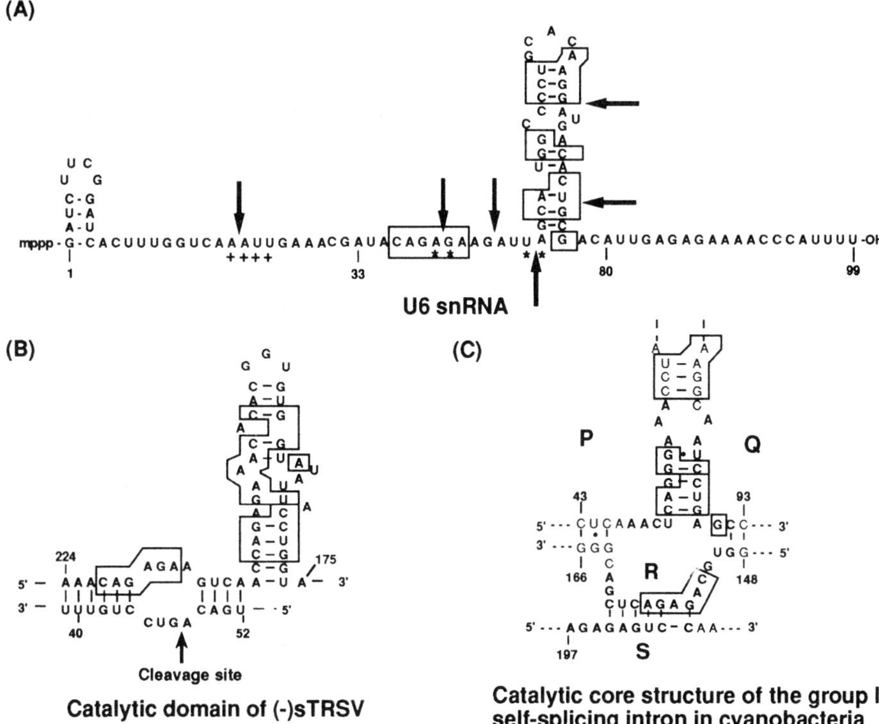

Fig. 11. A: The possible secondary structure of *S. pombe* U6 snRNA and positions of spliceosomal introns in U6 snRNA. Arrows denote the positions of spliceosomal introns found in *S. pombe, R. dacryoidum* and *R. hasegawae* U6 genes. The homologous sequences between U6 snRNA and the conserved core sequence of the group I self-splicing intron are boxed. Asterisks denote the nucleotides essential for the second step of the splicing reaction, and + represents the region involved in the aberrant branch formation between U6 snRNA and the substrate RNA in the *C. elegans* in vitro system [30]. B: The structure of the catalytic domain of (-)sTRSV. The arrow indicates the cleavage position. The sequences homologous with U6 snRNA are boxed. The numbering of nucleotides is according to Hampel and Tritz [25]. C: The conserved core structure of the group I self-splicing intron in the cyanobacteria. The boxes delineate nucleotide sequences identical to the *S. pombe* U6 snRNA sequence.

which is likely to be involved in the binding of guanosine essential for self-splicing reaction [27], and has sequences homologous with those of U6 snRNA in the stem region (Fig. 11C).

In addition, the chimeric RNA between *S. pombe* U6 snRNA and the catalytic RNA of (-)sTRSV, which is 49 bases in length and has an 83.3% sequence identity with the conserved region of *S. pombe* U6 snRNA, can cleave the substrate RNA with the GU sequence at the cleavage site [24]. These results suggest that the conserved region, including the second hairpin structure of U6 snRNA, may function as one of the components of the catalytic domain in pre-mRNA splicing [14].

The U6 introns are found near the nucleotides essential for the pre-mRNA splicing reaction

Interestingly, the U6 introns in *S. pombe* and *R. dacryoidum* are located immediately adjacent to the nucleotides essential for the second step of the splicing reaction [22,23]. The nucleotide adjacent to the *R. dacryoidum* U6 intron is also involved in the interaction between U6 snRNA and the 5′ splice site of a pre-mRNA in the spliceosome [28]. In addition, one of the U6 introns in *R. hasegawae*, which is present at +24, is located within one of the regions required for efficient splicing and spliceosome assembly in the in vitro splicing system using *Xenopus* oocytes [29]. These results support the hypothesis that the U6 intron occurred by the insertion of an intron from a pre-mRNA into a catalytic site in the U6 snRNA. It is noteworthy that the intron site at +24 is also involved in the branch formation between the splicing substrates and U6 snRNA observed in the nematode in vitro splicing system [30]. The product with this aberrant branch structure might be an intermediate of the intron insertion.

The U2 intron is also present in the region important for the splicing reaction

The two U2 regions upstream of the branch site recognition domain were shown to base pair with U6 snRNA [31,32]. The helix I is formed between the U2 sequence immediately upstream of the branch site recognition domain and the highly conserved region of U6 snRNA, and the helix II is between the 3′-terminal region of U6 snRNA and the 5′-terminal region of U2 snRNA. In addition to those base pairings, we found a third potential base pairing between the highly conserved sequence of the U6 snRNA immediately upstream of the indispensable ACAGAGA sequence and the conserved sequence of the U2 snRNA where the intron is present [20] immediately downstream of the branch site recognition domain (Fig. 12A). This potential base pairing is phylogenetically conserved among species, except for *Trypanosoma* [20]. Interestingly, the proposed novel base pairing can bring the branch site and the 5′ splice site of pre-mRNA close to the invariant conserved sequence ACAGAGA, which is presumed to be a catalytic site in U6 snRNA, in combination with helices I and II (Fig. 12B). In this model, U2/U6 base pairing might disturb the interaction between the branch site and U2 snRNA and that between the 5′ splice site and U6 snRNA. Thus, only one of the two alternative base pairings may form in different steps of the splicing reaction, or, alternatively, they might form a triple helix resembling the Holliday-like structure [33]. It is probable that the U2/U6 snRNA complex constitutes a catalytic domain in the splicing reaction. On that point, the spliceosomal intron in the U2 snRNA gene is also found within the region important for the splicing reaction.

Recently, we have cloned other spliceosomal snRNA genes, the U1, U4 and U5 snRNA genes, in *R. hasegawae* (Y. Takahashi, T. Tani and Y. Ohshima, unpublished results). U1 and U5 snRNA genes have spliceosomal introns, whereas the U4 snRNA gene has no introns. Interestingly, spliceosomal introns in the U1 and U5 genes are

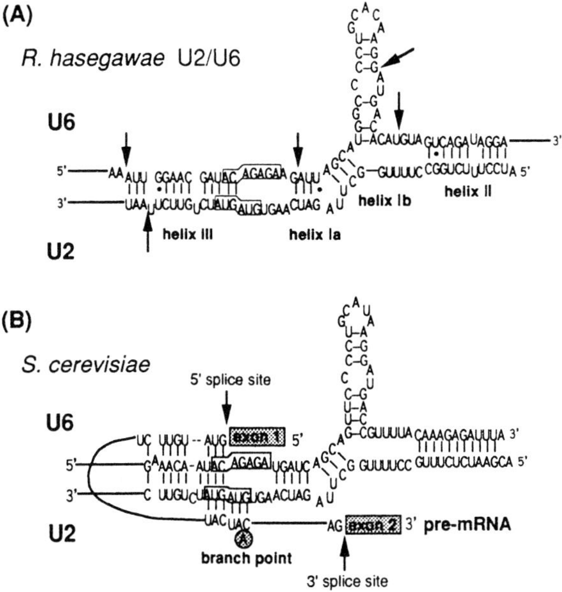

Fig. 12. A: Presumed base pairings between U2 and U6 snRNAs [20]. The novel base pairing that we propose (helix III) is shown together with helix I [32] and helix II [31]. The positions of spliceosomal introns found so far are indicated by arrows. B: Possible interactions between U6 snRNA and the 5′ splice site of a pre-mRNA and those between U2 snRNA and the branch site of the pre-mRNA are shown with the base pairing between U2 and U6 snRNAs.

also located within or immediately adjacent to the sequences essential for the splicing reaction, that is, the 5′ splice site recognition sequence in U1 snRNA and the conserved loop region which interacts with 5′ and 3′ exons in U5 snRNA. It is also interesting that the U4 snRNA gene has no spliceosomal introns, because U4 snRNA is the only spliceosomal snRNA that is released or destabilized from the spliceosome prior to the cleavage at the 5′ splice site.

Evolutionary origin of the spliceosomal introns in the spliceosomal snRNA genes

We examined the U6 snRNA genes of 52 organisms using PCR analysis, and found that the U6 snRNA genes in most of the organisms analyzed had no introns. In addition, the positions of the spliceosomal introns found in the U6 snRNA genes in three genera of yeasts are totally different from one another. Therefore, it is likely that all spliceosomal introns found in the U6 snRNA genes were not present in the ancestral U6 snRNA gene and arose independently during the evolution, although we cannot exclude an unlikely possibility that at least six introns were present in the ancestral U6 snRNA gene and they were removed in most organisms except for *S. pombe, R. dacryoidum* and *R. hasegawae.*

All of the spliceosomal introns found in the spliceosomal snRNA genes are located within or immediately adjacent to the regions functionally important for pre-mRNA

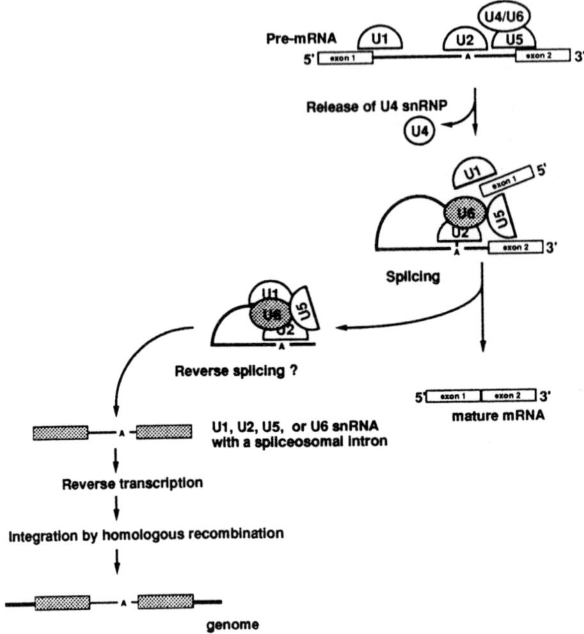

Fig. 13. A possible model of the origin of a spliceosomal intron in a spliceosomal snRNA gene.

splicing. We propose that the U2/U6 snRNA complex is a component of the catalytic site in the spliceosome. We also propose that spliceosomal introns in the spliceosomal snRNA genes arose through reverse splicing of an intron excised from an mRNA precursor into the snRNAs forming a catalytic site or located closely to the catalytic site in the spliceosome, followed by replacement of their cDNAs with the chromosomal genes during evolution (Fig. 13).

Acknowledgements

The authors are grateful to all those who have been involved in this work. This research was supported by grants from the Ministry of Education, Science and Culture of Japan.

References

1. Reddy R, Busch H. Small nuclear RNAs: RNA sequences, structure, and modifications. In: Birnstiel ML (ed) Structure and Function of Major and Minor Small Nuclear Ribonucleoprotein Particles. Berlin: Springer-Verlag, 1988;1–37.
2. Maniatis T, Reed R. The role of small nuclear ribonucleoprotein particles in pre-mRNA splicing. Nature 1987;325:673–678.
3. Green MR. Biochemical mechanisms of constitutive and regulated pre-mRNA splicing. Ann Rev Cell

Biol 1991;7:559–599.
4. Moore MJ, Query CC, Sharp PA. Splicing of precursors to mRNA by the spliceosome. In: Gesteland RF, Atkins JF (eds) RNA World. New York: Cold Spring Harbor Laboratory Press, 1993;303–358.
5. Kass S, Tyc D, Steitz JA, Sollner-Webb B. The U3 small nucleolar ribonucleoprotein functions in the first step of preribosomal RNA processing. Cell 1990;60:897–908.
6. Lamond AI, Konarska MM, Grabowski PJ, Sharp PA. Spliceosome assembly involves binding and release of U4 small nuclear ribonucleoprotein. Proc Natl Acad Sci USA 1988;85:411–415.
7. Blencowe BJ, Sproat BS, Ryder V, Barabino S, Lamond AI. Antisense probing of the human U4/U6 snRNP with biotinylated 2′-OMe RNA oligonucleotides. Cell 1989;59:531–539.
8. Sawa H, Shimura Y. Association of U6 snRNA with the 5′ splice site region of pre-mRNA in the spliceosome. Genes Dev 1992;6:244–254.
9. Sawa H, Abelson J. Evidence for a base-pairing interaction between U6 snRNA and the 5′ splice site during the splicing reaction in yeast. Proc Natl Acad Sci USA 1992;89:11269–11273.
10. Singh R, Reddy R. γ-monomethyl phosphate: a cap structure in spliceosomal U6 small nuclear RNA. Proc Natl Acad Sci USA 1989;86:8280–8283.
11. Kleinschmidt AM, Pederson T, Tani T, Ohshima Y. An intron-containing *Schizosaccharomyces pombe* U6 RNA gene can be transcribed by human RNA polymerase III. J Molec Biol 1990;211:7–9.
12. Maniatis T, Fritsch EF, Sambrook J. Molecular cloning: a laboratory manual. New York: Cold Spring Harbor Laboratory Press, 1989.
13. Domdey H, Apostol B, Lin R-J, Newman A, Brody E, Abelson J. Lariat structures are in vivo intermediates in yeast pre-mRNA splicing. Cell 1984;39:611–621.
14. Tani T, Ohshima Y. mRNA-type introns in U6 small nuclear RNA genes: implications for the catalysis in pre-mRNA splicing. Genes Dev 1991;5:1022–1031.
15. Tani T, Ohshima Y. The gene for the U6 small nuclear RNA in fission yeast has an intron. Nature 1989;337:87–90.
16. Mertins P, Gallwitz D. Nuclear pre-mRNA splicing in the fission yeast *Schizosaccharomyces pombe* strictly requires an intron-contained conserved sequence element. EMBO J 1987;6:1757–1763.
17. Potashkin J, Frendewey D. Splicing of the U6 precursor is impaired in fission yeast pre-mRNA splicing mutants. Nucl Acid Res 1989;17:7821–7831.
18. Brow DA, Guthrie C. Splicing a spliceosomal RNA. Nature 1989;337:14–15.
19. Tani T, Ohshima Y. The gene for U6 small nuclear RNA in *Tetrahymena thermophila* are repeated in tandem. Nucl Acid Res 1991;19:2495.
20. Takahashi Y, Urushiyama S, Tani T, Ohshima Y. An mRNA-type intron is present in the *Rhodotorula hasegawae* U2 small nuclear RNA gene. Molec Cell Biol 1993;13:5613–5619.
21. Singh R, Gupta S, Reddy R. Capping of mammalian U6 small nuclear RNA in vitro is directed by a conserved stem-loop and AUAUAC sequence: conversion of a noncapped RNA into capped RNA. Molec Cell Biol 1990;10:939–946.
22. Fabrizio P, Abelson J. Two domains of yeast U6 small nuclear RNA required for both steps of nuclear precursor messenger RNA splicing. Science 1990;250:404–409.
23. Madhani HD, Bordonné R, Guthrie C. Multiple roles for U6 snRNA in the splicing pathway. Genes Dev 1990;4:2264–2277.
24. Tani T, Takahashi Y, Ohshima Y. Activity of chimeric RNAs of U6 snRNA and (-)sTRSV in the cleavage of a substrate RNA. Nucl Acid Res 1992;20:2991–2996.
25. Hampel A, Tritz R. RNA catalytic properties of the minimum (-)sTRSV sequence. Biochemistry 1989;28:4929–4933.
26. Hampel A, Tritz R, Hicks M, Cruz P. "Hairpin" catalytic RNA model: evidence for helices and sequence requirement for substrate RNA. Nucl Acid Res 1990;18:299–304.
27. Michel F, Hanna M, Green R, Bertel DP, Szostak JW. The guanosine binding site of the *Tetrahymena* ribozyme. Nature 1989;342:391–395.
28. Sontheimer EJ, Steitz JA. The U5 and U6 small nuclear RNAs as active site components of the spliceosome. Science 1993;262:1989–1996.
29. Wolff T, Bindereif A. Reconstituted mammalian U4/U6 snRNP complements splicing: a mutational

analysis. EMBO J 1992;11:345–359.
30. Yu Y-T, Maroney PA, Nilsen TW. Functional reconstitution of U6 snRNA in nematode *cis*- and *trans*-splicing: U6 can serve as both a branch acceptor and a 5′ exon. Cell 1993;75:1049–1059.
31. Hausner T, Giglio LM, Weiner AM. Evidence for base-pairing between mammalian U2 and U6 small nuclear ribonucleoprotein particles. Genes Dev 1990;4:2146–2156.
32. Madhani HD, Guthrie C. A novel base-pairing interaction between U2 and U6 snRNAs suggests a mechanism for the catalytic activation of the spliceosome. Cell 1992;71:803–817.
33. Steitz JA. Splicing takes a Holliday. Science 1992;257:888–889.

Tracing Biological Evolution in Protein and Gene Structures.
M. Gō and P. Schimmel, editors.

The *Tetrahymena* ribozyme tolerates diverse activator domains that replace P5abc

Yoshiya Ikawa, Hideaki Shiraishi and Tan Inoue
Department of Chemistry, Faculty of Science, Kyoto University, Kyoto, Japan

Abstract. P5abc is a conserved RNA structure within the subclass C of the group I introns. The self-splicing group IC intron of *Tetrahymena thermophila* (*T. thermophila*) possesses such a region consisting of 69 nucleotides. It has been shown that this intron ribozyme lacking the P5abc domain is hardly capable of performing a self-splicing reaction. P5abc provides an activation function to the remainder of the ribozyme. This domain was presumably inserted to an ancestral intron in the course of evolution. We have reported successful reconstruction of the P5abc activator domain in that the in vitro selection technique is employed. Under the strict selection conditions very limited numbers of newly constructed activator domains were obtained. We discussed that the result could be attributed to either the nature of the ribozyme itself or the selection conditions. In order to further examine this issue, we attempted the selection experiment under mild conditions. In contrast to the previous result, a variety of new constructs were obtained. The result established that many domains consisting of a variety of sequences are able to substitute a particular functional domain in the ribozyme.

Key words: group I intron, in vitro selection, RNA.

Introduction

The introns categorized as groups I and II have been shown to function as ribozymes (RNA enzymes) that specifically cut and join RNAs [1–3]. Several other examples of ribozymes have also been known as self-cleaving RNAs and an RNA component (P RNA) of RNAase P [2,4].

It has been demonstrated that the group I intron ribozyme consists of several functionally independent components [5,6]. In the group I intron, the RNA components are covalently attached to each other. This molecular design resembles the functional organization of protein enzymes in that structural domains and subunits are associated. On the basis of this theory, it is intriguing to investigate the structural and functional relationship among the components in the group I intron RNA.

Despite fundamental differences between RNA and protein folding, P5abc, a portion of the group IC intron ribozyme, makes a particularly apt comparison with protein domains that in some cases may also have been units of evolution. A close P5abc homolog is found in only about one-sixth of the known group I introns, but

Address for correspondence: Dr Yoshiya Ikawa, Department of Chemistry, Faculty of Science, Kyoto University, Kyoto 606-01, Japan.

some others bear an extension from the L5 loop in which at least an equivalent to the A-rich bulge feature of P5abc has been discerned [7,8]. The most extensively studied P5abc in the *Tetrahymena* group IC intron ribozyme is a 69-nucleotide (nt) element, comprising one-sixth of the total molecular mass of the intron (Fig. 1). The ΔP5abc mutant *Tetrahymena* ribozyme, precisely deleted of P5abc, is essentially inactive under normal in vitro reaction conditions, but can be reactivated by adding spermidine and elevating the Mg^{2+} concentration [9,10]. Thus P5abc is neither part of the catalytic core nor essential for its folding, yet provides an activation function. This domain may have been acquired as one solution to a general problem of activating the core that has been solved differently in other subgroups [11,12].

The most striking evidence of its domain-like relationship to the rest of the *Tetrahymena* ribozyme is the fact that the separately synthesized P5abc molecule, if added in trans, can reactivate the ΔP5abc deletion mutant [13]. Since there is no

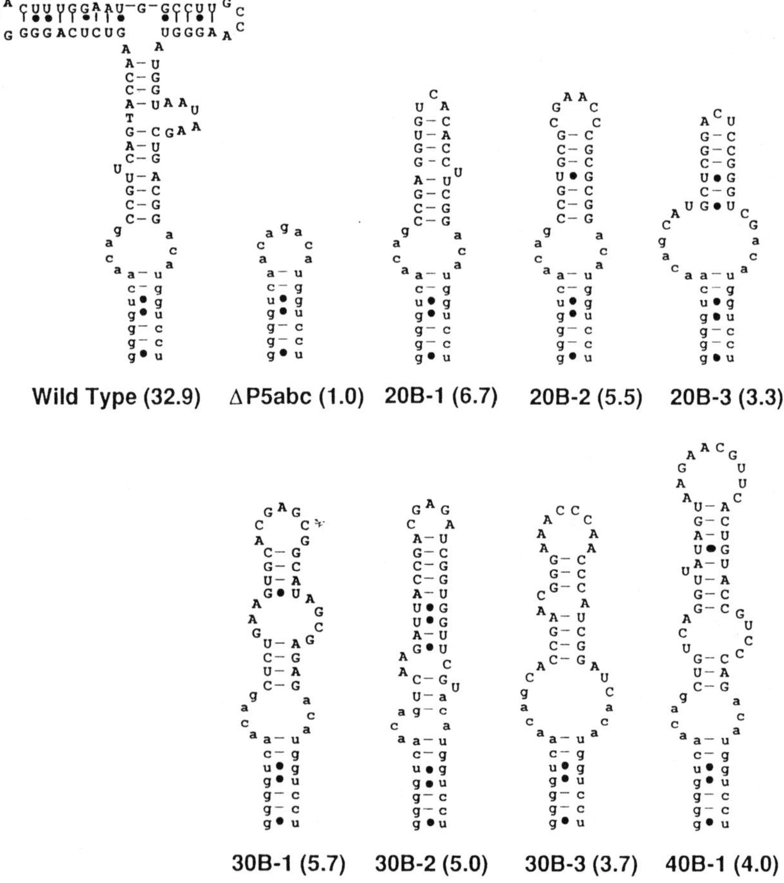

Fig. 1. Predicted secondary structures and relative activities of the wild-type (p5abc), ΔP5abc and the active selectants. The sequences corresponding to P5abc or the randomized region are in capital letters. Relative activities to ΔP5abc are shown in parenthesis.

apparent Watson-Crick base pairing, specific tertiary interactions should be involved [14–16]. In fact RNA-RNA gel mobility-shift assay revealed that the two RNAs form a specific complex [13].

To test the evolutionary hypothesis described above, we developed a new system for the in vitro selection of novel activator domains of the *Tetrahymena* ribozyme that replace P5abc [17]. A pool of a large number of sequence-varied RNAs, transcribed from a DNA template, is treated so as to select individual RNAs that exhibit a desired function. The selected individuals are then converted back into transcription templates by reverse transcription and PCR so that subsequent cycles of selection can proceed. Subcloning for analysis of the isolated individuals are performed in order to elucidate their sequences and structures.

We have reported the application of this system under particular selection conditions [17]. To select novel forms of the replaceable activator domain in the *Tetrahymena* ribozyme, P5abc domain was replaced by 20 or 40 nucleotides consisting of completely randomized sequences. For the two selection experiments the activity of the selectants nearly attained the level of that produced by the wild-type P5abc. Selectant DNAs were cloned and individual sequences were determined.

We performed the selection by employing the stringent conditions (conditions A, see Table 1) that were designated to select the RNAs with highest activity in a short period of time. Each pool was dominated by a particular sequence (and a few variant sequences) after the selection was performed for the 20 or 40 nucleotides pool. In addition to these, an unrelated sequence was also identified. At the primary structural level, these three selectant families bear no apparent relationship to P5abc (or each other).

In the previous report [17], we presented two possible interpretations for the reasons why the numbers of the selectants were so stringently limited. One interpretation was that the selection conditions were too strict to allow the others to survive. In this case the reason could have been that three selectants happened to survive. The other is that the ribozyme is not tolerant in its ability to use a variety of sequences as an activator. Which interpretation is likely to be the case remained unresolved.

Table 1. The selection conditions.

	Conditions A[a]	Conditions B
Randomness	20N, 40N	20N, 30N, 40N
Selection conditions	Stringent	Mild
Mg^{2+}	5 (mM) × 6 (round)	15 × 1 → 12.5 × 1 → 10 × 1 → 5 × 3
Time	5 min × 1 → 1 min × 5	60 min × 5 → 10 min × 1
Temp	30°C × 6	37°C × 5 → 30°C × 1
Character of the 6th-round pool		
Activity[b]	12–18 times	2–3 times
Varieties	2–3	>10

[a]See reference [17]; [b]relative activity to ΔP5abc.

In order to resolve the issue the second selection experiment was attempted under the mild conditions (conditions B, see Table 1). For the selection under conditions B the strategy was to eliminate bit by bit the RNAs that hardly exhibit the activity (instead of selecting the very best in the pool).

Material and Methods

Outline of the selection under the mild conditions (conditions B)

Differences of the selection conditions between conditions A and B are summarized in Table 1. In the selection under the mild conditions (conditions B), bimolecular ligation of the first round was carried out under a high Mg^{2+} concentration (15 mM) with a higher temperature (37°C) and longer reaction time (60 min) to recover all active variants. After active variants were amplified, the stringency of the selection conditions was increased gradually at each cycle by reducing the Mg^{2+} concentration in the reaction buffer, the reaction time and the temperature as indicated in Table 1. Finally, a ligation reaction for the 6th round was carried out under similar conditions (5 mM Mg^{2+}, 30°C, 10 min) as used in the selection under the previously employed stringent conditions.

Procedure of selection cycle

The bimolecular ligation reaction and the preparation of the lower-half RNA and the initial pool of upper-half RNA containing 20 or 40 randomized nucleotides were described previously [17]. The initial pool of upper-half RNA containing 30 randomized nucleotides was prepared accordingly. Oligonucleotides used in this experiment are as follows: D6 (GGC-AAG-ATC-TGT-GTT-GAC-TTA-GGA-CTT-GG), D8 (**TAA-TAC-GAC-TCA-CTA-TA**G-GGA-GCT-CTC-TAA-AAG-AAT-TTG-GAG-GGA-AAA-GTT-ATC-AGG; T7 promoter in boldface) [17].

Schematic representation of the selection cycle is shown in Fig. 2. After denaturing by heating to 70°C, upper-half RNA (100 pmol) was hybridized with the lower-half RNA (200 pmol) in 90 μl of H_2O by cooling slowly to room temperature. Ligation reaction was started by addition of 10 μl of 10 X reaction buffer (1 X buffer contained 50 mM EPPS, pH6.5, 200 mM NH_4OAc and various concentrations of $MgCl_2$ as listed in Table 1). The sample was incubated for various periods of time and at different temperatures as listed in Table 1. The reaction was terminated by addition of cold ethanol. RNA was recovered by ethanol precipitation. The RNA mixture was electrophoresed on a 5% denaturing polyacrylamide gel in order to separate the ligated RNAs from the unligated upper-half RNA. The ligated RNAs were converted to cDNAs with Mu-MLV reverse transcriptase using a primer D6 that is complementary to the 3' terminus of upper-half RNA. Then RNA was degradated by alkaline hydrolysis. To generate the template DNA of upper-half RNA, resulting cDNA was amplified by PCR using the primer D6 and the primer D8 which consists

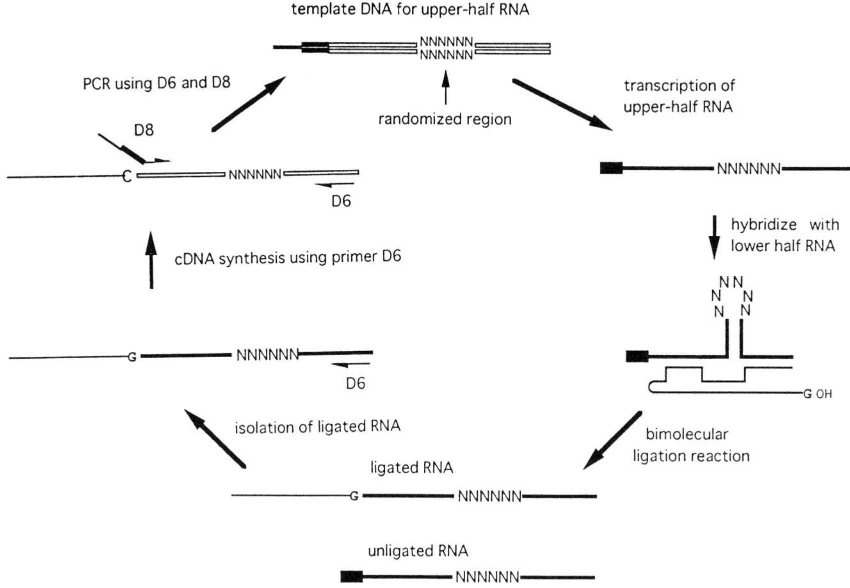

Fig. 2. Schematic representation of the selection cycle.

of T7 promoter sequence followed by the 5' terminal sequence of the upper-half RNA. To minimize the introduction of undesired base substitutions during PCR, high fidelity thermostable DNA polymerase (*Pfu* or Ex *Taq*) and small thermal cycle numbers (15–25 cycle) were used in the PCR reaction. The reaction mixture of PCR was directly used in transcription with T7 RNA polymerase to yield upper-half RNA for the next round of selection.

Bimolecular ligation assay

The activity of the individual selectants was analyzed by bimolecular ligation assay under the same conditions as those of the 6th selection cycle. The upper half RNA of individual selectants were prepared as follows. Inserts of the individual subclones were amplified by PCR using the primers D6 and D8 under high fidelity conditions as described in the preceding section. The PCR products were transcribed with T7 RNA polymerase in the presence of [α-^{32}P]GTP and the resulting RNAs were purified by electrophoresis on a 5% denaturing polyacrylamide gel. After terminating the bimolecular ligation reaction, the reaction mixtures were electrophoresed on a 5% denaturing polyacrylamide gel. The extent of the reaction was quantitated by using Bio-imaging Analyzer Fuji BA-100. Reactivity of the RNA pools recovered from each selection cycle was also quantitated in the same manner.

Results

The selection experiment under the mild conditions were performed as described in Material and Methods. The concentration of magnesium ion was gradually decreased during the course of the selection. To preserve large numbers of the active clones, the selection was started in the presence of a relatively high concentration of magnesium ions (15 mM), which presumably helps to stabilize the structure of RNA (Table 1). The selection was completed using 5 mM of magnesium ions, which is comparable to the previously reported selection conditions.

From the 6th-round pools in which 20, 30 and 40 nucleotides were randomized, 19, 21 and 18 clones were isolated, respectively. The ligation activity of the individual clones are compared and summarized in Fig. 3. As shown in Fig. 3, the activity of the selected pools and that of the clones are reasonably comparable. As anticipated, the clones with mediocre activity are dominant in the pool and the numbers of clones with either eminently lower or higher activity are small.

Nucleotide sequences of the clones recovered from the 6th round of selection under conditions B were determined. Each pool provided selectants consisting of more than 10 different variable sequences. Selectants which are more than 3 times as active as ΔP5abc are shown in Fig. 1. The averaged activity of these selectants is approximately one-half of that of the ones selected previously under stringent conditions. However, whether the selected pools contain the clones whose activity is as high as or even higher than the previously selected ones is uncertain.

The Zuker free-energy minimization algorithm was used to predict secondary structures for the selected sequences that are responsible for high activity (Fig. 1) [18]. From the selection under conditions B, we found that varieties of sequences are

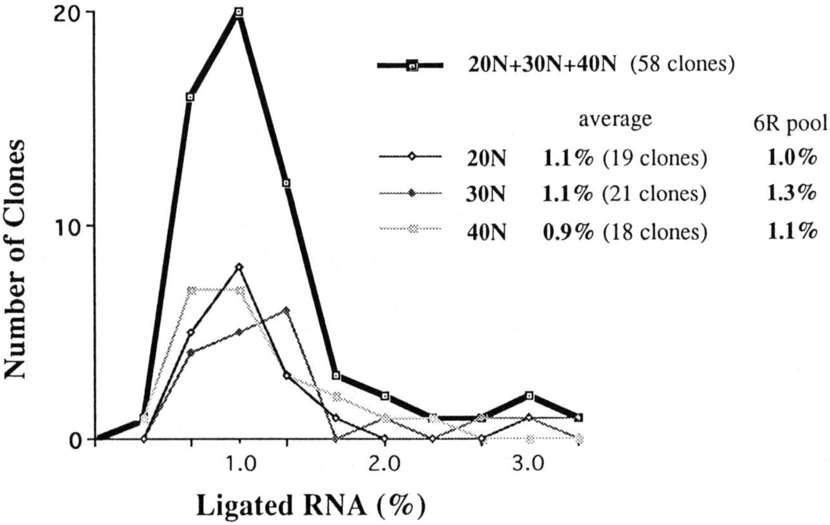

Fig. 3. The efficiency of the ligation reaction by the individual clones from the 6th round pools. The thin lines correspond to each pool and the thick line shows the total of all 58 clones.

capable of activating the ribozyme. As in the case of the previous selection under the stringent conditions, we were unable to discern clear similarities among the new sequence families and P5abc. For example, the A-rich bulge is again not properly represented among the selectants. Thus, this result emphasizes a prominent tolerance of the ribozyme in its ability to accept activation function from various RNA sequences.

To monitor enrichment of active variants during selection cycles, the activity of the selected pools (0, 3rd, 6th round) containing 20 randomized nucleotides was compared in the presence of 5 mM of magnesium ions. As shown in Fig. 4A, the activity of the pool was increased more than 10 times after three rounds of the selection when compared with the starting pool. The next three rounds of the selection enhanced the activity twice as high as that of the 3rd round pool. Thus the six rounds selection enhanced the activity of the pool to 30 times as high as that of the initial pool. Compared with the ΔP5abc ribozyme, the activity of the 6th round pool is doubled (Fig. 4A). However, no further improvement was observed by attempting the 7th round selection. A comparable relationship between the enrichment of the activity and the selection cycles was also observed for the pools containing the 30 or 40 randomized nucleotides. Contrary to these results, the activity of the 3rd round pool was at its maximum when the reaction was performed in the presence of 10 mM of magnesium ions (Fig. 4B). The RNAs that exhibit the highest activity only in the presence of 5 mM of magnesium ions might have been lost in earlier rounds of the selection with higher magnesium ions.

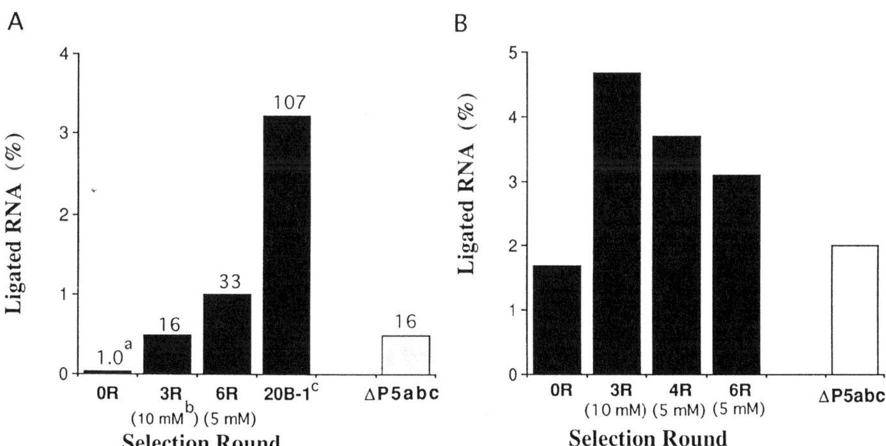

Fig. 4. The ligation activity of the selected pools containing 20 randomized nucleotides. A: Enrichment of the ligation under 5 mM Mg^{2+} conditions. Assay conditions not described in Material and Methods were as follows: in the presence of 5 mM Mg^{2+} at 30°C for 10 min. B: Transition of the ligation activity under 10 mM Mg^{2+} conditions. Assay conditions not described in Material and Methods were as follows: in the presence of 10 mM Mg^{2+} at 37°C for 60 min. [a]Relative activity to the starting RNA pool. [b]Mg^{2+} concentration in the ligand reaction of each selection cycle. [c]20B-1 is one of the selected clones in conditions B as listed in Fig. 3.

Conclusions

We stated in the previous paper two possible reasons why very limited variants had been selected as new domains in the *Tetrahymena* ribozyme from our selection experiments. The issue was whether this is due to the nature of the ribozyme or to the conditions for the selection. We attempted to evaluate the issue by performing the selection under mild selection conditions. As a consequence, the possibility that the ribozyme is merely tolerant in its ability to use apparently dissimilar RNA sequences is denied. It is now clear that, under the mild conditions, many distinctively different selectants are attainable. In the previously reported selection unintended selective pressure acted to unbalance a larger set of active sequences that might have survived crucial selective steps. We conclude here that the group I intron ribozyme is not indiscriminate in its ability to tolerate different activator domains.

The 6th and 7th rounds of selection were performed under the stringent conditions that were employed in the previous experiment [17]. The 6th-round pool contained a variety of RNAs with divergent activity. However, no apparent increase in activity by the 7th-round selection was observed. Perhaps this is attributed to an unfavorable selection pressure due to deplorable properties of the selection system. For example, if some types of poorly active RNAs were exceptionally well amplified by RT/PCR, these RNAs could survive even after the selection under very stringent conditions. Unfortunately, this class of RNAs cannot be eliminated in this system and in consequence the pool has to become equilibrated at a certain stage of the selection. In the previous selection the intolerant selection steps might have happened to discard this class of RNAs.

Through the sets of experiments, it became clear that a particular in vitro selection system can bring very dissimilar consequences by simply altering the selection conditions. The nature of the selectants will be investigated and published elsewhere.

Acknowledgements

A part of this work was supported by a Grant-in-Aid for Scientific Research from the Ministry of Education, Science and Culture, Japan.

References

1. Cech TR. Ann Rev Biochem 1990;59:543–568.
2. Cech TR. In: Gesteland R, Atkins J (eds) The RNA World. Cold Spring Harbor, New York: Cold Spring Harbor Press, 1993;239–269.
3. Saldanha R, Mohr G, Belfort M, Lambowitz AM. FASEB J 1993;17:15–24.
4. Altman S, Kirsebom L, Talbot S. FASEB J 1993;17:7–14.
5. Williams KP, Fujimoto DN, Inoue T. Proc Natl Acad Sci USA 1992;89:10400–10404.
6. van der Horst G, Inoue T. J Molec Biol 1993;229:685–694.
7. Collins RA. Nucl Acid Res 1983;16:2705–2715.
8. Michel F, Westhof E. J Molec Biol 1990;216:585–610.

9. Joyce GF, van der Horst G, Inoue T. Nucl Acid Res 1989;17:7879–7889.
10. Beaudry AA, Joyce GF. Biochemistry 1990;29:6534–6539.
11. Doudna JA, Szostak JW. Molec Cell Biol 1989;9:5480–5483.
12. Green R, Szostak JW. Science 1992;258:1910–1915.
13. van der Horst G, Christian A, Inoue T. Proc Natl Acad Sci USA 1991;88:184–188.
14. Murphy FL, Cech TR. Biochemistry 1993;32:5291–5300.
15. Murphy FL, Cech TR. J Molec Biol 1994;236:49–63.
16. Murphy FL, Wang Y, Griffith JD, Cech TR. Science 1994;265:1709–1712.
17. Williams KP, Imahori H, Fujimoto DN, Inoue T. Nucl Acid Res 1994;22:2003–2009.
18. Jaeger JA, Turner DH, Zuker M. Meth Enzymol 1990;183:281–306.

Tracing Biological Evolution in Protein and Gene Structures.
M. Gō and P. Schimmel, editors.

Aminoacyl tRNA synthetase and polymerase: modular design and evolution

Thomas A. Steitz, Laura F. Silvian and Virginia L. Rath
Department of Molecular Biophysics and Biochemistry, Howard Hughes Medical Institute, Yale University, New Haven, Connecticut, USA

Abstract. While aminoacyl tRNA synthetases and polymerases must have been among the earliest enzymatic activities in evolution, the surprising diversity among the present-day structures of these enzymes leads one to wonder whether the protein enzymes showing these activities were also early features. Structural studies of mutant $tRNA^{Gln}$ and aminoacyl adenylate analogues are establishing the ways in which the four protein domains work together to recognize the substrates. While one of these domains is common to half of the synthetases, the other three appear to be completely different thus far. Comparison of the structures of the four known polymerases shows a surprising diversity among the three subdomains of their polymerase domains, implying that, like the synthetases, all polymerases did not evolve from a common precursor. Could it then be that protein subdomains replaced the presumed ribozyme predecessor counterparts relatively late in evolution and in different ways for different organisms?

Key words: Klenow fragment, reverse transcriptase.

We consider here two classes of enzymes, aminoacyl tRNA synthetases and polymerases, whose activities must have existed very early in evolution. One imagines that the ability to replicate the genome, RNA or DNA, must have existed at the very beginning of evolution. And, of course, the translation of the genetic code must have coincided with the arrival of proteins on the scene. Now whether, in fact, the proteins catalyzing these activities were also early is a little less clear, since both classes of protein enzymes were presumably preceded by ribozymes that may have persisted well into the introduction of proteins. I shall first discuss a number of aspects of Gln-tRNA synthetase and then some aspects of polymerases, such as reverse transcriptase. Of particular interest is the polymerase mechanism as derived from the crystal structures and other data from four different polymerases. These four polymerases provide examples of not only divergent evolution, but of domain swapping and convergent evolution as well.

Address for correspondence: Thomas A. Steitz, Department of Molecular Biophysics and Biochemistry, Howard Hughes Medical Institute, Yale University, 266 Whitney Avenue, P.O. Box 208114, New Haven, CT 06520-8114, USA.

Glutaminyl-tRNA synthetase

The co-crystal structure of glutaminyl-tRNA synthetase (GlnRS) with tRNA and ATP shows that the enzyme has four domains, the largest of which contains the active site, has a Rossmann or dinucleotide fold, and interacts with the acceptor stem of the tRNA (Fig. 1) [1,2]. There is an insertion in this domain that is also involved in recognition of the acceptor stem, and there are two β-barrel domains that are involved in recognition of the anticodon. Superimposing one of these two β-barrel domains on the other shows they are, in fact, structurally similar to each other. There is a long arm extending from one β-barrel domain that stretches out toward the active site.

There are two aspects to substrate recognition required for translation of the genetic code by this class of enzymes: one aspect is its recognition of its cognate tRNA. Of particular interest here is its recognition of the anticodon and how that recognition is communicated to the active site; secondly, this enzyme has to recognize the correct amino acid. GlnRS belongs to a family of 10 enzymes (the class

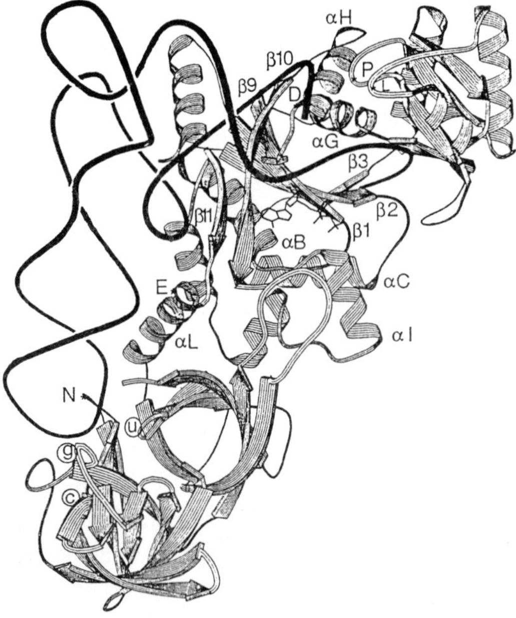

Fig. 1. A schematic drawing of the *E. coli* glutaminyl-tRNA synthetase complexed with its cognate $tRNA^{Gln}$ and ATP. α-Helices are represented by coiled ribbons, β-sheets by arrow-shaped ribbons. The Rossmann fold of the active site domain is composed of α-helices αB, αG and αH, and β-strands β1, β2, β3, β9, β10 and β11. The acceptor-binding domain is inserted in the Rossmann fold between β3 and αG. The two β-barrel domains that recognize the anticodon region follow αL. Asterisks indicate the positions of key residues proline 181 (P), leucine 136 (L), aspartate 235 (D), glutamate 323 (E) and asparagine 370 (N). The HIGH and MSK sequences conserved in all of the class I aminoacyl tRNA synthetases lies at the N-terminus of αB, and on the loop between β10 and αI, respectively. The approximate locations of the three anticodon bases C34, U35 and G36 are indicated by c, u, and g, respectively. For a complete labeling scheme see [1,2].

I synthetases), all of which have the same overall structural fold for the domain containing the active site, but all of which recognize different amino acids. Finally, we shall discuss the structural basis for why the yeast GlnRS enzyme does not charge the *Escherichia coli* (*E. coli*) tRNA and the *E. coli* GlnRS synthetase does not charge the yeast tRNA.

Why are the two signature sequences, KMSK and HIGH, conserved among the class I synthetases? Even though all of the enzymes of the class I synthetases have the Rossmann fold forming the active site, there are almost no conserved residues at all except for these two sequences which are, in fact, conserved among all of them. The HIGH and KMSK sequences are in the ATP-binding region interacting with the triphosphate moiety and with each other [3]. The first histidine of the HIGH is interacting with the α-phosphate of ATP, as is lysine 270 which is also interacting with the ATP phosphates [3]. Hence their primary role is the positioning of the phosphates of the ATP and stabilizing the pentacovalent phosphate transition state in the first step of the reaction. The histidine and the lysine are making interactions with oxygen atoms of the α-phosphate that we believe will stabilize this pentacovalent intermediate. As the O-P-O bond angle between the apical and equatorial oxygens is 90° in the pentacovalent intermediate, the lysine may be stabilizing that angle by simultaneously interacting with bridging and nonbridging oxygens, and the histidine could likewise be helping to stabilize this intermediate. Thus, it is highly unlikely that these conserved residues are doing something completely different in the homologous enzymes. Surely they are playing an identical and essential catalytic role, possibly what we proposed here.

Now in the preprotein ribozyme, what functional group might have been playing the same role as histidine and lysine in stabilizing the pentacovalent phosphate? Perhaps a magnesium ion could function very similarly as the epsilon amino group of Lys. We do not see any other catalytic residues in this active site. Thus, as long as something, a magnesium ion would be the most likely, can stabilize this pentacovalent intermediate this reaction would work perfectly well in a ribozyme world.

The location of the binding site for the glutamine was initially predicted by model building [3]. We have been unable to bind glutamine to the crystalline GlnRS, tRNA, ATP complex; when glutamine is diffused into the crystal, the ATP hydrolyzes to AMP. In order to experimentally determine the binding site for Gln we used an analogue of the aminoacyl adenylate similar to that done by Stephen Cusack et al. in their studies of substrate binding to the serRS [4].

The analogue of the aminoacyl adenylate intermediate that we have used was synthesized by Brian Sproat and contains a sulfamoyl instead of a phosphorate ester. It was co-crystallized with GlnRS and $tRNA^{Gln}$ and binds with its adenosine moiety in the same place as that of ATP. As in the case of the serine enzyme, the pyrophosphate of ATP that has to leave in the reaction is on one side of the α-phosphate and the attacking amino acid carboxylate is on the other side, perfectly placed for a displacement reaction. Having dragged it into the active site attached to the adenosine moiety, how does the amino acid actually interact with the active site? At the resolution of our present preliminary data (2.8 Å, V.L. Rath, unpublished) the binding site for glutamine in GlnRS is largely the same as predicted earlier [3].

To answer the question of how the binding of glutaminyl intermediate is related to the binding of the tyrosyl-AMP to tyrRS we superimposed the corresponding α-carbon atoms of the Rossmann folds of the tyrRS and the glnRS (L. Silvian, unpublished). The AMP moieties bound to the two enzymes superimpose reasonably well, but the tyrosine side chain goes off in one direction and the glutamine side chain goes off in another. The GlnRS side chains that are providing the specificity for tyrosine come from a different part of the Rossmann fold than the protein side chains that are specifying specificity for glutamine. Here, the difference in specificity for the amino acid side chains is not as simple as the case of the serine proteases where chymotrypsin and trypsin have the same substrate specificity pocket and changing the serine side chain to aspartate changes the specificity of chymotrypsin and trypsin from a large, hydrophobic side chain to the basic side chain of lysine or arginine [5]. Clearly, there is a much more complex answer to the differences in amino acid specificity in group I synthetases, at least in the case of a few of these enzymes.

How do the identities of the bases in the anticodon loop and in the acceptor stem effect catalysis some distance away? The identity elements in the acceptor stem and in the anticodon loop were originally identified from the co-crystal structure [1,2,6] and when subsequently mutated were found to affect catalysis [7], the biggest affect on k_{cat}/K_M being as much as 10^4. Although most of the effect was found to be on k_{cat}, the kinetic experiments need to be reinvestigated because the substrate concentration was not in excess of the enzyme concentration. Preliminary results with two $tRNA^{Gln}$s mutated in the anticodon indicate that most of the kinetic consequence is on Km (L. Silvian, unpublished). The structural problem we have addressed is how the changes in the anticodon loop affect catalysis? We have changed bases in the anticodon region, co-crystallized the mutated tRNAs with the enzyme and then determined their co-crystal structures (L. Silvian, unpublished). To date, two have been analyzed. One is the mutation of C34 to G, another is a mutation of G36 to U. Both the RNA structure in the anticodon loop and the protein structure interacting with the loop change in response to both of these mutations. However, they change in different ways. In the U36 to G mutant tRNA the hydrogen bond between U33 and U37 is broken and a stacking of anticodon loop bases on one side occurs that does not occur in the wild-type complex. Such that, the whole anticodon loop structure changes. By contrast, the mutation of C34 to G alters the base stacking on the other side.

These two anticodon mutations also affect the protein structure. Perhaps most prominent among these changes is the observation that the distal anticodon-binding β-barrel shifts. In the U36-containing mutant tRNA a cleft between the distal β-barrel and the rest of the protein opens up, while when the C at 34 is changed to a G this cleft closes down. The binding site for anticodon base 34 is affected by a mutation at position 36, causing the base at 34 to be disordered. How or whether this change is transmitted to the active site is too early for us to say. We had hypothesized earlier [2] that a conformational change might go through a long loop that extends from the β-barrels and tickles the active site through the KMSK sequence. We do not know whether that is going to be the case. There are other routes through this part of the protein and, of course, it may be that the mutated tRNA simply binds less tightly.

Our last report on the glutamine tRNA synthetase was introduced to us by Paul Schimmel: how did the $tRNA^{Gln}$s from yeast and *E. coli* co-evolve with their glutaminyl-tRNA synthetases to yield two GlnRS enzymes with specificities for two $tRNA^{Gln}$s containing different identity elements? Among the differences between $tRNA^{Gln}$ from *E. coli* and yeast are differences in the base pair at the end of the acceptor stem (1-72) and in base pair 3-70. There is also an important change in G10 that we will not discuss.

By modelling the sequence differences in the yeast GLnRS into the structure of the complex we can easily understand what differences in the yeast GlnRS enzyme make it now recognize the yeast $tRNA^{Gln}$ and no longer recognize the *E. coli* $tRNA^{Gln}$. We observed [1] that Asp235 from α-helix H recognizes base pair 3-70 through a specific interaction. Further, a protein loop is breaking apart the terminal base pair, 1-72. The yeast GlnRS differs from *E. coli* GlnRS in both the side chain of 235 and the nature of this loop. In *E. coli* GlnRS the carboxylate of aspartic acid 235 coming off of helix H interacts with the N2 of G3 (Fig. 2). In the yeast enzyme this aspartic acid is a leucine and the 3-70 base pair is an A-U. A model-built leucine superimposes reasonably well on the position of the aspartic acid. This hydrophobic side chain presumably may not discriminate between a U-A or A-U at 3-70. On the other hand, if the 2-amino group of G were sticking out at this position, there might well be a steric clash. At the terminal base-pair of the acceptor stem, the *E. coli*

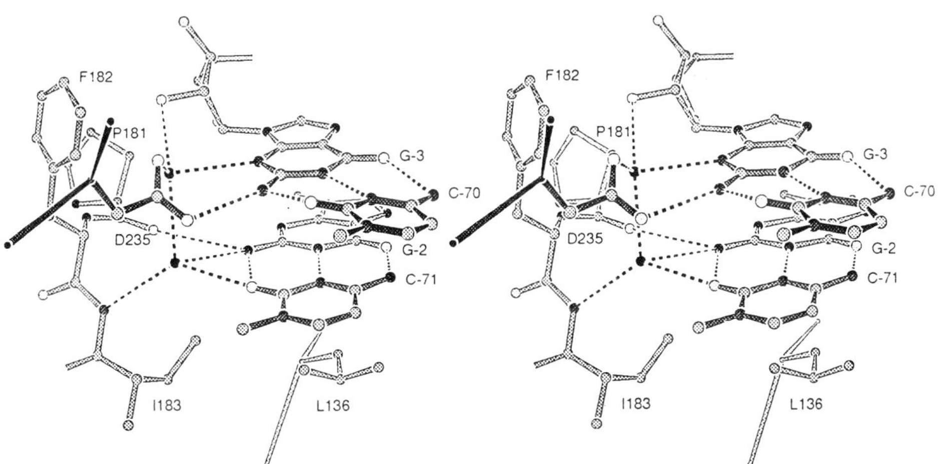

Fig. 2. Stereo view of the recognition interface of base pairs G2-C71 and G3-C70. Aspartate 235 directly bonds to the 2-amino group of guanine 3 via the minor groove. The backbone carbonyl of proline 181 is rigidly directed to hydrogen bond to the 2-amino group of guanine 2. A network of water molecules between the protein and minor groove of the tRNA, only two of which are shown here, appears to enforce a requirement for base pairs at these positions. The hydrophobic environment formed by the proline, phenylalanine, isoleucine, and the underside of the ribose sugars enhances the strength and specificity of these direct and water-mediated hydrogen bonds.

enzyme presents an antiparallel β-turn derived from the Rossmann fold insertion domain. A leucine coming off one end of the turn packs between the bases of two nucleotides (A_{72} and C_{71}) and disrupts base pair 1-72 (Fig. 3). In the yeast enzyme, a lysine replaces the leucine and the loop is much longer. How or why in the predecessors of yeast or *E. coli* one or the other Gln synthetase and $tRNA^{Gln}$ systems co-evolved to give rise to these two divergent synthetase specifications is not clear.

Polymerases

Most polymerases involved in the synthesis of nucleic acids catalyze the addition of a (deoxy)nucleoside triphosphate to the 3′ OH of the primer strand in a template-dependent fashion. The reaction catalyzed is the same whether it is an RNA-dependent or DNA-dependent RNA or DNA polymerase. One might expect that they all have exactly the same reaction mechanism and one might even have expected that

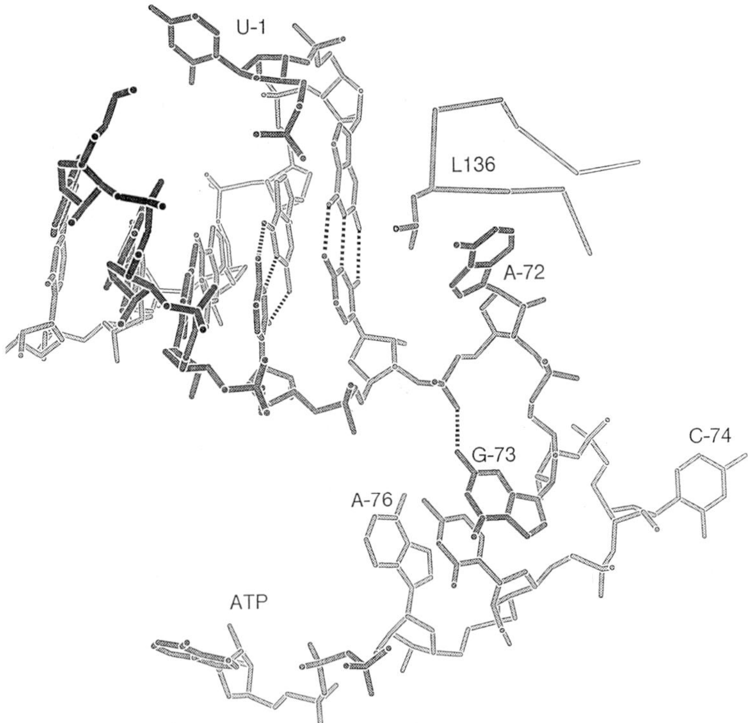

Fig. 3. Conformation of the end of the acceptor stem and the 3′ strand. The expected base pair between U-1 and A-72 is broken by leucine 136, which packs against the guanine of the G2-C71 base pair. The 2-amino group of guanine 73 hydrogen bonds to the phosphate backbone, stabilizing the hairpin conformation of the 3′ strand into the active site. Cytosine 74 binds into a tight pocket in the protein, allowing the bases of nucleotides 73, 75, and 76 to stack.

they all evolved from a common precursor enzyme which would result in all polymerases having the same active site structure. Until recently that has been the prevailing prejudice in the field; in fact, it is apparently not true that all polymerases are homologous [8–11]. Not all polymerases have the same active site structures which leads to some interesting consequences to consider.

There are four polymerases whose structures are known: the Klenow fragment (KF) was the first polymerase whose structure was determined [12] which was followed by reverse transcriptase (RT) from HIV [13], and then the RNA polymerase from T7 [14]. A fourth polymerase recently determined is the structure of DNA polymerase β (pol β) from rats [8,9]. All of these polymerases share something in common, a big cleft and an overall anatomy that resembles a right hand; thus, we named the polymerase subdomains "thumb", "palm", and "fingers" [13]. The thumb subdomain is a different protrusion in all four polymerases though it may be doing the same thing in all four. Three of the four polymerases have different fingers structures, while two of them, T7 RNA polymerase and Klenow fragment, have the same structure. The palm subdomain is where the relationships among the polymerases get interesting; the palm contains the catalytic residues, the heart of catalysis. Reverse transcriptase and Klenow fragment and the RNA polymerase have a similar palm structure, but pol β does not.

A comparison of the polymerase domain structures of RT and KF shows that their thumbs have different structures, and they occur in different places in the gene sequences of RT and KF. In the case of KF the thumb precedes the palm and finger sequences and in RT it follows it. The fingers subdomains have completely different structures in these two enzymes.

Some years ago, Dino Moras's group in Strasbourg looked for sequence similarities among all the polymerases [15]. A characteristic of all the polymerase sequences known were two carboxylate groups found in a motif called C, and one carboxylate group located in motif A. There are very few other sequence similarities that were common to all polymerase sequences. The KF sequence is not similar at all to any of the eukaryotic DNA polymerases, except for these two putative motif similarities. Even in this alignment, a D-E or D-D sequence is not obviously the same as, or homologous to, a D-T-D sequence. In the heart of the catalytic site what does it mean to insert an extra residue? Furthermore, both the T3 and T7 RNA polymerases have a second carboxyl group missing from motif C. Nevertheless, the HIV RT has a palm structure containing the D-D sequence which looks pretty much like that of KF containing a D-E sequence which encouraged us to think that many of the sequence alignments were correct. The superposition of the palm subdomains of RT and KF is pretty good [13]. The three β-strands and two α-helices from the palm superimpose reasonably well as do the corresponding three catalytic carboxylates, which might convince one that these two domains are similar and have evolved from a common precursor.

Eddy Arnold's group co-crystallized reverse transcriptase with a DNA primer template and solved this co-crystal structure [16]. The 3′ end of the primer strand lies near these conserved carboxylates, consistent with their playing an important role in catalysis. The template strand interacts with the fingers and the thumb lies in the minor groove. The co-crystal structure of duplex DNA bound to the Klenow fragment has its 3′ terminus in the exonuclease active site. However, in this case the DNA has to bend to get the 3′ end into the active site. Nevertheless, the primer terminus approaches the catalytic carboxyls from the same direction in both of these polymerases.

DNA polymerase β (pol β) complexed with DNA primer-template, a dideoxynucleoside triphosphate, and two metal ions was solved recently [10]. This structure confirms an earlier proposal [17] that polymerases utilize a two-metal-ion mechanism of phosphoryl transfer. It also showed that the palm subdomain of pol β is not homologous to the other polymerases but shows some analogous similarities [9]. To compare pol β with the other known polymerase structures, Kraut et al. superimposed the palm subdomain pol β on those of the other polymerases in a way that led them to conclude that all the other enzyme-substrate complexes had to be reinterpreted. They have superimposed the analogous palm subdomain structures containing β-strands, α-helices, and the catalytic carboxyl groups. Although this is a pleasing superposition, one of the superimposed β-strands runs in the opposite direction and is not co-linear with the rest of the superimposed structure in the sequence. Superimposed this way the DNA bound to reverse transcriptase is approaching the carboxylates in the cleft from one direction and the DNA complexed with pol β from the opposite direction.

By contrast, it has been suggested that with these nonhomologous enzymes it is more appropriate to superimpose the DNA substrates and the two catalytic metal ions, assuming that the enzymes use the same reaction mechanism [11]. The two palm subdomains are not homologous, but are presenting catalytic carboxylate groups to bind two metal ions. The only requirement for effective catalysis is that the two metal ions are properly oriented relative to the substrates. The two metal ions are required to stabilize pentacovalent intermediate, however, the way in which the carboxylates are arranged to ligate these metal ions is of no consequence. In fact, we have made the point that the metals could be presented to the substrate by an RNA enzyme [17].

It will be interesting to see the structures of a few more DNA and RNA polymerases because it seems likely that we have not seen the full range of their diversity yet. Quite possibly, the replicating eukaryotic polymerase, e.g., the DNA polymerase α, may have yet another structure. The question arises: why are there so many polymerase structures? How did evolution proceed? In going from one generation to the next, how is it possible to switch polymerases?

Were polymerases and synthetases late protein enzymes?

While there are no doubt many possible explanations for the surprising diversity of structures seen in aminoacyl-tRNA synthetases and in polymerases, one possibility is that the presumed ribozyme precursors to the protein enzymes persisted well into evolution, and in the case of polymerases perhaps even beyond the divergence of eukaryotes and prokaryotes. If the domains of the presumed predecessor RNA enzyme playing the roles of fingers, thumbs and palm were replaced by protein domains at different times in various diverged organisms, then the diversity of domain structures seen could be explained. In telomerase, a bound RNA forms the template for DNA synthesis. The RNA sequence forming the template is presented to the catalytic site by an even larger part of the RNA — an early fingers domain? Could the telomerase RNA be in part a molecular fossil, harking back to an era when the fingers domains of polymerase were made of RNA? Additional structures of diverse polymerases may prove instructive in this regard.

References

1. Rould MA, Perona JJ, Söll D, Steitz TA. Structure of *E. coli* glutaminyl-tRNA synthetase complexed with $tRNA^{Gln}$ and ATP at 2.8 Å resolution: implications for tRNA discrimination. Science 1989;246:1135–1142.
2. Rould MA, Perona JJ, Steitz TA. Structural basis of anticodon loop discrimination by glutaminyl-tRNA synthetase. Nature 1991;352:213–218.
3. Perona JJ, Rould MA, Steitz TA. Structural basis for transfer RNA aminoacylation by *E. coli* glutaminyl-tRNA synthetase. Biochemistry 1993;32:8758–8771.
4. Belrhali H, Yaremchuck A, Tukalo M, Larsen K, Berthet-Colominas C, Leberman R, Beijer B, Sproat B, Als-Nielsen J, Grübel G, Legrand J-F, Lehmann M, Cusack S. Crystal structures at 2.5 angstrom resolution of seryl-tRNA synthetase complexed with two analogues of seryl adenylate. Science 1994;263:1432–1436.
5. Steitz TA, Henderson R, Blow D. Structure of crystalline α-chymotrypsin III. Crystallographic studies of substrates and inhibitors bound to the active site of α-chymotrypsin. J Molec Biol 1969;46:337–348.
6. Rould MA, Steitz TA. Structure of the glutaminyl-tRNA synthetase-$tRNA^{Gln}$-ATP complex. In: Eckstein F, Lilley D (eds) Nucleic Acids and Molecular Biology, 6. Heidelberg: Springer-Verlag, 1992;225–245.
7. Jahn M, Rogers MJ, Söll D. Anticodon and acceptor stem nucleotides in $tRNA^{Gln}$ are major recognition elements for *E. coli* glutaminyl-tRNA synthetase. Nature 1991;253:258–260.
8. Davies JF II, Almassy RJ, Hostomska Z, Ferre FA, Hostomsky Z. 2.3 Å crystal structure of the catalytic domain of DNA polymerase β. Cell 1994;76:1123–1133.
9. Sawaya MR, Pelletier H, Kumar A, Wilson SH, Kraut J. Crystal structure of rat DNA polymerase β: evidence for a common polymerase mechanism. Science 1994;264:1930–1935.
10. Pelletier H, Sawaya MR, Kumar A, Wilson SH, Kraut J. Structure of ternary complexes of rat DNA polymerase β, a DNA template-primer, and ddCTP. Science 1994;264:1891–1903.
11. Steitz TA, Smerdon SJ, Jäger J, Joyce CM. A unified polymerase mechanism for nonhomologous DNA and RNA polymerases. Science 1994;266:2022–2025.
12. Ollis D, Brick P, Hamlin R, Xuong NG, Steitz TA. Structure of the large fragment of *E. coli* DNA polymerase I complexed with dTMP. Nature 1985;313:762–766.
13. Kohlstaedt LA, Wang J, Friedman JM, Rice PA, Steitz TA. Crystal structure at 3.5 Å resolution of HIV-1 reverse transcriptase complexed with an inhibitor. Science 1992;256:1782–1790.
14. Sousa R, Chung YJ, Rose JP, Wang BC. Crystal structure of bacteriophage T7 RNA polymerase at 3.3 Å resolution. Nature 1993;364:593–599.
15. Delarue M, Pock V, Tordo N, Moras D, Argos P. An attempt to unify the structure of polymerases. Protein Eng 1990;3:461–466.
16. Jacobo-Molina A, Ding J, Nanni RG, Clark AD Jr, Lu X, Tantillo C, Williams RL, Kamer G, Ferris AL, Clark P, Hizi A, Hughes SH, Arnold E. Crystal structure of human immunodeficiency virus type 1 reverse transcriptase complexed with double-stranded DNA at 3.0 Å resolution shows bent DNA. Proc Natl Acad Sci USA 1993;90:6320–6324.
17. Steitz TA. DNA- and RNA-dependent DNA polymerases. Curr Opin Struc Biol 1993;3:31–38.

Tracing Biological Evolution in Protein and Gene Structures.
M. Gō and P. Schimmel, editors.

Thermodynamics of RNA/DNA hybrid formation

Naoki Sugimoto
Department of Chemistry, Faculty of Science, Konan University, Higashinada-ku, Kobe, Japan

Abstract. The energetic behavior of eight pairs of RNA/DNA hybrid duplexes with identical nearest neighbors have been investigated by UV melting analysis. In the pairs with identical nearest-neighbor pairs the melting curve traces at the same strand concentration were very similar. The average difference in stabilization energy of these pairs was 4%, which was about the same expected from experimental errors. These results indicate that the nearest-neighbor model is valid for predicting the stability of the RNA/DNA hybrid duplexes as well as RNA/RNA and DNA/DNA duplexes. Thermodynamic values for 68 different hybrid sequences (136 single-stranded oligonucleotides) with 5—13 nucleotide length were also reported.

Key words: melting temperature, nearest-neighbor model, nondenaturing gel electrophoresis, RNA/DNA hybrid, stability prediction, thermodynamic parameter.

Abbreviations. CD — circular dichroism; UV — ultraviolet; HPLC — high-performance liquid chromatography; T_m — melting temperature; MOPS — 3-(N-morpholino)propanesulfonic acid; Na_2EDTA — disodium ethylenediaminetetraacetate; Tris — tris(hydroxymethyl)aminomethane.

Introduction

RNA/DNA hybrids often play an important role in biological systems such as transcription [1] and reverse-transcription [2]. When the information of DNA sequence is transcribed into RNA, it is considered that an about 12mer RNA/DNA hybrid is formed while an about 17mer DNA double helix bubble is located in the complex with RNA polymerase [3]. The RNA/DNA hybrids are also important in an antisense therapy [4]. In the regulation of a protein function of this therapy, the protein-coding mRNA should be targeted by chemically synthesized oligonucleotide [5]. To make a complementary association of an invaded oligodeoxyribonucleotide with the target site in mRNA inhibits formation of the characteristic secondary structure of the RNA and a translation into peptide. In a further antisense method, it was reported recently that the activity of a retrovirus like a human immunodeficiency virus (HIV) could be inhibited by cleaving the specific site of mRNA with a DNA-RNA chimeric-hammerhead ribozyme [6]. In this science it is necessary that one knows the hybrid stability quantitatively. However, little is known about the

Address for correspondence: Naoki Sugimoto, Department of Chemistry, Faculty of Science, Konan University, 8-9-1 Okamoto, Higashinada-ku, Kobe 658, Japan. Tel.: +81-78-435-2497. Fax: +81-78-435-2539.

thermodynamics of RNA/DNA hybrid formation and the parameters predicting its exact stability.

Thermodynamic parameters based on the nearest-neighbor model were determined using stabilities of DNA/DNA and RNA/RNA duplexes [7,8], and were also reported for formation of mismatches, dangling ends, internal loops, bulges, and hairpin loops in RNA [9—14]. The parameters were applied successfully to predict stable secondary structures and active centers of nucleic acids [15—19]. The nearest-neighbor prediction may be considered useful in determining the stability of RNA/DNA hybrids. But, if the parameters were to be applied to the prediction of stability of RNA/DNA hybrids, it is very important to know in advance whether the nearest-neighbor model is valid for RNA/DNA duplexes.

Therefore, in this paper thermodynamic behaviors have been investigated for eight pairs of RNA/DNA hybrid duplexes (16 hybrids) with identical nearest neighbors but different sequences. The thermodynamic values were also obtained for two pairs of RNA/DNA hybrid duplexes (four hybrids) with identical sequences between RNA and DNA strands but different nearest neighbors. Comparison of these results provides a direct test of the nearest-neighbor model for RNA/DNA hybrids. Here, thermodynamic values for 68 different hybrid sequences (136 single-stranded oligonucleotides) with 5 to 13 nucleotide length were also reported.

Materials and Methods

Materials

Ribonucleotides and deoxyribonucleotides used here were chemically synthesized on a solid support using phosphoramidite procedures and purified with HPLC after deblocking operations. These oligomers were further purified and desalted with a C-18 Sep-Pak cartridge column. Final purity of the oligomers was confirmed by HPLC and was greater than at least 98%. The length of the examined hybrid strands was at least five nucleotides and consisted of four nearest-neighbor base pairs and one initiation factor, and at most 13 nucleotides possessing 12 nearest-neighbor base pairs and an initiation factor.

All experiments were conducted in a buffer containing 1 M NaCl/10 mM Na_2HPO_4/1 mM Na_2EDTA, pH 7.0. Single-strand concentrations of the oligonucleotides were determined by measuring the absorbance of 260 nm at a high temperature, as described previously [9]. Single-strand extinction coefficients were calculated from mononucleotide and dinucleotide data by using a nearest-neighbor approximation [20]. RNA strand and its complementary DNA strand were mixed with the concentration ratio of 1:1 to obtain each RNA/DNA hybrid.

CD and nondenaturing gel electrophoresis measurements

CD spectra were obtained on a JASCO J-600 spectropolarimeter equipped with a temperature controller and interfaced to a NEC PC-9801 computer. The experimental temperature was 5.0°C. The cuvette-holding chamber was flushed with a constant stream of dry N_2 gas to avoid water condensation on the cuvette exterior. All CD spectra were measured from 350–200 nm in a 0.1 cm path-length cuvette. The concentration of the samples was 70 μM in a 1 M NaCl-phosphate buffer.

Electrophoresis experiments were conducted in gels containing 20% polyacrylamide (19% acrylamide/1% bisacrylamide) with an Atto AEP-500 apparatus cooled with circulating water to provide a running temperature of 5.0 ± 0.5°C. The hybrid samples were diluted into a 5 μl buffer containing 10 mM MOPS/1 mM Na_2EDTA/1 M NaCl. The solutions were heated to 90°C for 2 min, cooled slowly, incubated at 5.0°C for 24 h, and run at 5.0°C in 1 × TBE buffer (0.09 M Tris, 0.09 M boric acid, and 0.2 mM EDTA, pH 8.3) for 3 h. Gels were stained in 0.5 μg/ml ethidium bromide solution. The pictures of gel were taken with Fuji instant black and white film FP-3000 B.

UV measurements

Absorbance measurements in the UV region were made on Hitachi U-3200 or U-3210 spectrophotometer. Melting curves (absorbance vs. temperature curves) were measured at 260 nm, with these spectrophotometers connected to a Hitachi SPR-7 or SPR-10 thermoprogrammer. The heating rate was 0.5 or 1.0°C/min. The water condensation on the cuvette exterior at the low temperature range was avoided by flushing with a constant strcam of dry N_2 gas.

The melting curves of all the hybrids under several strand concentrations showed normal melting behaviors similar to that of DNA/DNA and RNA/RNA double helixes [15,21]. Melting data were collected and fitted with an NEC PC-9801 computer. The melting temperatures were obtained with the curve-fitting procedure described previously [22]. The thermodynamic parameters ($\Delta H°$, $\Delta S°$ and $\Delta G°$) for hybrid formation were determined by the van't Hoff method and analyzed as described in the Results and Discussion.

Calculation with nearest-neighbor parameters

According to the nearest-neighbor model, a free energy change ($\Delta G_{37}°$) of the non-self-complementary double-helix formation consists of two terms: 1) a free energy change for helix initiation to form a first base pair in the double helix; and 2) a free energy change for helix propagation which is the sum of each subsequent base pair [23]. In a free energy change for helix propagation there are 16 nearest-neighbor sequences for RNA/DNA hybrids, while there are only 10 nearest-neighbor pairs for RNA/RNA or DNA/DNA double helixes. With the nearest-neighbor parameters about RNA/RNA and RNA/DNA duplexes as reported previously [3,8], the $\Delta G_{37}°$ values of the RNA/DNA duplex formation were calculated as the sum of helix initiation and propagation.

Results and Discussion

Structure of RNA/DNA hybrid duplexes

It is well known that the CD spectrum of the DNA/DNA double helix has a positive peak around 280 nm and an intense negative peak of around 250 nm at the wavelength range from 220 to 350 nm. This means that the structure is a typical B-form conformation. On the other hand, the CD spectrum of RNA/RNA double helix has one positive peak around 270 nm and a relatively weak and negative peak around 230 nm at the same wavelength range as DNA/DNA, indicating a normal A-form structure [24].

Figure 1 shows CD spectra of r(CACGGCUC)/d(GAGCCGTG) and r(CUCAC-GGC)/d(GCCGTGAG) hybrid duplexes at 5.0°C. This pair of RNA/DNA hybrid duplexes consist of identical nearest neighbors:

r(CACGGCUC)/d(GAGCCGTG)
= rCA/dTG + rAC/dGT + rCG/dCG + rGG/dCC + rGC/dGC + rCU/dAG + rUC/dGA
= r(CUCACGGC)/d(GCCGTGAG).

The spectra of both hybrids with identical nearest neighbors in Fig. 1 have similar shapes: they have an intense positive peak around 270 nm and a small negative peak

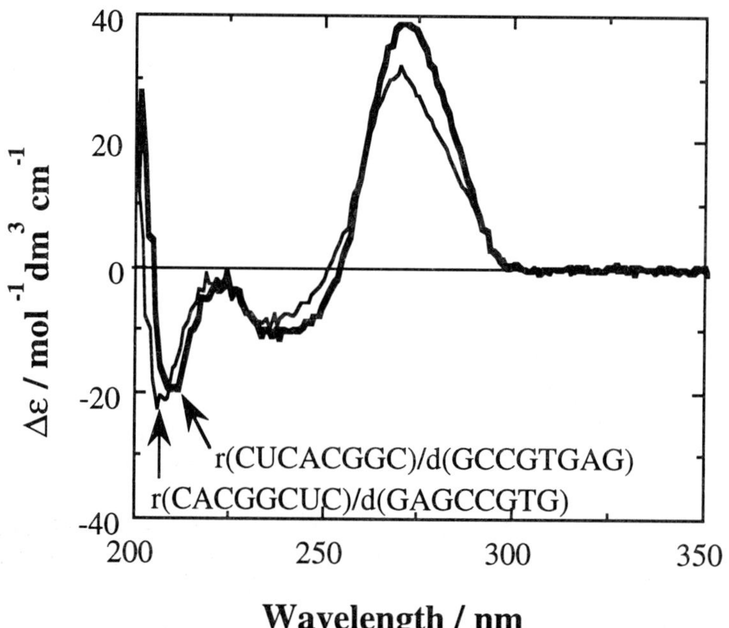

Fig. 1. CD spectra of r(CACGGCUC)/d(GAGCCGTG) (thin line), and r(CUCACGGC)/d(GCCGTGAG) (thick line). The concentration of the samples was 70 μM and measurements were done in 1 M NaCl-phosphate buffer at 5.0°C.

around 235 nm, though the intensities of the peak at 270 nm are slightly different. The same tendency was found in the other pairs of RNA/DNA hybrids used in the UV melting experiments. It suggests that the RNA/DNA hybrids with identical nearest neighbors have a similar structure which is closer to an A-form conformation than a B-form one.

Conformational similarity of the hybrid duplexes was also monitored by polyacrylamide gel electrophoresis under nondenaturing conditions. It is known that the mobility of gel is sensitive to the conformation of nucleic acid [25]. The result showed that r(CACGGCUC)/d(GAGCCGTG) seems to have almost the same mobility as r(CUCACGGC)/d(GCCGTGAG) [26]. This behavior is also found in the case of r(CAGCCGUG)/d(CACGGCTC) and r(GCCGUGAG)/d(CTCACGGC) with identical nearest neighbors [26]. The same mobility of two duplexes in each pair shows that the duplexes with identical nearest neighbors have a similar conformation, which was suggested by the CD spectra described above.

Melting behavior of RNA/DNA hybrid duplexes

Typical melting curves of the r(CACGGCUC)/d(GAGCCGTG) and r(CUCACGGC)/d(GCCGTGAG) hybrid duplexes at the same concentration of the oligomers are shown in Fig. 2. The shapes of the melting curves of these oligomers are very similar to each other. The values of the melting temperature (T_m) are very close, 44.3 and 45.4°C, respectively. These results suggest that the hybrids with identical nearest

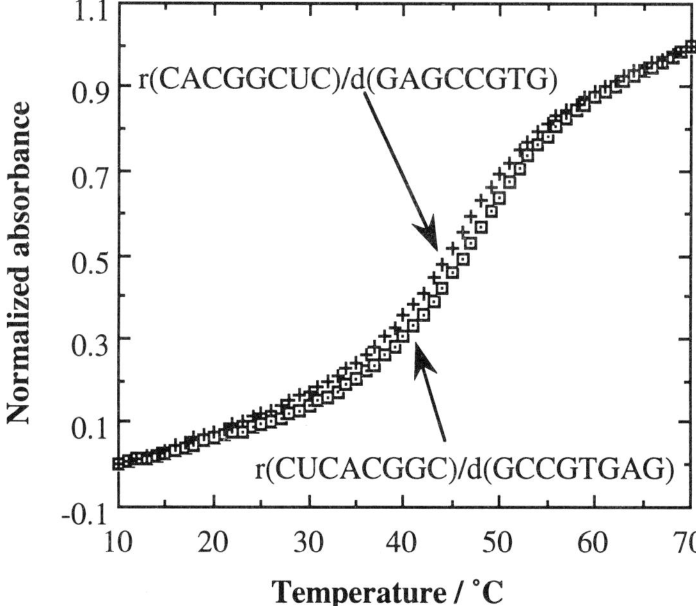

Fig. 2. Normalized melting curves of r(CACGGCUC)/d(GAGCCGTG) (+), and r(CUCACGGC)/d(GCCGTGAG) (□). The concentration of the samples was 10 μM in 1 M NaCl-phosphate buffer.

neighbors have similar melting behavior and stability.

If the melting of the RNA/DNA hybrid duplexes is the two-state transition from a double-helix state to a single-strand state, the following equation can be applied in order to obtain thermodynamic parameters [12]:

$$T_m^{-1} = (2.30R/\Delta H°)\log(C_t/4) + (\Delta S°/\Delta H°) \quad (1)$$

where T_m is the melting temperature of a duplex, $\Delta H°$ and $\Delta S°$ are enthalpy and entropy changes for duplex formation, R is the gas constant, and C_t is the total strand concentration. Plots of T_m^{-1} vs. $\log(C_t/4)$ for both hybrids are shown in Fig. 3. In Fig. 3 the plots of Eqn. 1 for these oligomers are very similar, and the linearity of the plots suggests the melting is due to the two-state transition, i.e., double-helix to single-strand transition.

Nearest-neighbor model for RNA/DNA hybrid duplexes

Thermodynamic parameters for the duplex formation were obtained by Eqn. 1 in the above section and Eqn. 2:

$$\Delta G°_{37} = \Delta H° - T\Delta S° \quad (2)$$

where $\Delta G°_{37}$ is the free-energy change for duplex formation at 37°C and T is 310 K.

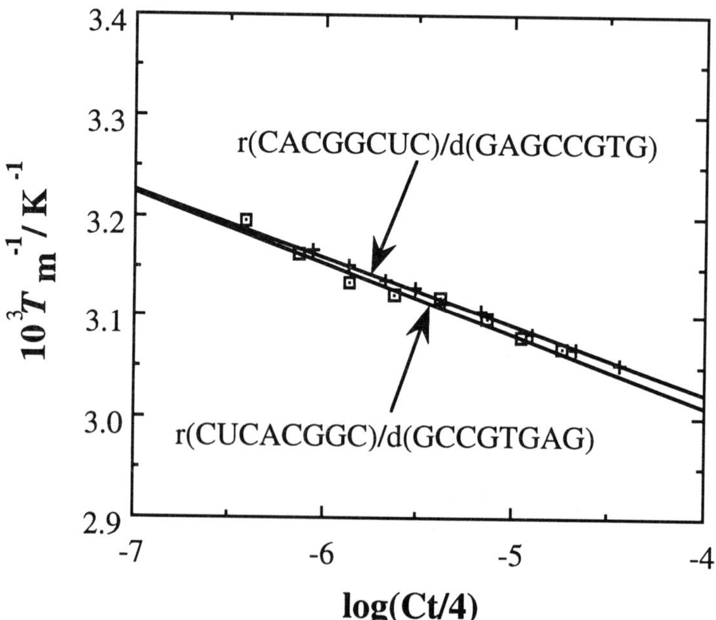

Fig. 3. Van't Hoff plots of r(CACGGCUC)/d(GAGCCGTG) (+), and r(CUCACGGC)/d(GCCGTGAG) (□). Measurements were done in 1 M NaCl-phosphate buffer at different oligomer concentrations.

The values of $\Delta H°$, $\Delta S°$, and $\Delta G°_{37}$ obtained from the plots of T_m^{-1} vs. $\log(C_t/4)$ like Fig. 4 are listed in Table 1 with those of T_m for eight pairs of the RNA/DNA hybrid duplexes (16 hybrids) with identical nearest neighbors but different sequences:
[A] r(GAGCCGUG)/d(CACGGCTC) and r(GCCGUGAG)/d(CTCACGGC);
[B] r(CACGGCUC)/d(GAGCCGTG) and r(CUCACGGC)/d(GCCGTGAG);
[C] r(AAGCGUAG)/d(CTACGCTT) and r(AGCGUAAG)/d(CTTACGCT);
[D] r(ACCGCA)/d(TGCGGT) and r(GCACCG)/d(CGGTGC);
[E] r(CACGGC)/d(GCCGTG) and r(GGCACG)/d(CGTGCC);
[F] r(AGUCCUGA)/d(TCAGGACT) and r(CUGAGUCC)/d(GGACTCAG);
[G] r(CUACGCUU)/d(AAGCGTAG) and r(CUUACGCU)/d(AGCGTAAG); and
[H] r(AAUGGAUUACAA)/d(TTGTAATCCATT) and r(AUUGGAUACAAA)/d(TTT-GTATCCAAT).

The thermodynamic values are also listed in Table 1 for two pairs of RNA/DNA hybrid duplexes (four hybrids) with identical sequences between RNA and DNA strands but different nearest neighbors; [I] r(AGCUUCA)/d(TGAAGCT) and r(UGAAGCU)/d(AGCTTCA), and [J] r(AAUGGAUUACAA)/d(TTGTAATCCATT) also contained in pair [H] and r(UUGUAAUCCAUU)/d(AATGGATTACAA). Each pair of the hybrid duplexes in pairs [I] and [J] do not consist of identical nearest neighbors; for example, in the case of pair [I]:

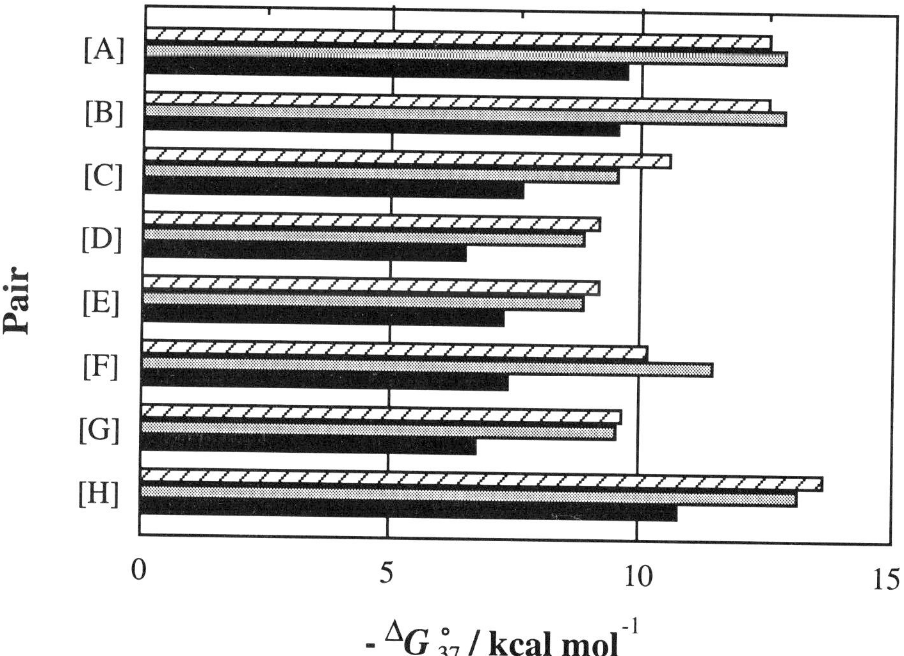

Fig. 4. Comparison of free energy changes for hybrid duplex formation measured in the present study (filled bars), and predicted with Freier's (stippled bar) and Yager and von Hippel's nearest-neighbor parameters (hatched bar). The free energy changes were calculated as average values of the oligomers with identical nearest-neighbor sequence pairs listed in Table 1.

Table 1. Thermodynamic values of duplex formation of RNA/DNA hybrids with identical nearest-neighbor sequences[a].

Pair #	Hybrid duplex	$\Delta H°$/kcal/mol	$\Delta S°$/cal/mol/K	$\Delta G°_{37}$/kcal/mol	$T_m{}^b$/°C
[A]	r(GAGCCGUG)/d(CACGGCTC)	–67.2	–186	–9.6	51.9
	r(GCCGUGAG)/d(CTCACGGC)	–69.1	–192	–9.7	51.8
[B]	r(CACGGCUC)/d(GAGCCGTG)	–69.0	–193	–9.1	48.8
	r(CUCACGGC)/d(GCCGTGAG)	–68.0	–187	–9.9	53.1
[C]	r(AAGCGUAG)/d(CTACGCTT)	–66.6	–190	–7.7	42.7
	r(AGCGUAAG)/d(CTTACGCT)	–61.5	–174	–7.5	42.1
[D]	r(ACCGCA)/d(TGCGGT)	–45.4	–126	–6.5	36.6
	r(GCACCG)/d(CGGTGC)	–47.9	–134	–6.4	36.3
[E]	r(CACGGC)/d(GCCGTG)	–47.3	–129	–7.1	40.5
	r(GGCACG)/d(CGTGCC)	–50.0	–138	–7.3	41.6
[F]	r(AGUCCUGA)/d(TCAGGACT)	–55.7	–158	–6.8	38.7
	r(CUGAGUCC)/d(GGACTCAG)	–59.0	–165	–7.8	43.8
[G]	r(CUACGCUU)/d(AAGCGTAG)	–52.1	–146	–6.8	38.9
	r(CUUACGCU)/d(AGCGTAAG)	–52.7	–149	–6.6	37.4
[H]	r(AAUGGAUUACAA)/d(TTGTAATCCATT)	–94.1	–269	–10.7	51.4
	r(AUUGGAUACAAA)/d(TTTGTATCCAAT)	–92.9	–265	–10.7	51.8
[I]	r(AGCUUCA)/d(TGAAGCT)	–54.1	–159	–4.8	25.1
	r(UGAAGCU)/d(AGCTTCA)	–41.8	–112	–6.9	40.0
[J]	r(AAUGGAUUACAA)/d(TTGTAATCCATT)	–94.1	–269	–10.7	51.4
	r(UUGUAAUCCAUU)/d(AATGGATTACAA)	–78.2	–226	–8.1	43.6

[a]All experiments were conducted in a buffer containing 1 M NaCl/10 mM Na_2HPO_4/1 mM Na_2EDTA, pH 7.0. Estimated errors in $\Delta H°$, $\Delta S°$, $\Delta G_{37}°$, and T_m are ±4, ±4, ±8, and ±2%, respectively. [b]Melting temperatures are calculated at the total oligomer concentration of 100 μM.

r(AGCUUCA)/d(TGAAGCT)
= rAG/dCT + rGC/dGC + rCU/dAG + rUU/dAA + rUC/dGA + rCA/dTG
≠ rUG/dCA + rGA/dTC + rAA/dUU + rAG/dCT + rGC/dGC + rCU/dAG
= r(UGAAGCU)/d(AGCTTCA).

The nearest-neighbor model in which the effect of nearest-neighbor base pair is very important for determining duplex stability is confirmed to be reasonable for most DNA/DNA and RNA/RNA double helixes [7,8]. The nearest-neighbor hypothesis predicts that the pairs with identical nearest neighbors will have identical melting behaviors and identical thermodynamic parameters for duplex formation. The values of $\Delta H°$, $\Delta S°$, $\Delta G°_{37}$ and T_m for each duplex in pair [A] in Table 1 differ by only 1.1%, 2.0%, 3.0%, and 1.5°C, respectively. For the other pairs in Table 1, the average differences in $\Delta H°$, $\Delta S°$, $\Delta G_{37}°$ and T_m are 3.9%, 4.5%, 4.1%, and 1.7°C, respectively, and they are within experimental errors. On the contrary, for each duplex in pairs [I] and [J] with identical sequences but different nearest neighbors, the average differences in $\Delta H°$, ΔS,° $\Delta G_{37}°$ and T_m are 22.1%, 26.1%, 31.8%, and 11.4°C, respectively, and they are much larger than those in pairs [A]–[H]. These results indicate that the nearest-neighbor model is valid for the RNA/DNA hybrid duplexes as well as RNA/RNA and DNA/DNA duplexes.

Comparison between measured and predicted thermodynamic stabilities for RNA/DNA hybrid formation

Nearest-neighbor parameters for predicting the stability of RNA double helixes were reported by Freier et al. [8]. As the structures of the RNA/DNA hybrids with identical nearest neighbors were A-form-like conformations in RNA/RNA duplexes described above, we examined whether the parameters can reproduce the measured $\Delta G_{37}°$ values for the RNA/DNA hybrids in Table 1. To our knowledge, almost no promising nearest-neighbor parameters of free energy change were determined previously for RNA/DNA hybrid duplex formation except Yager and von Hippel's parameters [3]. Unfortunately, in obtaining his parameters many hypotheses had to be used because there were not enough experimental data about the stability to exactly determine hybrid parameters at that time. For example, the values for some nearest-neighbor sequences had to be calculated as an average of DNA/DNA [7] and RNA/RNA parameters [8]. As several nearest-neighbor pairs had to be regarded as possessing the same free energy change, his parameters were given for 11 nearest-neighbor sequences, though the different parameters were necessary for 16 nearest-neighbor sequences. We also applied the parameters of Yager and von Hippel to predict the measured $\Delta G_{37}°$ values for the RNA/DNA hybrids in Table 1. The results are shown in Fig. 4.

Comparing the measured $\Delta G_{37}°$ values for the hybrids with the predicted values from Freier's and Yager and von Hippel's nearest-neighbor parameters, the average differences are 16 and 18%, respectively. The largest difference is 34% for the RNA/RNA parameters and 24% for the RNA/DNA ones. These differences are significantly larger than expected. Therefore, it is suggested that the nearest-neighbor model is valid in RNA/DNA duplexes as well as RNA/RNA and DNA/DNA ones, but the thermodynamic parameters for the nearest neighbors should be improved or newly determined in order to predict more precisely the stability of RNA/DNA hybrid duplexes.

Thermodynamic parameters for other RNA/DNA hybrid duplexes

Thermodynamic parameters of stabilities of DNA/DNA and RNA/RNA duplexes based on the nearest-neighbor model were determined [7,8], and they made it possible to predict stable secondary structures and active centers of nucleic acids [15–19]. The nearest-neighbor model and parameters for DNA/DNA and RNA/RNA duplexes can be described as follows: the duplex stabilities are due to the 10 nearest-neighbor sets such as dAA/dTT, dAT/dAT, dCG/dCG, dCT/dAG, dGA/dTC, dGC/dGC, dGG/dCC, dGT/dAC, dTA/dTA, dTG/dCA for DNA/DNA and RNA/RNA duplexes, though we have to describe U for T in the latter duplexes. Moreover, there are initiation and symmetry factors. The initiation factor corresponds to essential energy to form a first

Table 2. Thermodynamic values for hybrid formation in 1 M NaCl phosphate buffer[a].

RNA sequence[b]	$\Delta H°$/kcal/mol	$\Delta S°$/cal/mol/K	$\Delta G°_{37}$/kcal/mol	$T_m{}^c$/°C
AGCCG	–41.6	–117	–5.4	29.0
CGGCU	–44.9	–130	–4.7	24.9
CGUGC	–39.3	–111	–5.0	25.0
GCACG	–42.0	–121	–4.4	22.3
GGUGG	–45.3	–130	–5.0	26.7
ACCGCA	–45.4	–126	–6.5	36.6
CAAUCG	–45.9	–138	–3.2	16.2
CACGGC	–47.3	–129	–7.1	40.5
CGAUUG	–48.2	–147	–2.6	13.5
CGGUGC	–48.5	–133	–7.2	41.4
CGUGCC	–47.6	–133	–6.5	36.5
GCACCG	–47.9	–134	–6.4	36.3
GGCACG	–50.0	–138	–7.3	41.6
AAUACCG	–51.8	–152	–4.7	26.4
ACGUAUG	–57.9	–171	–4.9	29.9
AGCUUCA	–54.1	–159	–4.8	25.1
CACGGCU	–52.3	–141	–8.6	49.7
CAUACGU	–54.1	–160	–4.5	25.6
GGACUUA	–43.0	–122	–5.0	26.5
UAAGUCC	–44.9	–127	–5.5	30.2
UGAAGCU	–41.8	–112	–6.9	40.0
AAAAAAAA	–51.7	–155	–3.7	21.1
AAGCGUAG	–66.6	–190	–7.7	42.7
AAUCCAGU	–58.2	–168	–6.1	34.8
AAUGUCGC	–65.1	–186	–7.3	40.5
ACCUAGUC	–54.7	–154	–7.1	40.2
ACGACCUC	–53.1	–143	–8.6	49.6
ACUGGAUU	–55.9	–157	–7.3	41.3
AGCGUAAG	–61.5	–174	–7.5	42.1
AGUCCUGA	–55.7	–158	–6.8	38.7
CAACAGCA	–50.9	–139	–8.0	45.9
CACGGCUC	–69.0	–193	–9.1	48.8
CGCUGUAA	–58.0	–164	–7.1	40.2
CUACGCUU	–52.1	–146	–6.8	38.9
CUAGUGGA	–63.6	–179	–8.3	44.7
CUCACGGC	–68.0	–187	–9.9	53.1
CUGAGUCC	–59.0	–165	–7.8	43.8
CUUACGCU	–52.7	–149	–6.6	37.4
GACUAGGU	–54.8	–151	–8.0	45.5
GAGCCGUG	–67.2	–186	–9.6	51.9
GAGGUCGU	–72.0	–201	–9.5	50.4
GCCAGUUA	–61.0	–173	–7.3	40.7
GCCGUGAG	–69.1	–192	–9.7	51.8
GCGACAUU	–58.5	–164	–7.6	42.8
UAACUGGC	–62.6	–173	–8.9	49.2
UCCACUAG	–65.9	–189	–7.1	39.5

(*continued*)

Table 2. Continued.

RNA sequence[b]	ΔH°/kcal/mol	ΔS°/cal/mol/K	$\Delta G°_{37}$/kcal/mol	T_m[c]/°C
UGCUGUUG	−50.6	−141	−6.7	38.1
UGUUCGAC	−69.1	−199	−7.4	40.9
UUACAGCG	−60.1	−171	−6.9	39.0
UUGGCACC	−57.0	−156	−8.7	49.5
AUAACUGGC	−62.7	−173	−9.0	49.9
AUCUAUCCG	−59.8	−170	−6.9	39.1
CAACAGCAA	−63.2	−177	−8.5	47.0
CAACAGCAU	−71.9	−203	−9.1	48.4
CGCUGUUAC	−73.9	−213	−8.0	43.1
CGCUGUUAG	−67.5	−192	−8.0	43.9
CUAACAGCG	−72.0	−203	−8.9	47.6
GCCAGUUAA	−65.6	−187	−7.8	43.0
GUAACAGCG	−77.7	−221	−9.2	48.0
UUAACUGGC	−65.5	−183	−8.9	48.4
ACGUAUUAUGC	−96.7	−284	−8.8	44.4
GCAUAAUACGU	−95.1	−276	−9.6	47.3
AAUGGAUUACAA	−94.1	−269	−10.7	51.4
AUUGGAUACAAA	−92.9	−265	−10.7	51.8
GUCAGGAAUCUG	−81.3	−223	−12.0	59.5
UUGUAAUCCAUU	−78.2	−226	−8.1	43.6
UUUGUAUCCAAU	−113.0	−334	−9.4	45.3
AGCCUAACUCAGC	−125.0	−356	−14.5	58.1

[a]All experiments were conducted in a buffer containing 1 M NaCl/10 mM Na_2HPO_4/1 mM Na_2EDTA, pH 7.0. Estimated errors in $\Delta H°$, $\Delta S°$, $\Delta G_{37}°$, and T_m are ±4, ±4, ±8, and ±2%, respectively. [b]A hybrid duplex consists of the denoted RNA strand and its complementary DNA strand. [c]Melting temperatures are calculated at the total oligomer strand concentration of 100 μM.

base pair and the enthalpy change in the factor has been assumed to be zero though this is not true [7,8]. The symmetry factor has to be considered only in the case of self-complementary sequence as an entropy effect. Thus, the predicted stability of a complementary double helix is calculated as the sum of all the nearest-neighbor sets about the helix propagation, the initiation factor, and the symmetry factor.

There exist 16 nearest-neighbor sets for RNA/DNA duplexes as follows: rAA/dTT, rAC/dGT, rAG/dCT, rAU/dAT, rCA/dTG, rCC/dGG, rCG/dCG, rCU/dAG, rGA/dTC, rGC/dGC, rGG/dCC, rGU/dAC, rUA/dTA, rUC/dGA, rUG/dCA, and rUU/dAA. The initiation factor has to be estimated too. This factor was regarded as having no enthalpy effect because forming a first base pair would produce no stacking effect [8]. But, as the first base pairing needs to form hydrogen bonding, it would be necessary to break the stacking in the state of the single strands. It would also need solvent release. They may give the initiation factor as a positive enthalpy effect. On the other hand, entropy effects on the duplex initiation would be negative because of the helix association. In the case of DNA/DNA or RNA/RNA duplex formation, we also have to consider the symmetry factor which reflects the entropy decrease for self-complementary strands though not for non-self-complementary strands. However, for hybrid formation it is not necessary to consider the factor because hybrids always consist of non-self-complementary strands. Thus, in future, we will have to obtain 16 nearest-neighbor sets and the initiation factor to predict the stability of RNA/DNA hybrid duplexes.

To determine the nearest-neighbor parameters in the near future, the thermodynamics of 68 different hybrid sequences, including the duplexes with identical nearest neighbors described above, were investigated. The obtained thermodynamic parameters by using Eqns. 1 and 2 are listed in Table 2. The melting temperatures at the strand concentration of 100 μM are also listed in the Table. We plan to get the optimizing values for 16 nearest-neighbor sets and one initiation factor with these thermodynamic results.

Acknowledgements

I thank M. Sasaki, T. Sugano, A. Matsumura, K. Honda, M. Katoh, H. Kawai, T. Ohmichi, H. Nakamuta, M. Yoneyama, and S. Nakano for helpful advice and technical assistance. This work was supported in part by grants from the Ministry of Education, Science and Culture, Japan, Hyogo Science and Technology Association, Inamori Foundation, and Hirao Science Foundation.

References

1. von Hippel PH, Bear DG, Morgan WD, McSwinggen JA. Ann Rev Biochem 1984;53:389–446.
2. Varmus HE. Science 1982;216:812–820.
3. Yager TD, von Hippel PH. Biochemistry 1991;30:1097–1118.
4. Stein CA, Cheng Y-C. Science 1993;261:1004–1012.
5. Blake KR, Murakami A, Miller PS. Biochemistry 1985;24:6132–6138.
6. Shimayama T, Nishikawa F, Nishikawa S, Taira K. Nucl Acids Res 1993;21:2605–2611.
7. Breslauer KJ, Frank R, Blöcker H, Marky LA. Proc Natl Acad Sci USA 1986;83:3746–3750.
8. Freier SM, Kierzek R, Jaeger JA, Sugimoto N, Caruthers MH, Neilson T, Turner DH. Proc Natl Acad Sci USA 1986;83:9373–9377.
9. Sugimoto N, Kierzek R, Freier SM, Turner DH. Biochemistry 1986;25:5755–5759.
10. Sugimoto N, Kierzek R, Turner DH. Biochemistry 1987;26:4554–4558.
11. Sugimoto N, Kierzek R, Turner DH. Biochemistry 1987;26:4559–4562.
12. Turner DH, Sugimoto N, Freier SM. In: Saenger WR (ed) Nucleic Acids, Vol. 1c. Berlin: Landolt-Bornstein, 1990; chapter 3.6.
13. Longfellow CE, Kierzek R, Turner DH. Biochemistry 1990;29:278–285.
14. Peritz AE, Kierzek R, Sugimoto N, Turner DH. Biochemistry 1991;30:6428–6436.
15. Sugimoto N, Sasaki M. Chem Lett 1991;345–348.
16. Sugimoto N, Matumura A, Sasaki M. Chem Exp 1993;8:309–312.
17. Pyle AM, Murphy FL, Cech TR. Nature 1992;358:123–128.
18. LeCuyer KA, Crothers DM. Biochemistry 1993;32:5301–5311.
19. Connell GJ, Yayus M. Science 1994;264:1137–1141.
20. Richards EG. Handbook of Biochemistry and Molecular Biology. In: Fasman GD (ed) Nucleic Acids, 3rd Edn., Vol. 1. Cleveland, Ohio: CRC Press, 1975;597.
21. Sugimoto N, Tanaka A, Shintani Y, Sasaki M. Chem Lett 1991;9–12.

22. Sugimoto N, Shintani Y, Tanaka A, Sasaki M. Bull Chem Soc Jpn 1992;65:535–540.
23. Borer PN, Dengler B, Tinoco I Jr, Uhlenbeck OC. J Molec Biol 1974;86:843–853.
24. Saenger WR. Principles of Nucleic Acid Structure. Berlin: Spring-Verlag, 1984.
25. Williamson JR, Raghuraman MK, Cech TR. Cell 1989;59:871–880.
26. Sugimoto N, Katoh M, Nakano S, Ohmichi T, Sasaki M. FEBS Lett 1994;354:74–78.

Tracing Biological Evolution in Protein and Gene Structures.
M. Gō and P. Schimmel, editors.

Using sequence variation to model RNA three-dimensional structures

Robert Cedergren
Department of Biochemistry, University of Montreal, Montreal, Quebec, Canada

Abstract. Nucleic acid and protein sequence variation has been used extensively for the analysis of evolutionary relationships among organisms, since the number of sequence differences between homologous genes of different organisms are approximately related to the time of divergence from their most recent common ancestor. RNA sequence variation (or positional covariation) of a homologous sequence collection from a variety of organisms has also been used to establish secondary structural (base pairing) patterns given the rules of base complementarity: a paired nucleotide, when mutated, requires a compensatory change in the identity of its partner, i.e., its partner must be converted to the Watson-Crick complement.

We extend the use of sequence variation to three-dimensional modeling of RNA in two ways. First, modeling of the Rev-binding element (RBE) of HIV-1 has been achieved using structural constraints inferred from the pattern of sequence covariation in RNA molecules selected from a random pool [1]. These data used in the form of postulated canonical and noncanonical pairs in an internal bulge of the RNA were analyzed by the constraint-satisfaction algorithm implemented in MC-SYM. Models satisfying the constraints predict that the base pairing in the internal loop results in an opening of the RNA major groove sufficient to accommodate docking of the alpha helical arginine rich peptide of the Rev protein. The model of the peptide/RBE is consistent with chemical and genetic data on the binding site.

Second, the notion of structural compensation is applied directly to the maintenance of three-dimensional features of RNA [2]. We show that the proposed secondary structural pattern of unusual mitochondrial tRNAs having long anticodon stems (up to nine base pairs) can be fitted to the L form of tRNA by a structural compensation involving a reduction in the number of base pairs in the D-stem, the coaxial partner of the anticodon stem, and the number of nucleotides serving to connect the two helical domains of the tRNA L form. Furthermore, these mitochondrial tRNA models suggest new structural rules which help to define tRNA molecules. Both studies show that in the case of a well-defined collection of RNA molecules having the same activity or binding ability, reasonable (i.e., experimentally useful) three-dimensional models can be inferred.

Key words: mitochondria, Rev, Rev-binding element, tRNA.

Introduction

Evolutionary theory as applied to molecules suggests that as organisms distance themselves from a common ancestor and cease to exchange DNA, their gene sequences diverge progressively from each other. This phenomenon is rooted in the surprisingly accurate, but basically error-prone, process of replicating and transmitting

---

Address for correspondence: Dr R. Cedergren, Université de Montréal, Département de biochimie, C.P. 6128, Succersale centre-ville, Montréal, Québec, Canada H3C 3J7.

genetic information to progenitors. The "hazard" of Jacques Monod is manifested in the diversity of gene sequences that populate present-day organisms [3]. "Necessity" emerges at the molecular level as the maintenance of required functions in the face of this natural mutational pressure. A strict definition of homologous molecules (necessary for our purpose) is that they possess not only a common ancestor but also a common function.

The notion that sequence differences in homologous molecules could reveal the evolutionary relationship between the organisms encoding them was first put forward by Zuckerkandl and Pauling [4]. To take this step, the notion of time has to be assimilated with sequence variation, since only then could changes be interpreted as showing temporal distance from an ancestor. In spite of the fact that the evolutionary clock is running at widely different tempos for various genes and even for the same gene in different branches of the evolutionary tree, it has been possible to define much more conclusively than ever before the relationship between organisms using sequence variation (e.g., [5]). However, the hypothesis that the behaviour of a single gene, whether an RNA or a protein gene, faithfully represents the evolutionary history of an entire organism is rapidly losing support; just in time, I might add, for the tidal wave of sequences originating from genomic sequencing projects and the resulting possibility to compare and analyze simultaneously many homologous gene collections.

The analysis of sequence variation in collections of homologous RNA sequences in order to propose secondary structures became an important approach to RNA structure as more sequences became available. This process can best be illustrated by the determination of the tRNA cloverleaf structure: in the paper describing the sequence of the first tRNA (yeast $tRNA^{Ala}$) three possible base-pairing patterns were proposed [6]; however, when the second structure, a $tRNA^{Tyr}$, shared one and only one of the patterns of $tRNA^{Ala}$, the notion of the generality of the tRNA secondary structure pattern was born [7]. The concept of the cloverleaf structure of tRNA is so powerful that it has dominated the tRNA structural paradigm for more than 25 years. tRNA genes are identified almost exclusively now by the ability of a DNA fragment to fold into the cloverleaf structure. Only a few mitochondrial (mt) tRNAs have dared to breach this tradition, and because of this, these mostly animal mt $tRNA^{Ser}$s owe their discovery to the tenacity of researchers in their quest for that particular tRNA and the fact that the small size of the mt genome permitted the folding of every micrometer of it, until something vaguely resembling a tRNA could be recognized [8]. I address the unusual tRNAs at a later stage in this report to illustrate how sequence variation in their proposed structures led to a novel view of tRNA tertiary structure. RNA structure by sequence variation, rather covariation, could not be discussed without recognizing the great legacy of the Woese school which turned the determination of the complex folding pattern of ribosomal RNAs into an art form [9].

The pioneering publication of Levitt [10] lifted the veil for the first time on the possible use of sequence covariation in the hunt for tertiary interactions between nucleotides. His analysis of variation patterns in known tRNA sequences resulted in the first molecular model of tRNA. Aside from the relative juxtaposition of the two helical domains of tRNA, the Levitt model was spectacular in its similarity to the X-ray crystallographic model produced several years later.

Recently the massive ribosomal RNA database created by Woese and Gutell has provided considerable depth in the understanding of ribosomal RNA tertiary interactions [11]. Part of the incisiveness of their approach has been provided by sophisticated covariance programs. In their detailed and insightful review on 16S and 23S rRNA structures, the authors conclude by asking what additional information about RNA structure can be inferred by using comparative methods [11]. Here, I illustrate two additional steps in the application of the comparative approach: first, covariation along with steric and geometric constraints is shown to provide enough data to propose a full three-dimensional structure for an RNA molecule, and second, the notion of compensatory structural changes is extended to include constancy in elements of three-dimensional structures.

Sequence variation is governed by the three-dimensional structure and function of RNA

Source of sequence-variation data

Experimentally derived data particularly from site-directed mutagenesis can be very useful in establishing nucleotides which may be in contact in three-dimensional space. However, up until now, the major source of sequence variation data in RNA structures has been that produced by the process of evolutionary sequence drift. Nevertheless, the utility of these evolutionarily related sequence collections is limited, ironically, by the fact that RNA sequences are involved: the four-letter alphabet of RNA renders the alignment of an RNA sequence collection extremely difficult, much more so than in the case of protein sequences. Known secondary structural elements are generally required for alignment. tRNA and rRNA sequences are, therefore, aligned by first superimposing their secondary structures. A problem arises when dealing with collections of RNA molecules whose secondary structure is unknown. In this case, a complete analysis of possible folding patterns of each RNA would be necessary in order to find a pattern of base pairing common to all molecules. Covariation analysis could then be used to confirm the secondary structure proposal.

A relatively new source of data is proving very powerful: collections of RNA sequences selected from random libraries [12]. The principle of selection from random pools of sequence is particularly significant, since no preconceptions are made with regard to sequence or structure expected. In principle then, the sequence landscape for a given function is much more completely explored than in phytogenetically derived collections. Since phylogenetic collections of molecules are homologous by definition, all members of the group must be related via biologically active ancestors. The requirement of functional maintenance is a major constraint, for it is possible, even likely, that other areas of sequence space could give rise to an identical function. The small amount of mutational change at each generation is probably not enough to cross the great barriers that might exist between regions of sequence that could be functional. That this explanation is plausible can be inferred from the sequence collections arising from random libraries possessing clearly unrelated secondary structures.

Of course, the advantage of full landscape exploration in experiments where molecules are selected from random libraries must be tempered by the problem of finding a subset of molecules similar enough in their secondary structure to be used in covariation analyses. A common secondary structure is not always evident in some of these collections. Although in the data presented here, the problem of relatedness of selected molecules has not been an issue, parsing of molecules into similarity classes will become a limiting factor in this strategy.

The structural constraints

As previously noted, Watson-Crick variation in a region generally denotes base-paired, double helical RNA. The mere existence of base pairing and double helicity impose severe conformational constraints on an individual nucleotide within a region of RNA. Succeeding nucleotides must have their nitrogen bases parallel and the sum of the six angles describing their ribophosphate backbone must produce the appropriate displacement between nucleotides such that the RNA A helix is obtained. Examination of a nucleotide conformation database reveals that there is only one combination of angles (within the precision of the modeling method) which responds to these criteria [13]. In a similar manner, all other secondary structure features impose constraints, some more than others [14]. In the case of the Rev-binding element, sequence variation allowed us to infer not only that noncanonical pairing was taking place, but also the precise way in which the nucleotides must be juxtapositioned in space. Other features such as loops and, particularly, pseudoknots impose many constraints on the shape of a molecule and the conformations of the nucleotides which compose it. Of course, even though these notions are self-evident, they are of little practical value unless an appropriate tool is available to exploit them.

A three-dimensional modeling scheme based on constraints

The multiplicity and complexity of constraints imposed on the three-dimensional structure of an RNA molecule by the secondary structure, steric hindrance and spatial proximity of nucleotides distant in the primary structure require analysis by either a very keen structuralist or some kind of automatic methodology. The lack of the former propelled our laboratory into the search for a computer-based procedure for modeling RNA structures.

A constraint satisfaction algorithm

Fortunately, the difficulty of treating mathematical or topological constraints is a known problem in the field of artificial intelligence, where algorithms had been developed to solve constraint satisfaction problems. Our implementation of this algorithm, called MC-SYM, involves defining the nucleotides as variables, the values of which are related to the coordinates of nucleotides in three-dimensional space [15]. The modeling problem is considered "solved" when each variable has been evaluated and all constraints are satisfied. Solutions are in the form of all-atom, three-dimensional models.

Models are built nucleotide by nucleotide; at each position all nucleotide conformations consistent with the imposed constraints are retained. The result of this operation is a tree-like diagram where at every node a value is assigned to a nucleotide and the number of branches at each node represents the number of different conformations which satisfy the constraints. Any node which gives rise to an assignment which is not consistent with the constraints is pruned from the tree.

In many ways, modeling with MC-SYM is an exercise in writing scripts expressing the structural information which is available on the RNA molecule [16]. A good deal of experience is necessary to write a script which will produce a convenient number of three-dimensional solutions. Factors such as distance constraint limits and the amount of base stacking desired in the final structures greatly affect the quantity and quality of solutions. As a general rule, we tend to start the modeling procedure with scripts describing the severest limits to the distance constraints and the maximum of base stacking in order to reduce the number of solutions. Slowly and systematically the constraints can be loosened to determine which have the greatest effect on the number of solutions.

In our implementation of the algorithm, two means are available to include experimental or theoretical data. The constraints serve as one method: most constraints used up until now are based on minimum and maximum distances in Å between two atoms. Note that although the variables are nucleotides, constraints apply to individual atoms. The second entry point for data, and perhaps the key element to our approach, is in the domain of values that the nucleotides are permitted to adopt [13]. Although the six backbone and the glycosidic angles whose values are necessary to describe a nucleotide conformation could in principle adopt almost any value, extensive sampling of different angles at this stage of modeling would provoke a computational explosion such that only very small molecules could ever be modeled. On the other hand, it is well known that the value of the backbone and glycosidic angles are not completely independent, so that the use of a number of discreet nucleotide conformations can be done without compromising unduly the precision in the final model.

A nucleotide-conformation database

The problem of selecting nucleotide conformations for models was addressed by first creating a database containing the coordinates of nucleotides from all nucleic acids whose three-dimensional structure had been determined by either X-ray crystallography or nmr studies. Classification of the conformations by hierarchical methods allowed the use of a number of discrete conformations commensurate with the precision desired in the resulting structure [13]. Normally, 20–50 different nucleotide conformations have been tried at each position.

When conformational information is known at a given position, a subset of the database is created which is composed of only the nucleotides in the bank having the desired conformation. For example, when a nucleotide is known to be stacked, classification involves a subset of the original database containing only "stacked" nucleotides. With the reduced set, only three conformations produce the same precision in modeling as that when 20 conformations are used in cases where no conformational information is available. A Watson-Crick pairing in a helical domain translates into the use of only a single nucleotide conformation. Further constraints limiting the number of permitted values in the domain of values are derived from postulates of noncanonical base pairs, syn-anti conformers, 2′- or 3′-endo conformations. The overall result of reducing the number of values permitted for a variable is to considerably decrease the combinational complexity of the problem.

The constraints

Perhaps the most useful constraint in modeling has proven to be based on loop closure, not only due to the prevalence of loop structures in RNA but also because loops can be inferred where interactions between nucleotides in space have been postulated. The first type of loop constraint comes from the fact that a loop must be closed, and the vector sum of the backbone angles from the nucleotides composing it must be consistent with a loop structure (Fig. 1A). Consider, after using 10 different nucleotide conformations at each of three positions in a tetraloop, that only conformations of the fourth nucleotide placing their 3′-hydroxyl close enough to the 5′-phosphate of the first stem nucleotide to form a bond can be added to the 1,000 intermediates generated up to this point. In retrospect, the loop closure constraint also limits the number of acceptable conformations at the third position, since the third nucleotide must be placed such that a single nucleotide could span the distance between it and the terminus of the stem. Similarly, the second and first positions are constrained, albeit to a lesser degree, such that the number of possible models of a

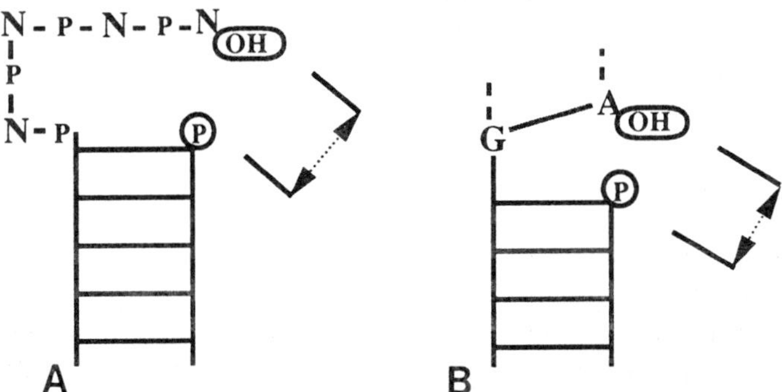

Fig. 1. Loop constraints at the terminus of a stem (A) and from base pairing in an internal loop (B).

loop is easily manageable.

The second type of loop closure comes from the interaction between nitrogen bases in space (Fig. 1B). When dealing with double-stranded RNA helices, the standard A conformation is used, and therefore, each loop closing due to the interaction of bases on different strands is automatic. On the other hand, when noncanonical base pairs are considered the use of only the A conformation is no longer justified and a full complement of conformations must be evaluated at each position. In this situation only conformations for the two base-paired nucleotides which allow the formation of internucleotide bonds of reasonable length and at the same time permit pairing can be used.

A modeling strategy: the structure of the Rev-binding element (RBE)

The availability of a collection of RNA molecules that bound the Rev protein of HIV-1 provided the first opportunity to test the MC-SYM program with nucleotide covariation data [17]. A cursory examination of the natural HIV sequences revealed little variation in the region of the RNA presumed to be responsible for the binding of the Rev protein; however, enough structural information could be gleaned to permit the expectation that a long stem containing an internal loop closed by a tetraloop could be used to probe binding. Thus, libraries of RBE composed of these features containing randomized positions in the internal loop were prepared. Analysis of the covariation in the collection of RNA which bound Rev (the aptamers) provided the first clues to an interaction between nucleotides on opposite sides of the internal loop [1,17]. Note that the selection experiments must have sampled much more sequence space than the biological molecules, since sequence variation was observed in the region completely conserved among the native sequences. This fact afforded a much more detailed analysis of noncanonical base-pairing constraints, such as the G-G pair in the "wild-type" structure.

One of the key elements during the modeling of the molecule was the use of the notion that base-base interactions permit the formulation of loop-closure constraints. Selecting which conformations simultaneously permitted the proper base-base interaction and were compatible with the requirement of continuous ribophosphate background cut deeply into the number of possible structures in the internal loop of RBE. In the prediction scheme, six different loop-closure constraints were used, five of which were formulated from the noncanonical interactions of the internal loop (Fig. 2).

The solution of the problem expressed in the form of constraints indicated in Fig. 2 produced 14 structures which distinguished themselves from each other by the conformation of two nucleotides: several conformations at U72 and two different pairing schemes for residues G50, A68 and C69. After the application of energy minimization techniques and molecular dynamics, 12 of the conformers merged into a single structure. The remarkable aspect of this structure was the formation of an uncharacteristically wide major groove in the helix formed in the internal loop region

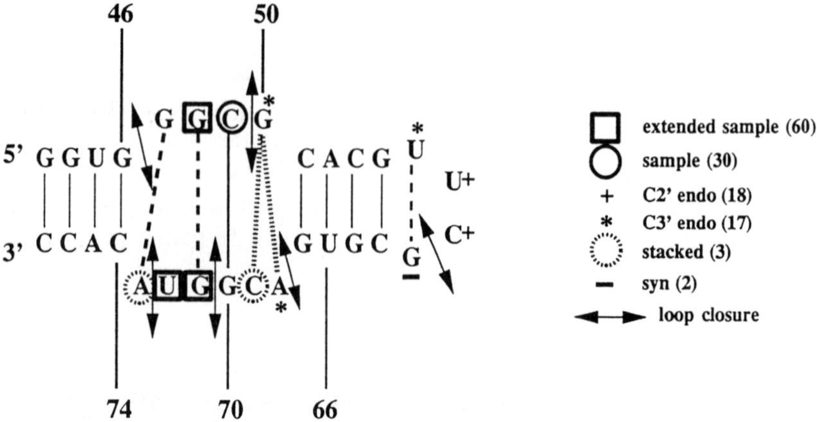

Fig. 2. Constraints inferred from aptamer data. The "wild-type" sequence and constraints used in modeling the RBE. The number of nucleotide conformations evaluated at each position is given in parenthesis. See [1] for further information.

containing the noncanonical base pairs. Equally remarkable was the fact that the groove was just wide enough to allow the docking of the α-helix of a protein. Previously it has been suggested that the region of the Rev protein which bound the RNA adopted the form of an α-helix upon interaction with the RNA [18]. When the arginine-rich motif of the binding region was docked, many charge-charge interactions became possible between arginine residues of the protein and the phosphate residues of the RNA backbone (Fig. 3). Although bases also interact with the protein, a striking correlation was found linking the binding capacity of aptamers

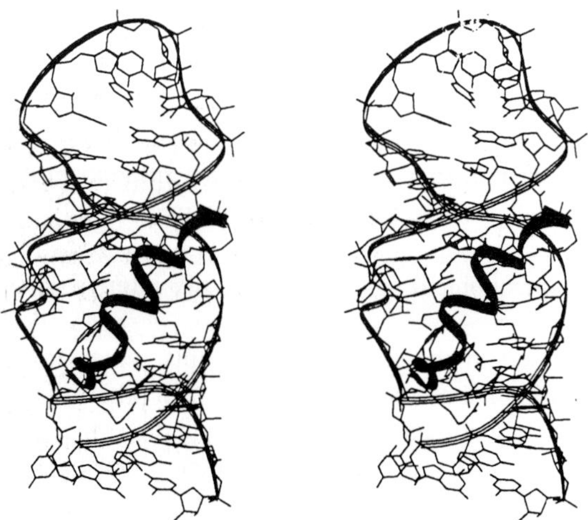

Fig. 3. The stereo drawing the Rev/RBE complex. The RNA backbone is represented by the lined ribbon and the peptide by the solid ribbon.

with the distance between key phosphates in the models, again suggesting that the helix groove opening is the significant aspect of Rev binding [1]. Finally, initial studies on the docking of neomycin B, an inhibitor of complex formation [19], revealed that the amino groups of neomycin bind to most of the same phosphates predicted to bind to the protein.

The ability of the RBE model to explain biochemical data and to act as a basis for continuing on the use of this interaction in drug design suggest that a combination of the modeling techniques which we have developed with a selection of RNA molecules from a random library may be a very powerful strategy for the study of RNA-protein interactions in the future. Here, I think it important to underline that the "correctness" of the model should not be the only criterion in judging these models, since equally important is the ability to plan new and incisive experiments based on the visualization of a structure.

Structural compensation in a new light

The mt tRNA story

To date, I have dealt only with the use of sequence covariation to produce secondary structural elements which constrain the tertiary structure of RNA. Clearly covariation or structural compensation might be interpreted directly in terms of a tertiary structural feature under conditions where the tertiary structure was well known. The unusual mt tRNA structures provided us with just such a case [2]. Although most mt tRNAs fold into recognizable cloverleaf patterns, a number of animal mt tRNAs are recalcitrant and have, therefore, become the orphans of the tRNA world since their relationship to other tRNAs has been impossible to infer.

In a prior modeling effort, we proposed the concept of a primary axis of tRNA defined as the distance from the anticodon to the aminoacceptor stem [20]. Further, we reasoned that the length of the tRNA primary axis must be common to all tRNAs in order that they all make the appropriate contact between the mRNA and the peptidyl transfer site of the ribosome during protein synthesis. The hypothesis is supported by the fact that some unusual tRNAs apparently function in the presence of the normal full length tRNAs. If a common length primary axis were true, how could this idea be applied to the shorter, unusual tRNAs? A clue to this dilemma was provided by a proposal of a seven-base pair anticodon stem for one of these tRNAs, even though its primary sequence is shorter than normal tRNAs [21]. If this tRNA was to maintain the length of the primary axis, some structural compensation would be necessary. Note that we count the noncanonical base pair at the base of the anticodon stem; therefore, a normal tRNA has six base pairs.

Maintenance of the length of the primary axis

The tertiary L form of tRNA is produced by the intersection of two helical domains, one formed by the coaxial stacking of the anticodon and D stems and the other by the T and aminoacceptor stems (e.g., [22]). The two domains are held together by two short connector regions between the acceptor and D stems and the anticodon and T stems (Fig. 4A). Inspection of the detailed three-dimensional structure of tRNA determined by X-ray crystallography suggested a compensatory strategy; by moving the junction point in the helical domain formed by the anticodon D stem closer to the acceptor T-stem domain and reducing the length of the connector regions, the tRNA could be virtually superimposed on the known three-dimensional structure of normal tRNAs (Fig. 4B). Subsequent examination of the tRNA database [23] demonstrated the presence of not just one, but several unusual tRNAs which could be folded in the same manner.

With this type of three-dimensional compensation in mind, it became obvious that other unusual tRNAs could perhaps be folded using the same strategy. In a similar manner, a series of eight-base pair anticodon stem tRNAs was found whose three-dimensional structure could be explained on the basis of moving the junction one position closer to the acceptor-/T-stem helix and reducing further the length of the connector regions. A nine-base pair class of tRNAs was also found which obeyed the stated rules. Finally, a group of tRNAs which folded into a pattern where only five-base pairs are present in the anticodon stem were discovered. As expected in this latter case, the junction point is moved away from the acceptor-/T-helical domain and the number of connector nucleotides is increased (Fig. 4C). Three-dimensional models now exist for each class of tRNA molecules described here. In addition, this analysis provides the framework for the evolutionary relationship of these unusual tRNAs with normal tRNAs because for the first time, they can all be aligned.

Considering the possibility that one or the other of these unusual forms could exist

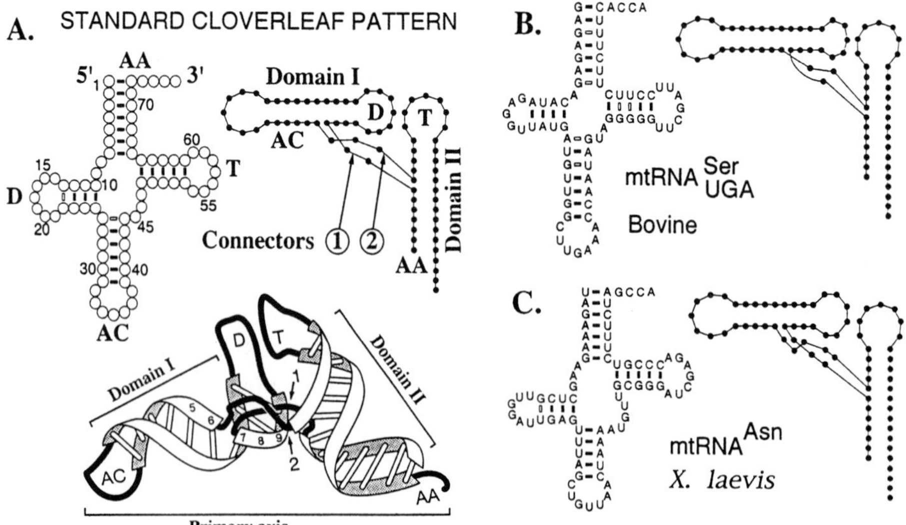

Fig. 4. tRNA cloverleaves. A: Normal cloverleaf pattern, note that the noncanonical base pair 26-44 is counted. B: 7-bp anticodon stem tRNA. C: 5-bp anticodon stem tRNA.

in nuclear or bacterial genomes, the new structural correlations inferred for the unusual mt tRNAs were used to search the entire sequence database. This study was facilitated by the use of the program RNAMOT, a utility developed in our laboratory, which searches databases for primary as well as secondary and tertiary structural features [24]. Scanning the database has demonstrated the presence of several unusual mt tRNA structures, but significantly the vast majority of selected genes are previously recognized tRNA genes (V. Bourdeau, unpublished work).

Applications to proteins

The question is inevitably posed as to why the strategy proposed here could not be applied to proteins. The response is both simple and profound. Simple in that the secondary structure of proteins cannot presently be determined by an analysis of positional covariation; profound in that why a covariation analysis cannot be applied to proteins is not clear.

In this laboratory we have consecrated a good deal of time searching for such correlations (M. Turcotte, unpublished work). Although it might be expected that amino acid variation could be explained better by covariation between classes of amino acids rather than simple covariation of amino acids, analyses of these types have not given encouraging results. Recently, reports of some correlated amino acid changes have appeared [25], and although correlations have not been perfect, i.e., there are false positives, it may be possible that proteins eventually will cede to the treatment we have reserved for RNA. The lack of a secondary structure prediction method analogous to the covariation of nucleotides has repercussions at the level of the tertiary structure of proteins as well. The only constraints possible at the present time would be the presence and position of disulphide bridges. Of course, several bridges would be necessary to model an entire protein. On this positive side, the fact that only two angles are required to pin down the peptide backbone would greatly simplify the combinatorial approach used in the MC-SYM program.

Conclusions

I have presented two examples of the three-dimensional modeling of RNA structures based on the analysis of sequence variation. Clearly the above cases are special in that extensive data of various types exist for both, so that we do not know to what extent this approach can be applied to other RNAs. Suffice to say that the present results are encouraging because they demonstrate a high potential for the rapid production of RNA tertiary structures. Again it should be realized that even though models close to the resolution of nmr or X-ray crystallography studies are the ideal, less precise models could be extremely useful in designing experimental programs concerned with the structure/function relation in RNA.

Acknowledgements

I would like to acknowledge and thank many co-workers and collaborators for their contribution to this work. First, the insightful work of François Major, Daniel Gautheret and Marcel Turcotte on the MC-SYM algorithm; Daniel Gautheret and François Major for the RNAMOT algorithm and Marcel Turcotte for covariation programs described in this report. I note with pleasure the superb and effective collaboration of Fabrice Leclerc and Andrew Ellington in the modeling of the Rev-binding element. Sergey Steinberg's intuition on the structure of tRNA is second to none and his collaboration is gratefully appreciated. This work was supported by NSERC and MRC of Canada. I am a fellow of the Evolutionary Biology Programme of the Canadian Institute for Advanced Research.

References

1. Leclerc F, Cedergren R, Ellington A. Nature Struc Biol 1994;1:293–299.
2. Steinberg S, Cedergren R. Nature Struc Biol 1994;1:507–510.
3. Monod J. Le Hazard et la Nécessité: essai sur la philosophie naturelle de la biologie moderne. Edition de Seuil, Paris, France, 1970.
4. Zuckerkandl E, Pauling L. J Theor Biol 1965;8:357–366.
5. Cedergren R, Gray MW, Abel Y, Sankoff D. J Molec Evol 1988;28:98–112.
6. Holly RW, Apgar J, Everett GA, Madison JT, Marquisee M, Merrill SH, Penswick JR, Zamir A. Science 1965;147:1462–1465.
7. Madison JT, Everett GA, Kung HK. Cold Spring Harbor Symp Quant Biol 1966;31:409–416.
8. Acari P, Brownlee GG. Nucl Acid Res 1980;8:5207–5212; de Bruijn MHL, Schreier PH, Eperon IC, Barrell BG, Chen EY, Amstrong PW, Wong JFH, Roe BA. Nucl Acid Res 1980;8:5213–5222.
9. Fox GE, Woese CR. Nature 1975;256:505–507.
10. Levitt M. Nature 1969;224:759–763.
11. Gutell RR, Larsen N, Woese CR. Microbiol Rev 1994;58:10–26.
12. Ellington AD. Curr Biol 1994;4:427–429.
13. Gautheret D, Major F, Cedergren R. J Molec Biol 1993;229:1049–1064.
14. Gautheret D, Konings D, Gutell RR. J Molec Biol 1994;242:1–8.
15. Major F, Turcotte M, Gautheret D, Lapalme G, Filion E, Cedergren R. Science 1991;253:1255–1260.
16. Major F, Gautheret D, Cedergren R. Proc Natl Acad Sci USA 1993;90:9408–9412.
17. Bartel DP, Zapp ML, Green MR, Szostak JW. Cell 1991;67:529–536.
18. Tan L, Chen L, Buettner JA, Hudson D, Frankel AD. Cell 1993;73:1031–1040.
19. Zapp ML, Stern S, Green MR. Cell 1993;74:969–978.
20. Steinberg S, Gautheret D, Cedergren R. J Molec Biol 1994;236:982–989.
21. Yokogawa T et al. Nucl Acid Res 1991;19:6101–6105.
22. Rich A, RajBhandry UL. Ann Rev Biochem 1976;45:805–860.
23. Steinberg S, Misch A, Spinzl M. Nucl Acid Res 1993;21:3011–3015.
24. Laferrière A, Gautheret D, Cedergren R. CABIOS 1994;10:211–212.
25. Göbel U, Sander C, Schneider R, Valencia A. Protein Struc Func Gen 1994;18:309–317.

Tracing Biological Evolution in Protein and Gene Structures.
M. Gō and P. Schimmel, editors.

Modules of barnase: the physicochemical basis for their structures

Tosiyuki Noguti and Mitiko Gō
Department of Biology, Faculty of Science, Nagoya University, Nagoya, Japan

Abstract. *Background.* Globular domains of proteins consist of structural units, modules. Modules are defined as the least extended or the most compact segments consisting of 10–40 amino acid residues in a globular domain. Boundaries of modules of several proteins are closely correlated with intron positions of genes encoding the proteins. It was considered that modules might be the original building blocks of proteins encoded by exons and might be recruited into contemporary proteins by exon shuffling.

Methods and Results. Barnase, a monomeric bacterial RNase from *Bacillus amyloliquefaciens*, was decomposed into at least six modules, M1 to M6. We analyzed the atomic interactions of barnase, on the bases of module organization, and found hydrogen bonds to be localized mainly within each of the modules. We synthesized polypeptide fragments corresponding to the modules and examined their solution structures by two-dimensional NMR. Secondary structures were observed at positions similar to those in the intact barnase.

Conclusions. From these observations it appears that the native conformations of modules of barnase are specified predominantly by interactions within the modules and that these conformations are further stabilized by association with other modules. Thus, compact structures similar to the original modules seem to be maintained as remnants in contemporary proteins.

Key words: exon shuffling, hydrogen bond, molecular evolution, peptide fragments, protein conformation, protein folding.

Introduction

Protein has a unique three-dimensional structure and carries out specific functions. A globular prn consists of one or more globular domains containing 50–200 amino acid residues, which behave more or less as independent folding units. A concept of modules was introduced from a purely structural point of view [1]. Modules are defined by partitioning a globular domain into small regions, each of which is a contiguous polypeptide segment of 10–40 residues with a compact conformation. A correlation has been found between boundaries of protein modules and intron positions of genes in globins [1], lysozyme [2,3], triose phosphate isomerase [4–7] and other proteins [8]. The presence of an additional intron was predicted on the basis of exon-module correspondence in the globin family [1] and the intron was found in the soybean leghemoglobin gene at the expected position [9,10]. Recently another intron was identified in the triose phosphate isomerase gene from a mosquito and the

Address for correspondence: Mitiko Gō, Department of Biology, Faculty of Science, Nagoya University, Chikusa-ku, Furo-cho, Nagoya 464-01, Japan. Tel.:+81-52-789-2976. Fax: +81-52-789-2977.

position of the intron corresponded to one of the module boundaries of the protein [7].

The correlation between modules and exons suggests that the former were encoded by corresponding exons in genes for ancestral proteins and functioned as evolutionary units; new proteins might be produced by combining various modules through exon shuffling or fusion [1,2,4,8,11–14]. It is essential that each module conserve its conformation in a compact form through the evolutionary process. The compact conformation of each module is specified mainly by interactions within the module and to some extent is independent of the remaining part of the protein [8]. To better understand the physicochemical bases for the role of modules in protein evolution and in protein folding, we studied modules of barnase, and asked whether compact conformations of modules are intrinsic characteristics of the corresponding polypeptide fragments. Barnase is an extracellular ribonuclease (RNase) with weak specificity and it is produced by *Bacillus amyloliquefaciens* [15]. It is a monomer of Mr 12,382 consisting of 110 amino acid residues [16]. Barnase is one of the smallest globular proteins containing no disulfide bonds and is a protein suitable for analysis of atomic interaction, other than the disulfide bonds, within and between modules.

Barnase is decomposed into six modules

Boundaries of modules are close to local centers of long contiguous segments [4], and are detected by considering contiguous segments of a certain length (a window length) and by locating their center. We locate their centers at minima of F_i, an index of centripetal character of ith residue within a contiguous segment of a window length of $(2k + 1)$ residues, which is the mean square distance between C^{α} atom of ith residue and C^{α} atoms of the residues in the contiguous segment of the polypeptide chain from $(i - k)$ to $(i + k)$th residues, i.e.,

$$F_i = \frac{\sum_{j_1 \le j \le j_2} r_{ij}^2}{(j_2 - j_1)}$$

where $j_1 = \max(1, i - k)$, $j_2 = \min(n, i + k)$, and n is the number of residues in the protein.

Using the X-ray coordinates of barnase [17], centripetal profiles of window lengths $(=2k + 1)$, 41, 51, 61, 71, 81 and 91 residues were drawn to detect residues at module boundaries of barnase (Fig. 1A). The local minima common to all the profiles are identified as module boundaries, at residues 24, 52, 73, 88 and 98 [18].

Barnase was decomposed into six modules, M1 to M6, M1: 1–24, M2: 25–52, M3: 53–73, M4: 74–88, M5:89–98 and M6:99–110. The boundaries of the modules are shown in the distance map that indicates residue pairs separated by more than 20 Å (Fig. 1B). As almost no residue pairs separated by more than 20 Å (Fig. 1B) are

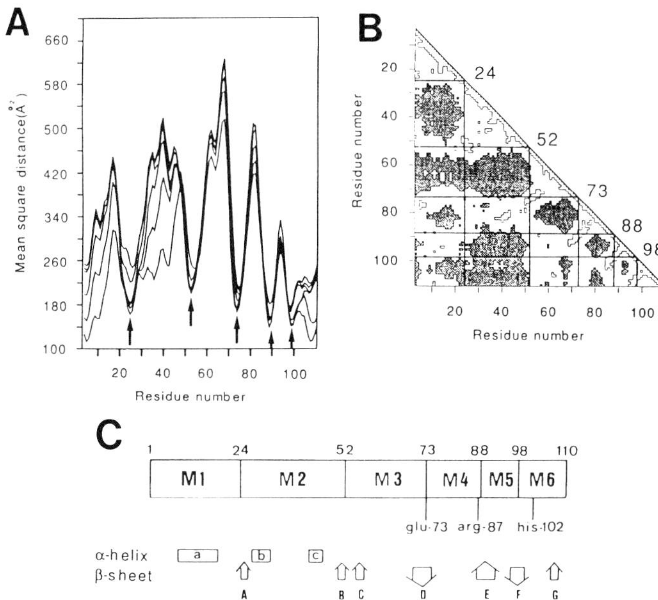

Fig. 1. Modules M1–M6 identified in barnase. A: The centripetal profile [4]. The vertical axis is the mean square distance between C^{α} atom of *i*th residue and C^{α} atoms of the residues which are within *k* (=20, 25, 30, 35, 40 and 45) residues along the polypeptide chain from *i*th residue. The module boundaries are indicated by arrows at the local minima of the curves. B: The distance map of barnase. Residue pairs of which C^{α} atoms are separated by more than 20 Å are marked. Solid lines indicate the module boundaries identified by the centripetal profiles. C: Schematic diagram of modules, catalytic sites, α-helices and β-strands. (Reproduced with permission from Noguti et al. [18] ©1993 Wiley-Liss, Inc.).

observed within each module, it is clear that each module is in a compact conformation in the protein.

Figure 1C shows modules, α-helices, β-strands and locations of the catalytic residues schematically. There are three α-helices (a–c); 7–17, 27–32 and 42–45, and seven β-strands (A–G): 24–25, 50–51, 55–56, 71–75, 87–91, 96–99 and 107–108. The first two β-strands (A and B) constitute a parallel sheet, and the other five (C–G) constitute an antiparallel sheet. Glu 73, Arg 87 and His 102 are the catalytic sites of barnase [19]. Glu 73 and Arg 87 are close to module boundaries.

All module boundaries in barnase are located on β-strands (Figs. 1C and 2a). Modules M2, M3, M4 and M6 form a shallow and wide cavity for the substrate binding, as seen in Fig. 2b. Modules M1 and M5 seem to support the other four modules and bring them together at positions so as to form a globular protein as an assembly of the six modules.

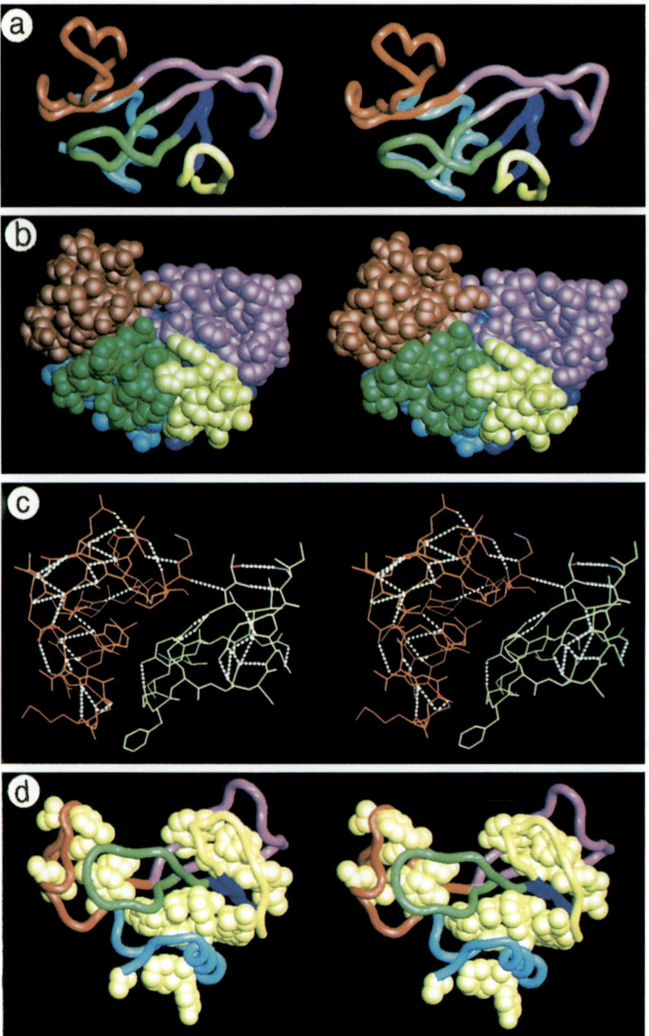

Fig. 2. Modules of barnase shown by different colors and in stereopairs. M1 is sky blue, M2 red, M3 magenta, M4 green, M5 blue and M6 yellow. (a) The folding of backbone is represented in tube model and shown from the direction of the catalytic sites. (b) Module organization in space-filling models viewed from the same direction as (a). (c) Hydrogen bonds (white dotted lines) in a pair of M2 and M4. The covalent bonds of modules M2 and M4 are in red and green, respectively. (d) Side-chain atoms of hydrophobic clusters shown in yellow space-filling models. Val, Ile, Leu, Phe, Trp and Tyr are shown as hydrophobic residues. The molecule is viewed from the bottom of (a). (Reproduced with permission from Noguti et al. [18] ©1993 Wiley-Liss, Inc.).

Hydrogen bonds are localized mainly within each module

Protein conformations are stabilized and specified by hydrophobic interactions, hydrogen bonds, electrostatic interactions, nonbonded interactions and others. Based on the three-dimensional structure of barnase [17], we investigated whether intra-molecular interactions, especially hydrogen bonds and hydrophobic interactions, are designed to stabilize structures of the modules [18].

If a donor and an acceptor of a hydrogen bond exist in the same module, we call it an intra-module hydrogen bond, and if not, it is called an inter-module hydrogen bond. Numbers of intra- and inter-module hydrogen bonds are presented in Table 1A. Most hydrogen bonds are strikingly intra-modular. Modules M2 and M4 are in contact in space as seen in Fig. 2b. Though 25 and 13 hydrogen bonds are formed within M2 and M4, respectively, only one hydrogen bond exists between them (Fig. 2c, Table 1A). However, this observation of intra-modular localization of hydrogen bonds might simply be because most hydrogen bonds are formed between residues close in sequence [20]. For elucidation, we defined "pseudomodules" as segments starting at a center of one module and ending at the center of the following module, and repeated analyses of hydrogen bond distribution for these hypothetical modules. These pseudomodules are similar to real modules in size. In Table 1B numbers of hydrogen bonds in and between these pseudomodules are given. Many hydrogen bonds form between adjacent pseudomodules. Hydrogen bonds are localized mainly within each module rather than within other segments of a similar size. Thus,

Table 1. Distribution of hydrogen bonds.

A: Number of intra- and inter-module hydrogen bonds

	M1	M2	M3	M4	M5	M6
M1	19	1	0	3	1	3
M2	1	25	1	1	0	0
M3	0	1	12	0	6	3
M4	3	1	0	13	0	6
M5	1	0	6	0	10	4
M6	3	0	3	6	4	11

Note: Numbers of intra- and inter-module hydrogen bonds are given at the diagonal and off-diagonal parts, respectively. M1 to M6: the actual modules of barnase.

B: Number of intra- and inter-"pseudomodule" hydrogen bonds

	m1	m2	m3	m4	m5	m6	m7
m1	5	8	0	2	0	0	3
m2	8	19	9	1	0	1	0
m3	0	9	9	6	0	2	0
m4	2	1	6	3	14	1	0
m5	0	0	0	14	7	14	0
m6	0	1	2	1	14	6	8
m7	3	0	0	0	0	8	1

Note: The pseudomodules, m1 to m7 consist of residues 3–12, 13–38, 39–62, 63–80, 81–93, 94–104 and 105–110, respectively. Hydrogen bonds are defined according to Baker and Hubbard [40]. (Reproduced with permission from Noguti et al. [18] ©1993 Wiley-Liss, Inc.).

this intra-module localization of hydrogen bonds is a general characteristic of module conformations.

Intra-module hydrogen bonds are observed in the helical parts and turns, and between both termini of each module or in their vicinities. These intra-module hydrogen bonds are formed in such a manner so as to direct each module to fold into a compact conformation.

Hydrogen bonds are localized mainly within each of the modules, thereby indicating that their locations are designed to primarily specify and stabilize conformations of individual modules [18].

Native-like secondary structures are formed in modules isolated in solution

Do modules maintain their native conformations, even when they are excised from the protein and isolated in solution? To search for an answer to this question, we studied the solution structures of isolated modules of barnase M1, M2 and M3 by two-dimensional NMR studies [21] (Takahashi et al., submitted).

One residue overlapped at each boundary of modules; i.e., M1: 1—24, M2: 24—52 and M3: 52—73 were chemically synthesized. Each of the modules M1, M2 and M3 was dissolved, independently, in a mixture of 50% d_3-TFE and 50% H_2O. All the NMR spectra were measured at 28°C. Sequence-specific resonance assignments of the modules were made according to the method described by Wüthrich and co-workers [22]. The distance geometry calculations were made for isolated modules M2 and M3 on the basis of distance constraints obtained from analysis of the NOESY spectra [23]. Details of experiments and results are described elsewhere [21] (Takahashi et al., submitted).

Module M1 in the crystal structure of barnase is made up from an N-terminal tail, the first α-helix (a in Fig. 1C) at the region Phe7-Tyr17, an extended strand folding back from the C-terminus of the α-helix and a turn near the C-terminus of the chain. In the solution structure of the isolated M1, an α-helix was found to form in a region corresponding to the first helix position of the intact barnase (Takahashi et al., submitted). In the crystal structure of intact barnase, M2 has a short stretch of the first β-strand (A in Fig. 1C) at Tyr24-Ile25, the second α-helix (b) at Lys27-Ala32, a turn-like structure at Ala37-Gly40, the third α-helix (c) at Leu42-Val45, a β-turn at Ala46-Lys49 and another short stretch of the second β-strand (B) at Ser50-Ile51. The two β-strands constitute a parallel β-sheet forming a bridge over N- and C-termini of the module. Structures similar to the second α-helix and the β-turn were located at the corresponding locations in the isolated M2 (Fig. 3). Module M3 in the crystal structure of intact barnase consists of two β-turns at Asn58-Gly61 and at Lys66-Arg69, and an antiparallel β-sheet bridging over N- and C-terminal regions of M3, Ile55-Phe56 (C) and Trp71-Glu73 (a part of the β-strand, D). The isolated M3 were found to form an extended structure at region Ile55-Ser57 and a β-turn at region Lys66-Arg69 [21]. In summary, about half of the secondary structures in the intact protein were formed in the isolated modules, M1, M2 and M3. The second helix was observed at regions Lys27-Ala32 and Thr26-Gly34 in the X-ray crystal structure and in the NMR solution structure [24] of the intact barnase, respectively. The corresponding helical region in the isolated M2 was Lys27-Val36. Except for this slight difference in boundaries of the helical regions, the secondary structures were found to correspond to those in intact barnase. Thus, the amino acid sequences of the three modules M1, M2 and M3 contain structural information that specifies at least about half of the secondary structures in each module [21].

Recent two-dimensional NMR studies revealed that peptide fragments of proteins in water or water/TFE mixtures take on secondary structures corresponding to those

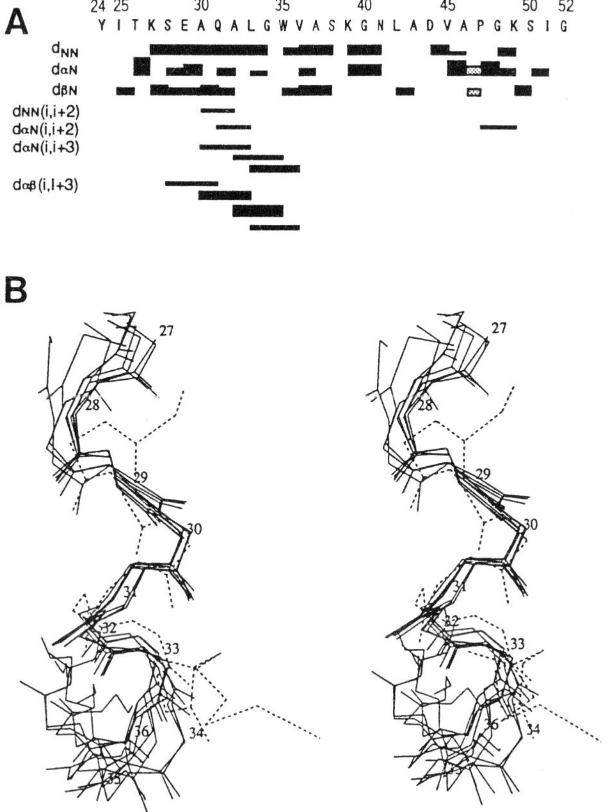

Fig. 3. Native-like secondary structures observed in isolated module M2. (A) Distance connectivities observed for module M2. The height of the bars indicates the magnitude of the NOESY cross-peaks. The patterns of the connectivities indicating α-helix and type II β-turn structures are observed in the regions of Lys27-Val36 and ALa46-Lys49, respectively. (B) Stereo view of superposition of the segments Lys27-Val36 in the 10 backbone conformations of module M2, consisting of nine conformations of M2 by NMR and distance geometry calculation (solid lines) and the X-ray structure of the corresponding segment in the intact barnase (a dashed line). The average of backbone root-mean-square displacements (RMSDs) in the nine NMR conformations is 1.2 Å for the segment. The backbone RMSD between the average NMR conformation and the X-ray structure is 1.0 Å for a segment of Ser28-Leu33. The residue numbers are those in the intact barnase. (Reproduced with permission from Ikura et al. [21], ©1993 Wiley-Liss, Inc.).

in the intact proteins [25–30]. Those structures are in rapid equilibrium with unfolded states and marginally stable so as to be detectable. In structures derived from the distance geometry calculation, the segment Lys26-Val36 of the isolated M2 took on stretched helical conformations, rather than such a typical α-helix as observed in the crystal structure (Fig. 3B). The apparent stretch of the helical conformations is possibly the result of rapid equilibrium between the α-helix and random coils [21].

As hydrogen bonds are formed mainly within modules [18], each module has the propensity within itself for taking on a native conformation. The observation of native-like secondary structures in the isolated modules is consistent with this propensity. Tertiary structures of the isolated modules M2 and M3 were, however, not uniquely determined by distance geometry calculation, therefore the isolated modules do not have the definite tertiary structures sufficiently stable to be detected. To better comprehend the stability of tertiary structures of isolated modules, an attempt should be made to discriminate between conformational stability and specificity [31]. Stability of a protein conformation depends on a delicate balance of free energy between native and denatured states. Therefore, one cannot conclude that an isolated module should take on its native conformation in solution, even though the module possesses the main part of interactions for conformational specificity. Isolated modules probably exist in a state where their secondary structures are in a dynamic equilibrium with unfolded states.

Kippen et al. divided barnase into two fragments, consisting of residues 1–22 and 23–110, respectively, and studied conformations of the isolated fragments and their mixture by CD and fluorescence spectroscopies [32]. The conformations of the isolated fragments were unstable in aqueous solution but the two fragments associated to form a fully native-like complex in the mixture. They also observed the isolated fragment 1–22 in water/TFE solution taking on the α-helix at a location corresponding to that in the intact barnase [30]. They concluded that barnase can fold by association of independently folded regions of structures. Interestingly, the N-terminal fragment 1–22 contains the most part of module M1, 1–24.

We also directed attention to remaining factors stabilizing the native conformation of modules in intact proteins, and for this we analyzed hydrophobic interactions in the intact barnase [18]. Hydrophobic side-chains of the protein form four clusters (Fig. 2d); the first one consists of M1 side-chains, the second mostly of M2 side-chains, the third of M3, M5 and M6 side-chains and the fourth of M1, M4 and M5 side-chains. Distribution of hydrophobic interactions within and between modules are given in Table 2. Many intra-module hydrophobic bonds exist within modules M1, M2 and M5. Localization of hydrophobic interactions are observed in M2, however, this module forms two hydrophobic bonds with M1 and M4, respectively. Module M5, on the other hand, forms 17 hydrophobic bonds with other modules. The degree of the localization of hydrophobic interactions in M1 is intermediate between M2 and M5. Modules M3, M4 and M6 do not have many hydrophobic bonds within themselves but do form more bonds with other modules, especially with M5. Oas and Kim studied the solution structure of two fragments of BPTI connected by a disulfide bond. They found that the complex of fragments formed a tertiary structure similar to the corresponding part of the native structure of the intact BPTI [33]. A hydrophobic core is present at the interface of the two fragments. Their results demonstrated that hydrophobic interactions between the fragments were essential for stability of the tertiary structure of the fragments. These hydrophobic interactions between modules of barnase are likely to contribute to maintaining the stability of the tertiary structure of each module.

Figure 4 shows a summary of all these discussions. Interactions within the module, hydrogen bonds, hydrophobic interactions and so forth, specify the native-like conformation of the module. In protein, interactions between modules stabilize compact conformations of individual modules. The excised and isolated module takes on secondary structures similar to those of the intact protein, but the tertiary structure is unstable because of a lack of stabilization by inter-module interactions. This interpretation can serve as one working hypothesis.

Based on the view mentioned above, inter-module interactions do not specify a compact conformation of a module, rather they mainly stabilize the conformation. If this view is tenable, then conformation of the module could also be stabilized by interactions, probably by hydrophobic interactions, with modules different from those which make contact with the module in the native protein. Such an example was found [34], as described in the next section.

Module M1 takes on a compact stable conformation during self-assembly

The peptide fragment of M1 was synthesized chemically, dissolved in aqueous solution and incubated at 5°C. CD spectra showed that the M1 fragment was in a predominantly random coil state immediately following the dissolution. However, formation of a helical structure began after 1 week and a plateau was reached in 20 days. The change in CD spectrum was accompanied by formation of filamentous structures present in the solution [34]. Therefore, the modules accumulated to constitute filamentous assemblies in which each of the modules assumed a stable

Table 2. Distribution of hydrophobic interactions.

	Number of intra- and inter-module hydrophobic bonds					
	M1	M2	M3	M4	M5	M6
M1	13	2	0	4	3	1
M2	2	9	0	2	0	0
M3	0	0	5	0	6	1
M4	4	2	0	2	4	0
M5	3	0	6	4	12	4
M6	1	0	1	0	4	1

Note: Hydrophobic bonds are defined here between a pair of hydrophobic residues, the distance between C^{β} atoms being less than 8 Å. Val, Leu, Ile, Phe, Trp, Cys, Met and Tyr are defined as hydrophobic residues.

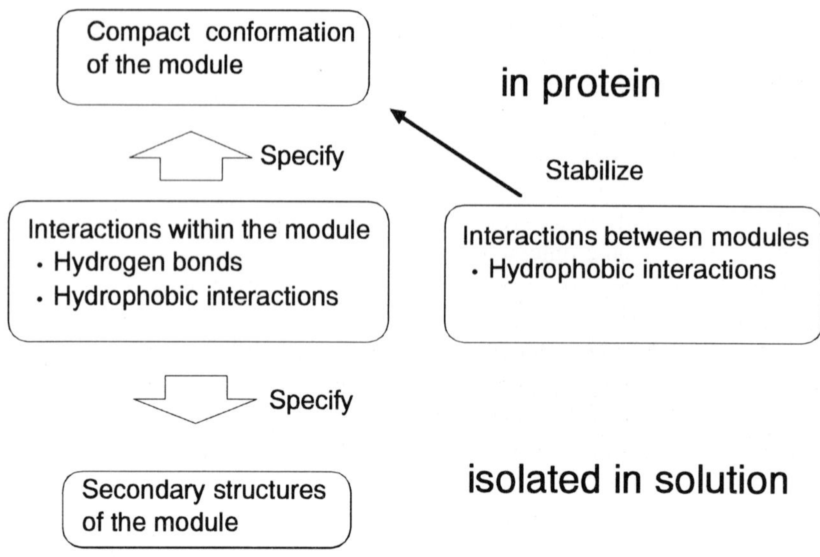

Fig. 4. Diagram illustrating how the conformation of modules is specified and stabilized.

conformation containing α-helical structures. NMR study of module M1 isolated in 50% TFE solution showed that it has the inherent propensity to form an α-helix at the location corresponding to that in the intact barnase (Takahashi et al., submitted). Therefore, the α-helix in M1 constituting the filamentous structure was probably formed at a location similar to that of the first α-helix in the barnase.

Scanning tunneling microscopy (STM) revealed the filamentous structures to be made up of protofilaments containing numerous disks apparently stacked on top of each other. From the high-resolution image of the protofilament, the size of the disk was about 2.0 nm in diameter and was about 1.5 nm thick [34]. It should be noted that M1 in the native barnase takes on a compact disk-like shape of about 2.0 nm in diameter and is about 1.5 nm thick with hydrophobic side-chains in the center and others in the margin of the disk-like shape (Fig. 5A). The shape and dimensions of the disks stacking up in the protofilament essentially conform to the native conformation of M1. Together with the α-helix formation, probably at the native-like location, our observations suggest that the M1 modules formed a native-like secondary and tertiary structure to assume the disk-like shape and stacked up through hydrophobic interactions between their side-chains. These interactions may thus facilitate the compact conformation of M1 and may be the driving force constituting the protofilament. A putative model for the protofilaments, consisting of stacked modules M1 in the native-like conformation is given in Fig. 5B [34].

Concluding discussion

To elucidate the role of modules in protein evolution and folding, we examined the

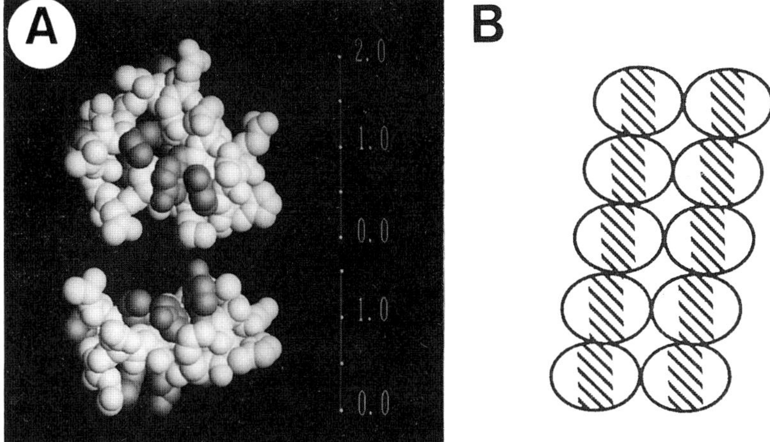

Fig. 5. (A) Disk-like shape of M1 in the native protein. Conformation of M1 is shown in the top and side views of the disk-like shape (upper and lower). Scales are represented by an nm unit. Hydrophobic sidechains are in dark tone. (B) A possible schematic model of the protofilaments. Each disk-like unit represents module M1 in the native-like conformation. The hydrophobic cores are hatched. ((B) reproduced with permission from Yoshida et al. [34], ©1993 the American Chemical Society).

structural stability and specificity of the modules of barnase. Hydrogen bonds are localized mainly within each of the modules, indicating that the native conformation of the module is specified by interactions within itself [18]. This view was further supported by observation of native-like secondary structures in the solution structures of synthesized modules M1, M2 and M3 [21] (Takahashi et al., submitted). The overall state does seem to be unstable. Module M1 assembled in aqueous solution [34]. NMR and CD spectra and STM image showed each M1 module in the assembly to take on a compact conformation similar to that in the intact barnase, and indicated that native conformations of modules which are specified by intra-module interactions are stabilized by the assembly of modules.

On the basis of the physicochemical propensity of modules, the following scenario is drawn for module recruitment into globular proteins through exon shuffling [8,13,35]. In early stages of evolution, selection pressure may have worked on the conformational stability of independent modules encoded by minigenes or of the assembly of modules. Conformations of some but not all modules may have been stable, but at least were likely to be specified by the interactions within themselves, If some modules did assemble, then their intrinsic conformations would be stabilized by interactions between them. Conformations of the modules may have been changed slightly by these interactions but nevertheless they retained shapes similar to the original. Some modules may have carried out specific primordial functions. In fact isolated modules M2, M3 and M6 of barnase were seen to have weak RNase activity [36]. More efficient functions may have derived from such an assembly of autonomous modules.

After the assembly of modules into a globular domain or protein, the selection pressure might then have been directed to stability of the assembly rather than of individual modules. Thus, interactions between modules, particularly hydrophobic interactions, have been developed by amino acid replacements. Particularly replacements by hydrophobic amino acid residues are likely to have occurred at the interface of the modules. This would explain why some contemporary modules, when excised from the protein, are unstable in solution.

The gene of barnase is not split by introns [37], as is also the case for other prokaryotic genes. It has been speculated that introns were lost from prokaryotic genes [5,8,13,38,39]. According to this view, module organizations observed in three-dimensional structures of contemporary proteins are vestiges of exon-intron organization of ancient genes, and if such is indeed the case, then the positions of these presumably lost introns in prokaryotic genes can be traced by identifying the module structures.

Acknowledgements

The studies reported here were carried out in cooperation with the groups of Drs N. Gō, F. Inagaki, H. Yanagawa, S. Iwanaga and K. Nagayama. Part of this work was supported by Grants-in-Aid from the Ministry of Education, Science and Culture of Japan to M.G. and T.N.

References

1. Gō M. Correlation of DNA exonic regions with protein structural units in haemoglobin. Nature 1981;291:90–92.
2. Gō M. Modular structural units, exons, and function in chicken lysozyme. Proc Natl Acad Sci USA 1983;80:1964–1968.
3. Isaacs NW, Machin KJ, Masakuni M. The three-dimensional structure of goose-type lysozyme from the egg white of the Australian black swan, *Cygnus atratus*. Austr J Biol Sci 1985;38:13–22.
4. Gō M, Nosaka M. Protein architecture and the origin of introns. Cold Spring Harb Symp Quant Biol 1987;52:915–924.
5. Gilbert W, Marchionni M. McKnight G. On the antiquity of introns. Cell 1986;46:151–154.
6. Gilbert W, Glynias M. On the ancient nature of introns. Gene 1993;135:137–144.
7. Tittiger C, Whyard S, Walker VK. A novel intron site in the triosephosphate isomerase gene from the mosquito *Culex tarsalis*. Nature 1993;361:470–472.
8. Gō M. Protein structures and split genes. Adv Biophys 1985;19:91–131.
9. Jensen EØ, Paludan K, Hyldig-Nielsen JJ, Jørgensen P, Marcker KA. The structure of a chromosomal leghaemoglobin gene from soybean. Nature 1981;291:677–679.
10. Blake CCF. Exons and the structure, function and evolution of haemoglobin. Nature 1981;291:616.
11. Gilbert W. Why genes in pieces? Nature 1978;271:501.
12. Straus D, Gilbert W. Genetic engineering in the precambrian: structure of the chicken triosephosphate isomerase gene. Molec Cell Biol 1985;5:3497–3506.
13. Blake CCF. Exons and evolution of proteins. Int Rev Cytol 1985;93:149–185.
14. Gilbert W. Genes-in-pieces revisited. Science 1985;228:823–824.
15. Nishimura S, Nomura M. Ribonuclease of *Bacillus subtilis*. J Biochem 1959;46:161–167.

16. Hartley RW, Barker EA. Amino acid sequence of extracellular ribonuclease (barnase) of *Bacillus amyloliquefaciens*. Nature New Biol 1972;235:15–16.
17. Mauguen Y, Hartley RW, Dodson EJ, Dodson GG, Bricogne G, Chothia C, Jack A. Molecular structure of a new family of ribonucleases. Nature 1982;297:162–164.
18. Noguti T, Sakakibara H, Gō M. Localization of hydrogen bonds within modules in barnase. Proteins 1993;16:357–363.
19. Hill C, Dodson G, Heinemann U, Saenger W, Mitsui Y, Nakamura K, Borisov S, Tischenko G, Polyakov K, Pavlovsky S. The structural and sequence homology of a family of microbial ribonucleases. Trends Biochem Sci 1983;8:364–369.
20. Stickle DF, Presta LG, Dill KA, Rose GD. Hydrogen bonding in globular proteins. J Molec Biol 1992;226:1143–1159.
21. Ikura T, Gō N, Kohda D, Inagaki F, Yanagawa H, Kawabata M, Kawabata S, Iwanaga S, Noguti T, Gō M. Secondary structural features of modules M2 and M3 of barnase in solution by NMR experiment and distance geometry calculation. Proteins 1993;16:341–356.
22. Wüthrich K, Wider G, Wagner G, Braun W. Sequential resonance assignments as a basis for determination of spatial protein structures by high-resolution proton nuclear magnetic resonance. J Molec Biol 1982;155:311–319.
23. Braun W, Gō N. Calculation of protein conformations by proton-proton distance constraints. A new efficient algorithm. J Molec Biol 1985;186:611–626.
24. Bycroft M, Ludvigsen S, Fersht AR, Poulsen FM. Determination of the three-dimensional solution structure of barnase using nuclear magnetic resonance spectroscopy. Biochemistry 1991;30: 8697–8701.
25. Wright PE, Dyson HJ, Lerner RA. Conformation of peptide fragments of proteins in aqueous solution: implications for initiation of protein folding. Biochemistry 1988;27:7167–7175.
26. Fry DC, Byler DM, Susi H, Brown EM, Kuby SA, Mildvan AS. Solution structure of the 45-residue MgATP-binding peptide of adenylate kinase as examined by two-dimensional NMR, FTIR, and CD spectroscopy. Biochemistry 1988;27:3588–3598.
27. Dyson HJ, Merutka G, Waltho JP, Lerner RA, Wright PE. Folding of peptide fragments comprising the complete sequence of proteins: Models for initiation of protein folding I. Myohemerythrin. J Molec Biol 1992;226:795–817.
28. Dyson HJ, Sayre JR, Merutka G, Shin H-C, Lerner RA, Wright PE. Folding of peptide fragments comprising the complete sequence of proteins: Models for initiation of protein folding II. Plastocyanin. J Molec Biol 1992;226:819–835.
29. Sancho J, Neira JL, Fersht AR. An N-terminal fragment of barnase has residual helical structure similar to that in a refolding intermediate. J Molec Biol 1992;224:749–758.
30. Kippen AD. Arcus VL, Fersht AR. Structural studies on peptides corresponding to mutants of the major α-helix of barnase. Biochemistry 1994;33:10013–10021.
31. Lattman EE, Rose GD. Protein folding – what's the question? Proc Natl Acad Sci USA 1993;90: 439–441.
32. Kippen AD, Sancho J, Fersht AR. Folding of barnase in parts. Biochemistry 1994;33:3778–3786.
33. Oas TG, Kim PS. A peptide model of a protein folding intermediate. Nature 1988;336:42–48.
34. Yoshida K, Shibata T, Masai J, Sato K, Noguti T, Gō M, Yanagawa H. Protein anatomy: spontaneous formation of filamentous helical structures from the N-terminal module of barnase. Biochemistry 1993;32:2162–2166.
35. Blake CCF. Do genes-in-pieces imply proteins-in-pieces? Nature 1978;273:267.
36. Yanagawa H, Yoshida K, Torigoe C, Park J-S, Sato K, Shirai T, Gō M. Protein anatomy: functional roles of barnase module. J Biol Chem 1993; 268:5861–5865.
37. Paddon CJ, Hartley RW. Cloning, sequencing and transcription of an inactivated copy of *Bacillus amyloliquefaciens* extracellular ribonuclease (barnase). Gene 1986;40:231–239.
38. Darnell JE, Doolittle WF. Speculations on the early course of evolution. Proc Natl Acad Sci USA 1986;83:1271–1275.
39. Kersanach R, Brinkmann H, Liaud M-F, Zhang D-X, Martin W, Cerff R. Five identical intron

positions in ancient duplicated genes of eubacterial origin. Nature 1994;367:387–389.
40. Baker EN, Hubbard RE. Hydrogen bonding in globular proteins. Prog Biophys Molec Biol 1984;44:97–179.

Tracing Biological Evolution in Protein and Gene Structures.
M. Gō and P. Schimmel, editors.

Mechanical stability of protein modules determined by molecular dynamics simulations

Ken-ichi Takahashi, Mitiko Gō and Tosiyuki Noguti
Department of Biology, Faculty of Science, Nagoya University, Nagoya, Japan

Abstract. *Background.* Exon shuffling would yield a stable globular protein if exons encode stable structural units. A module, a compact segment within a globular protein domain, is considered to be such a structural unit once encoded by an exon, since close correlation is observed between module and exon boundaries in various proteins. In order to elucidate the mechanism of protein evolution and the principle of protein architecture, we investigated the mechanical stability of excised modules.

Methods. We carried out 1 ns molecular dynamics simulations of six modules of barnase, a bacterial RNase from *Bacillus amyloliquefaciens*, starting with native conformations.

Results. We found that all but the C-terminal module retained native-like conformations through 1 ns simulations. Hence, the modules are mechanically stable. Molecular dynamics simulations with modified nonbonded interactions showed that the main contribution to the stability is due to van der Waals and nonionic partial charge interactions, indicating that compact conformation of the modules is essential for mechanical stability.

Conclusions. Our observations support the view that modules are suitable evolutionary building blocks of proteins.

Key words: barnase, exon shuffling, molecular evolution, protein folding.

Introduction

A globular domain of a protein is decomposed into compact structural units called modules (Fig. 1), which are contiguous segments as long as 10–40 amino acid residues [1,2]. Boundaries of modules are determined by the centripetal profile and extension profile [3] based on three-dimensional structures of proteins. In various proteins, module boundaries correspond to positions of introns in genes encoding the proteins [1–4]. This suggests that each module was encoded by an exon in early evolutionary periods and that exon shuffling or fusion [4–6] produced various combinations of modules that resulted in formation of proteins with various functions. The module organization observed in contemporary protein structures implies that each module has retained its primordial conformation through evolutionary processes of various combinations with other modules. If this is indeed the case, conformation of a module should be specified mainly by intramodule interactions (interactions between atoms within the module) rather than intermodule interactions (interactions

Address for correspondence: Mitiko Gō, Department of Biology, Faculty of Science, Nagoya University, Chikusa-ku, Furo-cho, Nagoya 464-01, Japan. Tel.: +81-52-789-2976. Fax: +81-52-789-2977.

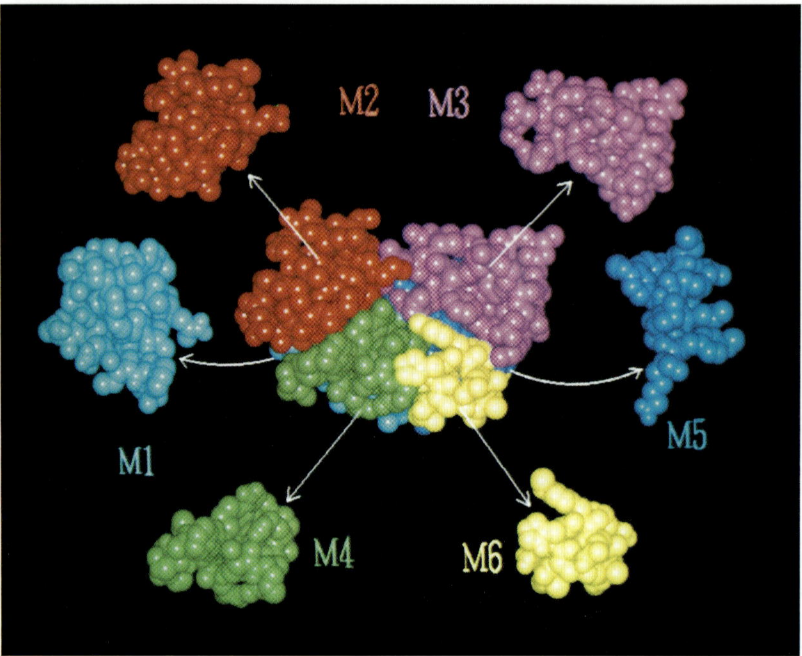

Fig. 1. Module structure of barnase. Barnase consists of a globular domain, which is decomposed into six modules, M1 (cyan), M2 (red), M3 (magenta), M4 (green), M5 (blue) and M6 (yellow).

between atoms in the module and those in other modules). This point cannot be given too much attention if one is to understand the evolution of proteins as well as the principle of protein architecture.

The following points suggest the autonomous nature of module conformations: localization of hydrogen bonds within rather than between modules [7]; formation of native-like secondary structures in solution structures of isolated modules [8]; and self-assembly of modules in which each module takes on a stable native-like compact shape [9].

To examine whether native conformations of modules in an existing protein are specified by intramodule interactions, we investigated the mechanical balance in intramodule interactions and the conformational stability of modules by molecular dynamics simulations. Conformation of each module is mechanically stabilized in the native protein, where all forces working on its atoms, including intra- and inter-module interactions and interactions with water molecules, are balanced. If forces from neighboring modules and water molecules are eliminated and only forces of intramodule interactions are left, what will happen to module conformation? This hypothetical state of the module isolated in vacuum can be simulated by molecular dynamics starting from its native conformation (Fig. 2). The isolated module will retain its native conformation in simulation if forces of intramodule interactions are

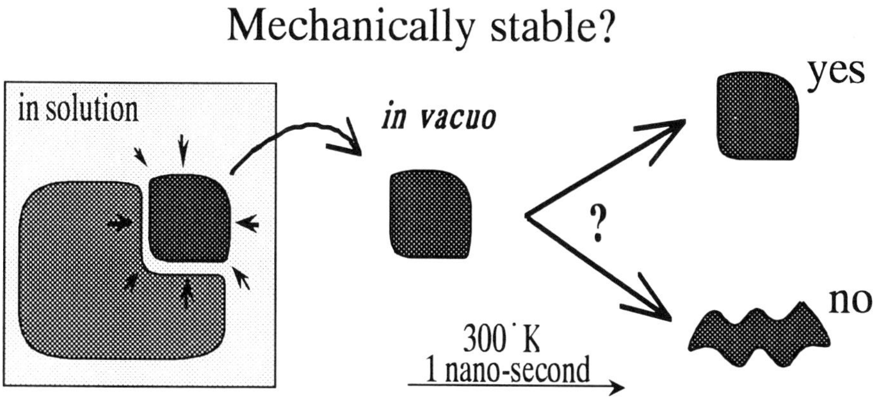

Mechanically stable?

in solution

in vacuo

?

yes

no

300˙K
1 nano-second

Fig. 2. Diagram illustrating the simulation study for mechanical stability of modules. In the native state of a protein in solution (left on the figure), a module (darker shaded square) interacts with the other part of the protein and solvent. These interactions (represented by short arrows) are eliminated by isolating the module in vacuum (middle). If the native conformation of the module is mechanically stable, the isolated module retains the structure. If not, it is deformed (right). Molecular dynamics simulation of the isolated module elucidates the problem.

balanced. Many mechanical properties of a protein molecule, such as conformational changes caused by ligand binding or by amino acid substitution, are characterized by dynamics in the time range of pico- to nanoseconds [10]. If conformation of the module remains similar to the native conformation for 1 ns in molecular dynamics simulation, then the native-like conformation of the module is mechanically stable.

Barnase is a bacterial RNase from *Bacillus amyloliquefaciens* [11]. Its tertiary structure has been elucidated by X-ray crystallography [12,13] and NMR spectroscopy [14]. Based on the tertiary structure, barnase was decomposed into six modules, M1 to M6 [7] (Fig. 1). We carried out molecular dynamics simulations of the modules of barnase for 1 ns and studied their mechanical stability. "Pseudomodules" of barnase were defined as segments starting at a center of one module and ending at the center of the following one (Fig. 3), and their mechanical stability was also studied by molecular dynamics. In contrast to compact conformations of modules, those of pseudomodules are extended forms in the native protein. We found that conformations of modules were mechanically stable, but those of pseudomodules were unstable (Takahashi et al., submitted for publication). We further examined the origin of the mechanical stability of modules by molecular dynamics simulations. Of the intramodule interactions, nonbonded interactions consisting of electrostatic interactions, hydrogen bonds, and van der Waals interactions stabilize module conformations. To investigate the role of these nonbonded interactions in the mechanical stability of modules, we simulated the isolated state of the modules by molecular dynamics calculations with hypothetically modified potential functions where ionic charges and interactions of hydrogen bonds were eliminated.

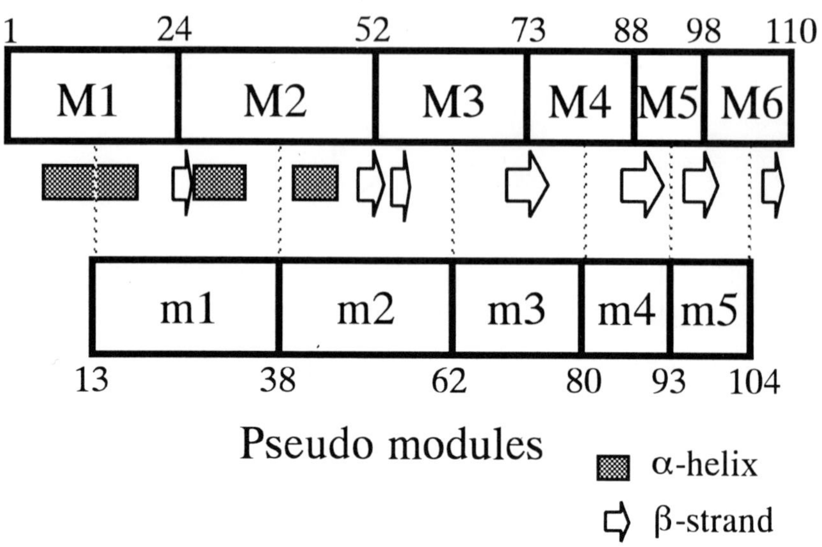

Fig. 3. Schematic diagram of modules, pseudomodules, and secondary structures of barnase. M1 to M6 and m1 to m5 are modules and pseudomodules, respectively. Residue numbers of boundaries of modules and pseudomodules are shown.

Materials and Methods

Modules of barnase

Barnase has three α-helices and seven β-strands (Fig. 3). The first and second β-strands form a parallel β-sheet and the remaining five form an antiparallel β-sheet. The module boundaries assigned by the centripetal profiles [3] are at residues 24, 52, 73, 88, and 98 [7] (Fig. 3). Thus, barnase is decomposed into six modules, M1 to M6; M1: 1–24, M2: 25–52, M3: 53–73, M4: 74–88, M5: 89–98, and M6: 99–110. The module boundaries are almost all located on one of the β-strands (Fig. 3). The N- and C-terminal parts of each main chain of modules, M2 to M5, are linked by hydrogen bonds, forming a part of the β-sheets. The first and two other α-helices are included in modules M1 and M2, respectively. We defined “pseudomodules” of barnase as segments starting at a center of one module and ending at the center of the following one. There are five pseudomodules, m1 to m5; m1: 13–38, m2: 39–62, m3: 63–80, m4: 81–93, and m5: 94–104 (Fig. 3). Each pseudomodule includes one or two β-strands.

Simulation methods and energy parameters

Conformations of modules and pseudomodules in the X-ray structure of barnase were energy-minimized independently. Each polypeptide chain of modules and pseudomodules was terminated at the peptide bond without charged or capped ends. Minimization was performed to find a conformation where the root mean square of atomic forces was less than 0.01 kcal/mol/Å. The energy-minimized conformations were taken for the initial conformations of the molecular dynamics (MD) simulations. The MD simulations were performed on modules and pseudomodules for 1 ns and 50 ps, respectively. Time steps were 1 fts unless mentioned otherwise. The temperature was maintained at 300 K by coupling (τ = 0.2 ps) to an external bath [15]. Structures were stored every 1 ps during the simulations for analysis. In all minimizations and MD simulations, a linear distance dependent dielectric function, $\varepsilon = r$, was applied and nonbonded interactions were all calculated, without truncation.

To study mechanical stability of each module or pseudomodule of barnase, we carried out two MD simulations, which we call "standard MD simulations", using the program and energy parameters of AMBER 3A [16] and those of BIOGRAF [17], respectively. The united-atom model was used in both simulations. Side chains of Lys, Arg, Asp, and Glu were ionized. Time steps of 0.25 fts were used. To examine the origin of conformational stability of the modules, we carried out the following two MD simulations for each of modules M1 to M5.

1. Charge-reduced MD simulation. In this simulation we used the program of AMBER 3A" and the reduced-charge, all-atom parameter set" of AMBER 4, in which the net charge of each Lys, Arg, Asp, and Glu was reduced to 0.2. We carried out this simulation to investigate effects of ionic charges on mechanical stability.
2. Chargeless, HB-less MD simulation. In this simulation we turned off interactions of hydrogen bonds and set net charge of each Lys, Arg, Asp, and Glu to be zero. The energy parameter and program of BIOGRAF for united-atom models were used. Effects of hydrogen-bond interactions on the mechanical stability were studied using this simulation.

Conformational comparison

Simulated conformations were averaged over the last half-trajectory of each MD simulation. We will hereafter call the averaged conformations "MD-averaged conformations". MD-averaged conformations and the corresponding X-ray structures of modules or pseudomodules were superimposed and compared. Besides the root mean square deviation of the C^{α}-atom positions (RMSD) between the superimposed structures, we used the root mean square deviation of the C^{α}-C^{α} distances (RMSΔD), defined in the following manner in order to measure conformational differences. The RMSΔD value between an MD-averaged conformation of a module and the corresponding X-ray structure is defined as

$$\mathrm{RMS}\Delta\mathrm{D} = \left[\sum_{i-j>4} \sum \frac{\left(d_M(i,j) - d_X(i,j)\right)^2}{n} \right]^{1/2},$$

where $d_M(i,j)$ and $d_X(i,j)$ are distances between C^α atom of the *i*th residue and that of the *j*th residue in the MD-averaged conformation and in the X-ray structure, respectively. The summation is taken over all of the pairs of residues that are separated by more than four residues ($i - j > 4$) and n is the number of such residue pairs. RMSΔD was more suitable than RMSD to measure differences in global main-chain folding patterns of modules.

Results

Molecular dynamics simulation without modification of nonbonded interactions (standard MD simulation)

Five of the six modules of barnase, M1 to M5, retained native-like conformations through the 1 ns molecular dynamics simulations (Takahashi et al., submitted for publication). The compact foldings of the main chains of the modules remained similar to those in the native conformation of barnase through the simulation. The RMSΔD values between the X-ray structures of the five modules and the corresponding MD averaged conformations were in the range of 2.0–3.6 Å (Table 1). The C-terminal module, M6, of which the native conformation is less compact than those of the other modules, folded into a substantially compact form. The RMSΔD value between the X-ray structure and the MD-averaged conformation of M6 was 6.9 Å. No pseudomodules could maintain their native conformation for even 50 ps; their extended conformations collapsed into compact forms. The RMSΔD values between the X-ray structures and the MD-averaged conformations were in the range of 8.8–13.6 Å (Table 1). These results mean that all modules, except for the C-terminal module M6, are mechanically stable, whereas all pseudomodules are mechanically unstable. The RMSD values of C^α atoms between the X-ray structures and the MD-averaged conformations were in the range of 2.3–4.4 Å for the five modules M1 to M5, 4.7 Å for M6 and in the range of 5.0–8.9 Å for the pseudomodules. Results of the standard MD simulations by the programs and energy parameters of BIOGRAF were essentially the same.

Molecular dynamics simulation with modified nonbonded interactions (charge-reduced MD simulation and chargeless, HB-less MD simulation)

All modules, except for M6, retained their native-like conformations through the standard MD simulations. Side chains of Asp, Glu, Arg, and Lys residues were treated as being fully ionized in the standard MD simulations. Numbers of the charged groups in the modules are given in Table 2. Effects of the charged groups on the conformational stability of the modules were eliminated by reducing their charges in the charge-reduced MD simulations. The five modules, M1 to M5, retained their native-like conformations throughout 1 ns in the charge-reduced MD simulations (Fig. 4). Their RMSΔD values between the MD-averaged conformations of the

Table 1. The root mean square deviation of the C^{α}-C^{α} distances (RMSΔD) in the MD-averaged conformation of each module or pseudomodule of barnase in the standard MD simulation from those in the corresponding X-ray structure.

Module	RMSΔD (Å)	Pseudomodule	RMSΔD (Å)
M1	2.0	m1	9.1
M2	3.1	m2	10.7
M3	3.6	m3	13.6
M4	2.6	m4	13.4
M5	2.9	m5	8.8
M6	6.9		

modules and the corresponding X-ray structures were in the range of 1.3—2.5 Å (Table 2). The RMSD values were 1.5—2.8 Å. The secondary structures in the X-ray structures were all observed at the corresponding locations in the MD-averaged conformations of the modules (Fig. 4). The hydrogen bonds of the α-helix in M1, the two α-helices and the parallel β-sheet in M2, and the parts of the antiparallel β-sheet formed in M3, M4, and M5 were maintained in the simulated conformations of individual modules. These results of the charge-reduced MD simulations show that the modules are mechanically stable even if charges of ionized groups are reduced.

Hydrogen bonds play essential roles in secondary structures and tertiary structures of proteins. Their contribution to the mechanical stability of module conformations were investigated by the chargeless, HB-less MD simulations in which both the hydrogen-bond interactions and electrostatic interactions of charged groups were hypothetically turned off. The result was surprising; four of the five examined modules, M1, M2, M4, and M5 retained their characteristic foldings of the native conformations throughout 1 ns (Fig. 5). Module M3 was largely deformed. The RMSΔD values between the MD-averaged conformations of the four modules and the corresponding X-ray structures were 2.1—3.8 Å and that of M3 was 4.9 Å (Table 2). The corresponding RMSD values of the four modules were in 2.4—5.1 Å and that of M3 was 5.0 Å. Some secondary structural features remained in the simulated conformations of the modules despite the absence of hydrogen-bond interactions, especially the turns or turn-like structures, which are critical to determine the compact

Table 2. The root mean square deviation of the C^{α}-C^{α} distances (RMSΔD) in the MD-averaged conformation of each module in the charge-reduced MD simulation and chargeless, HB-less MD simulation from those in the corresponding X-ray structure.

Module	No. of charged groups		RMSΔD	
	+	–	Charge-reduced MD (Å)	Chargeless, HB-less MD (Å)
M1	1	3	2.1	3.8
M2	3	2	1.7	3.0
M3	5	3	2.5	4.9
M4	2	2	2.2	2.7
M5	1	1	1.3	2.1

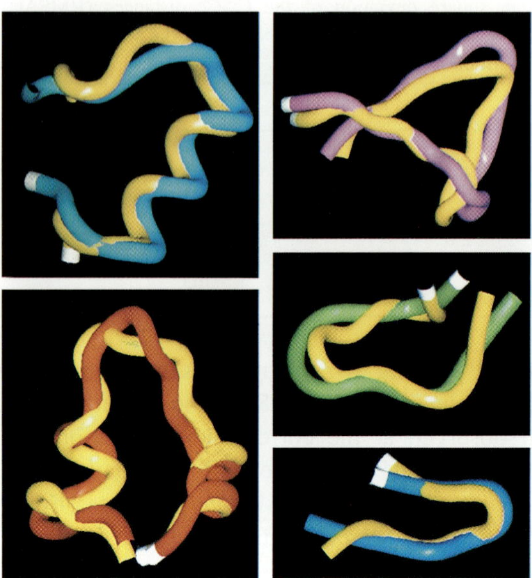

Fig. 4. Conformations of modules simulated by the molecular dynamics with reduced ionic charges (charge-reduced MD) and their corresponding X-ray structures. The trace of the main chain of each module is shown in a tube model. X-ray structures of modules are colored; M1: cyan, M2: red, M3: magenta, M4: green, and M5: blue. Conformations in the last 500 ps trajectory of each molecular dynamics simulation are averaged. The resultant conformation of each module is colored in ocher and superimposed on the corresponding X-ray structure. The N-terminal of each module is indicated in white.

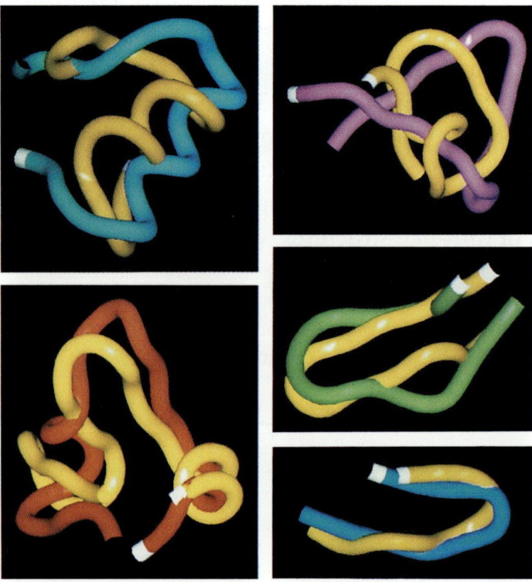

Fig. 5. Conformations of modules simulated by the molecular dynamics, without ionic charges and hydrogen-bond interactions (chargeless, HB-less MD) and their corresponding X-ray structures. Legend as in Fig. 4.

native conformations of the modules, were retained in M2, M4, and M5. The MD-averaged conformation of M5 is similar to the native conformation, an antiparallel β-sheet. Modules M1 and M2 formed native-like helical structures, although their length and location were not identical to the native ones. These results of chargeless, HB-less MD simulations show that most of the modules are mechanically stable even if hydrogen-bond interactions are eliminated.

Discussion

We simulated the hypothetical state of barnase modules isolated in vacuum by 1 ns molecular dynamics. All but the C-terminal module retained native-like conformations through the 1 ns simulations. This means that atomic interactions are balanced within each of the modules and thus conformations of the modules are mechanically stable (Takahashi et al., submitted for publication) and it also suggests that tertiary structures of modules are determined mainly by intramodule interactions.

We further examined roles of nonbonded interactions in the conformational stability of the modules by simulating their isolated state with modified potential functions where ionic charges and hydrogen-bond interactions are eliminated. We found that conformations of the modules remain stable without the hydrogen-bond interactions and electrostatic interactions of ionic charges. Therefore, other nonbonded interactions, the van der Waals interaction and the electrostatic interaction between partial chargcs of nonionic groups, mainly contribute to the mechanical stability. These nonbonded interactions are weak for each atomic pair and are only effective in case of a short distance. A sufficient number of atoms should be in contact to stabilize the conformations. In general, the more compact conformation has a larger number of atomic contacts. Modules are so compact as to have a large number of atomic contacts. Therefore, it is concluded that compact conformations of modules are essential for mechanical stability. In contrast to modules, pseudomodules are not compact and do not have a large number of atomic contacts, thus their conformations are mechanically unstable. Hydrogen bonds, however, play an important role in the mechanical stability of modules. Secondary structures are retained better in charge-reduced MD simulation (Fig. 4) than in chargeless, HB-less MD simulations (Fig. 5); where interactions of hydrogen bonds are turned off, hydrogen-bond interactions contribute to stabilization of local secondary structures.

To examine the mechanical balance in intramodule interactions, we simulated a hypothetical state of an isolated module in vacuum. In an actual situation, modules are integrated into a protein surrounded by water molecules; the surface of modules is considerably exposed to water. Electrostatic interactions and hydrogen bonds between atoms exposed to water are weakened by hydration effects. In vacuum simulations, effects of electrostatic interactions and hydrogen bonds between atoms on module surfaces might be more potent than in the actual situation. However, this does not seem to affect the results of our simulations. We found that most of the modules investigated are mechanically stable in the complete absence of ionic charges and hydrogen bonds (Fig. 5).

We carried out the same simulation of modules of four other proteins: myohemerythrin, immunoglobulin V_H domain, flavodoxin, and lysozyme, which are α-, β-, α/β- and $\alpha + \beta$-type proteins, respectively. Most of the modules were mechanically stable.

Correlation between intron positions and module boundaries [1–4] suggests that there was a one-to-one correspondence between exons and modules in the early evolutionary period and that exon shuffling or fusion [4–6] produced various combinations of modules, which resulted in the creation of stable proteins with various functions. It is essential that each module be structurally stable and maintain a compact conformation in various combinations of modules. We showed that most of the modules are mechanically stable. This result further supports the view that modules are indeed evolutionary building blocks of proteins. Assembly of mechanically stable modules are likely to result in mechanically stable proteins. Modification of a protein by exchanging modules is also likely to be successful. Alternative splicing produces several combinations of exons from a single copy gene, depending on cell types or stages of development. The resultant proteins often have slightly tuned functions. The mechanical stability of modules is thought to be important in this modification of protein functions.

Acknowledgements

We thank Drs G. Dodson and C. Hill for providing the X-ray atomic coordinates of barnase, Drs F. Hirata and A. Matsumoto for helpful comments on MD calculations, and M. Ohara for critical comments on the manuscript. Part of this work was supported by Grants-in-Aid from the Ministry of Education, Science and Culture of Japan to M.G. and T.N.

References

1. Gō M. Correlation of DNA exonic regions with protein structural units in haemoglobin. Nature 1981;291:90–92.
2. Gō M. Modular structural units, exons, and function in chicken lysozyme. Proc Natl Acad Sci USA 1983;80:1964–1968.
3. Gō M, Nosaka M. Protein architecture and the origin of introns. Cold Spring Harb Symp Quant Biol 1987;52:915–924.
4. Gō M. Protein structures and split genes. Adv Biophys 1985;19:91–131.
5. Gilbert W. Why genes in pieces? Nature 1978;271:501.
6. Blake CCF. Do genes-in-pieces imply proteins-in-pieces? Nature 1978;273:267.
7. Noguti T, Sakakibara H, Gō M. Localization of hydrogen-bonds within modules in barnase. Proteins 1993;16:357–363.
8. Ikura T, Gō N, Kohda D, Inagaki F, Yanagawa H, Kawabata M, Kawabata S, Iwanaga S, Noguti T, Gō M. Secondary structural features of modules M2 and M3 of barnase in solution by NMR experiment and distance geometry calculation. Proteins 1993;16:341–356.
9. Yoshida K, Shibata T, Masai J, Sato K, Noguti T, Gō M, Yanagawa H. Protein anatomy:

spontaneous formation of filamentous helical structures from the N-terminal module of barnase. Biochemistry 1993;32:2162–2166.
10. Karplus M, Petsko GA. Molecular dynamics simulations in biology. Nature 1990;347:631–639.
11. Nishimura S, Nomura M. Ribonuclease of *Bacillus subtilis*. J Biochem 1959;46:161–167.
12. Mauguen Y, Hartley RW, Dodson EJ, Dodson GG, Bricogne G, Chothia C, Jack A. Molecular structure of a new family of ribonucleases. Nature 1982;297:162–164.
13. Baudet S, Janin J. Crystal structure of a barnase-d(GpC) complex at 1.9 Å resolution. J Molec Biol 1991;219:123–132.
14. Bycroft M, Ludvigsen S, Fersht AR, Poulsen FM. Determination of the three-dimensional solution structure of barnase using nuclear magnetic resonance spectroscopy. Biochemistry 1991;30: 8697–8701.
15. Berendsen HJC, Postma JPM, van Gunsteren WF, DiNola A, Haak JR. Molecular dynamics with coupling to an external bath. J Chem Phys 1984;81:3684–3690.
16. Weiner SJ, Kollman PA, Case DA, Singh UC, Ghio C, Alagona G, Profeta S Jr, Weiner P. A new force field of molecular mechanical simulation of nucleic acids and proteins. J Am Chem Soc 1984;106:765–784.
17. Mayo SL, Olafson BD, Goddard III WA. Dreiding: a generic force field for molecular simulations. J Phys Chem 1990;94:8897–8909.

Tracing Biological Evolution in Protein and Gene Structures.
M. Gō and P. Schimmel, editors.

Helix-turn-helix module distribution and module shuffling

Kei Yura and Mitiko Gō
Department of Biology, Faculty of Science, Nagoya University, Nagoya, Japan

Abstract. *Background.* DNA-binding helix-turn-helix (HTH) motif exists in eukaryotic and prokaryotic gene-regulatory proteins. The motif consists of a single module, a compact unit of protein structure. The HTH modules were found in a repressor 2 to 3 times. The repetitive existence of HTH modules in repressors indicates that a DNA-binding HTH module was duplicated. The modules that do not compose HTH motif have not been reported to bind DNA. Correlation between module boundaries and intron positions observed in various proteins implies that the module was encoded by a single exon in ancient genes. If the module had been shuffled by exon shuffling, the HTH modules could be found in other proteins.

Methods. Three-dimensional structure, the best conserved quality during protein evolution, was compared between the HTH module and all the modules of the apparently non-DNA/RNA-binding proteins whose three-dimensional structures are available in the Protein Data Bank at Brookhaven National Laboratory.

Results. The HTH modules were also found in several proteins that have not been reported to interact with DNA/RNA.

Conclusions. The existence of the HTH modules in non-DNA-binding proteins suggests that the modules were recruited as a scaffold for other modules with functions other than binding DNA. The existence of HTH module as a DNA-binding module and as a foundation to appropriately allocate other modules gives supporting evidence for module shuffling.

Key words: exon shuffling, HTH motif, molecular evolution, repressor, three-dimensional structure.

Introduction

The HTH motif is one of the most frequently observed DNA-binding substructures in prokaryotic and eukaryotic transcription regulatory proteins such as repressors, homeodomains, POU domains and histone H5 [1–3]. The name of the motif comes from its three-dimensional structure; two helices connected by a turn in an angle of about 120°. DNA recognition is achieved by inserting the latter helix, called recognition helix, into a major groove. Side chains of the helix recognize specific bases in the groove. Amino acid sequence and DNA sequence analyses of the motif from several DNA-binding proteins suggests that the motif was derived from a common ancestor [4].

The HTH motif is composed of a single module [5]. A module is a contiguous polypeptide segment with a compact conformation in a globular protein [6]. The

---

Address for correspondence: Mitiko Gō, Department of Biology, Faculty of Science, Nagoya University, Furo-cho, Chikusa-ku, Nagoya 464-01, Japan. Tel.: +81-52-789-2576. Fax: +81 52-789-2977.

average length of a module is about 15 amino acid residues long. Compactness of the module indicates that it is a building block of proteins. Introns are located close to module boundaries [6–12]. Correspondence of module boundaries and intron positions suggests that a single module was once encoded by a single exon. Combined with the notion of exon shuffling [13], the module is considered to be a unit of protein evolution.

Correspondence of the HTH motif to a module suggests that a protein with an HTH motif might have been formed using the HTH module plus other modules. Module structure analysis of other parts of repressors revealed that HTH modules were used repetitively (Fig. 1). Phage 434 Cro [14] was made from three HTH modules and two other modules. Phage 434 repressor [15] was made from three HTH modules plus another module. Gamma repressor (Fig. 2) [16] was made from two HTH modules plus five other modules. The amino acid sequences of HTH modules in repressors showed a statistically significant similarity. This means that the repressors evolved by multiplying the HTH module. Presumably a functionally important DNA-binding HTH module was multiplied. HTH modules that were not

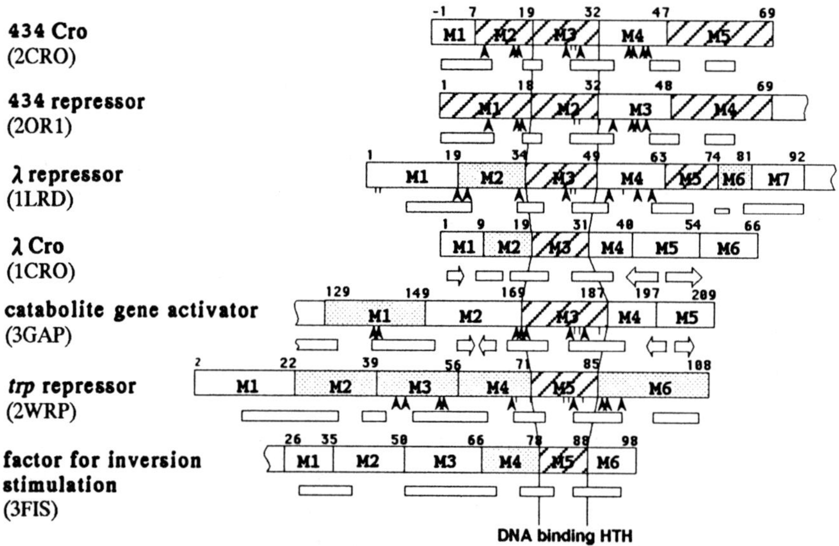

Fig. 1. Module organization and secondary structure of repressors. The name of each repressor is given on the left with the accession ID of Protein Data Bank in parentheses. A box with a number starting with M indicates a module. A rectangle below the modules is an α-helix, a small rectangle is a 3_{10} helix and an arrow is a β-strand. A striped box is the HTH module. A gray module has two helices on both termini of the module, but its overall structure is different from the HTH modules. Modules drawn at the center of the figure connected with vertical lines are DNA-binding HTH modules. A thin bar on the modules indicates a residue that interacts with a DNA base. A wedge indicates a residue that interacts with the DNA phosphate. Three-dimensional structure of 434 Cro was determined by Mondragon and Harrison [14], 434 repressor by Aggarwal et al. [25], λ repressor by Jordan and Pabo [16], λ Cro by Anderson et al. [26], catabolite gene activator by Schultz et al. [27], *trp* repressor by Otwinowski et al. [28] and factor for inversion stimulation by Yuan et al [29]. (Reproduced by permission, from [5]. Protein names are modified from the original figure.)

involved in DNA-binding function might have lost the function during the process of evolution [5]. Repetitive usage of HTH module in repressors suggests that the HTH module should be used in seemingly non-DNA/RNA binding proteins, if module shuffling is the likely mechanism of protein evolution. We, therefore, searched for the HTH module in apparently non-DNA/RNA-binding proteins in the Protein Data Bank at Brookhaven National Laboratory [17].

Material and Methods

Module boundary assignment

Modules were assigned by centripetal and extension profiles [10]. Module boundaries are located in the center of a local segment of a protein. The centripetal profile detects residues at the center of the local segment by calculating the mean square distance between the C^{α} atom of *i*th residue and the other C^{α} atoms within the range of $\pm k$ from *i*th residue. The values were plotted against *i*. The local minima of the profile are candidates for module boundaries. Compactness of the fragment between two boundaries was confirmed by extension profile [18].

Structure comparison procedure

Comparisons of modules were made based on their three-dimensional structures. C^{α} atoms of two modules were superimposed by the method of McLachlan [19], and root mean square deviations (RMSD) of C^{α} atoms were calculated. Two modules do not necessarily have the same length of amino acid sequences, so the shorter module was shifted along the longer one to acquire all the possible superimpositions. The least RMSD was assigned to the RMSD of the two.

All the proteins in the Protein Data Bank [17] that have not been reported to interact with DNA/RNA were divided into modules. As a representative of an HTH module, module M2 of 434 repressor [15] that composes the HTH motif was chosen. All modules were superimposed against 434 repressor module M2.

The criteria for similarity was set at 1.9 Å. The significance of criteria was justified statistically by the RMSD calculation of protein fragments. Superimposition of C^{α} atoms of fragments with 15 residues from proteins resulted in an average RMSD of 6.00 ± 1.45 Å. 1.9 Å is more than 2.8 σ below the average [20].

Sequence comparison procedure

Sequence comparison of the HTH modules was performed based on the three-dimensional structure alignment. Similarity score was calculated between the pairs of residues that correspond when two modules were superimposed. The significance of the similarity was assessed by the deviation of the score from the average of scores between one of the two modules and 10^4 randomized sequences of the other module. The deviation was scaled by the standard deviation [5].

Results and Discussion

Search for the HTH module in the proteins of the Protein Data Bank resulted in detection of a number of HTH modules in a wide range of proteins (Yura K et al., submitted). HTH modules were not limited to repressors, but found in proteins that have not been reported to bind DNA. Proteins that had the HTH module were sarcoplasmic calcium-binding protein [21], cytochrome P450cam [22], xylose isomerase [23] and so forth. Superimposition of the HTH modules revealed differences in length of helices on both termini, yet the packing angles of the two helices were the same (Fig. 3).

Amino acid sequence similarity among the HTH modules in repressors suggested that the DNA-binding and non-DNA-binding HTH modules had been derived from a common ancestor [5]. The same analysis was carried out between the DNA-binding HTH modules and the newly found HTH modules. The HTH module in cytochrome P450cam marked a high similarity score, a score equivalent to the ones found among DNA-binding HTH modules. The result suggests that the HTH module in cytochrome P450cam and DNA-binding HTH modules in repressors have a common ancestor. Although other HTH modules did not mark high similarity score, low similarity score could be the result of distant relationships. One cannot rule out the possibility, however, that the similarity of amino acid sequences arose from the preference of amino acid residues in the secondary structure formation.

The possibility that sarcoplasmic calcium-binding protein, cytochrome P450cam and xylose isomerase bind DNA by the HTH modules is low. Our analysis of DNA-binding HTH modules and non-DNA-binding HTH modules in repressors showed a difference in amino acid sequences between the two. The DNA-binding HTH module had a residue with a small side chain at the fifth position from conserved glycine on the turn, whereas non-DNA-binding HTH module had a residue with a bulky side chain at the corresponding position (Fig. 4). A large side chain at the position hampers DNA binding of the module. The residue at the fifth position was included in the recognition helix of DNA-binding HTH module. When the recognition helix is inserted into a major groove, the residue faces the DNA side. The fifth residue with a large side chain prevents the recognition helix from being inserted into a major groove, and thereby hampers interaction between residues and DNA phosphates or bases [5]. The HTH modules found in apparently non-DNA-binding proteins have bulky side chains at the fifth position. Sarcoplasmic calcium binding protein has a methionine, cytochrome P450cam has an aspartic acid, and xylose isomerase has a histidine side chain. The residues are bulky compared to the ones of DNA-binding HTH modules (Fig. 4). Binding energy of λ repressor module M3 to DNA was calculated on a computer. By binding to DNA, the energy reduced by 3.61 kcal/mol. When the side chain of the fifth residues was changed to aspartic acid, the energy reduced only by 1.08 kcal/mol. When the side chain is bulky, sufficient interaction between residues of the recognition helix and DNA phosphates or bases cannot be obtained [5].

The HTH modules in non-DNA/RNA-binding proteins seemed to serve as a

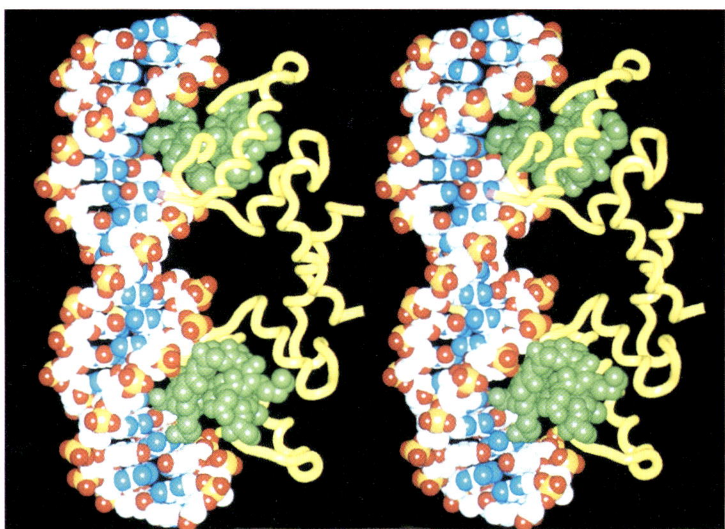

Fig. 2. Structure of λ repressor dimer in complex with a DNA [16]. The DNA-binding HTH module is displayed by space-filling model colored in green. Other parts of the repressor are depicted by the tube model colored in yellow. The NH_2-terminus is colored in pink. A carbon atom of DNA is colored in white, a nitrogen atom in blue, an oxygen atom in red, and a phosphorus atom in orange. (Reproduced with permission from [30].)

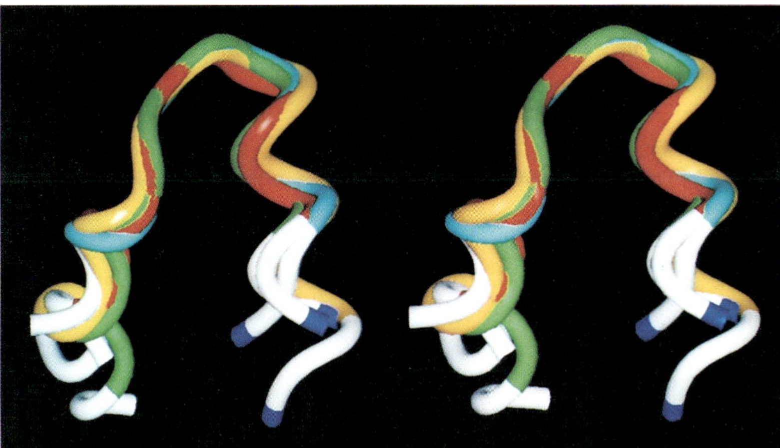

Fig. 3. Superimposition of the HTH modules. The red module is from 434 repressor, the green from sarcoplasmic calcium-binding protein, the orange from cytochrome P450cam and the sky blue from xylose isomerase. NH_2-terminus is colored in dark blue. The white regions of each module belong to the previous and next modules. RMSD of each module against 434 repressor M2 is less than 1.9 Å.

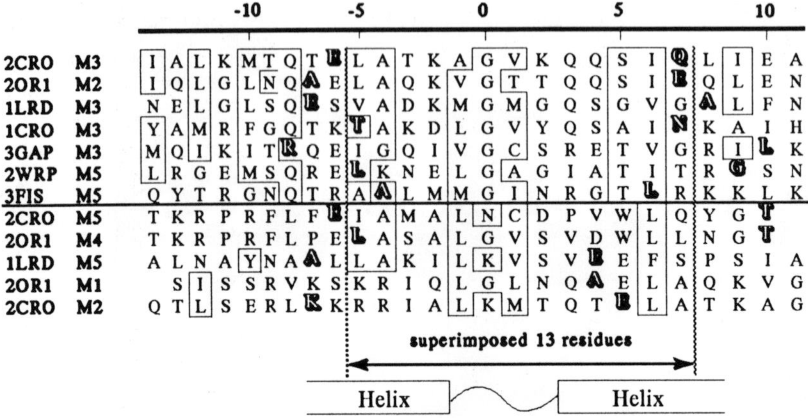

Fig. 4. Structural alignment of HTH modules in repressors. Accession ID in Protein Data Bank is given on the left. A module number is given next to the code starting with the letter M. An outlined letter is a module boundary. An arrow at the bottom shows the range of similar three-dimensional structures. Modules above the thick bar in the middle are DNA-binding HTH modules and below the bar are non-DNA-binding HTH modules. The similar residues throughout the alignment are boxed. The scale on the top is numbered as a glycine at the turn of the HTH modules becomes zero. The side chains of residues at the fifth position show a clear difference between DNA-binding HTH module and non-DNA-binding HTH modules. The residue for DNA-binding HTH module has a small side chain, whereas the one for non-DNA-binding HTH module has a bulky side chain. (Reproduced with permission from [5].)

scaffold for other modules involved in the function of each protein. In repressors, the non-DNA-binding HTH modules served as a scaffold to locate the DNA-binding HTH module so as to insert the recognition helix into a major groove [5]. The HTH modules found in this report have similar characteristics to the non-DNA-binding modules in repressors; the modules have residues with a large side chain at the fifth position and they do not have obvious functions. The similarities between the non-DNA-binding HTH modules in repressors and the newly found HTH modules in other proteins indicate that the HTH modules were used as a scaffold in other proteins as well as in the repressors.

The HTH modules were also found in DNA/RNA binding proteins other than repressors. Cases were found where the HTH modules appeared repeatedly in a single protein [5] (Yura K et al., submitted). The HTH modules were used in two different ways in a variety of proteins; they were used as a DNA-binding module and formed a DNA-binding protein, or as a scaffold for other modules of non-DNA-binding proteins.

The finding of HTH modules in a variety of proteins supports the notion that HTH modules were shuffled during protein evolution. The shuffling was plausibly carried out by exon shuffling [13], because correspondence of intron positions and module boundaries suggests that a module was encoded by a single exon. One cannot rule out the possibility that appearance of the HTH modules in many proteins have been the result of convergent evolution from different structures; originally different structures in each protein ended in the same HTH structure, because the structure was useful for each protein. However, the convergent evolution is less likely, unless strong preference acted on the structure of a module, and not on the function.

A new protein was likely to be formed using available modules; some of these modules were recruited for specific functions and others for proper allocation of modules involved in functions. Barnase, a bacterial ribonuclease, is composed of six modules. Studies on the excised modules revealed that three of the six modules had ribonuclease activity and the other three of them did not. The latter three modules were considered to be scaffold for the former three modules [24]. Modules for the scaffold did not require functions in themselves, yet needed to be stable enough to sustain other modules dealing with the function. Therefore, any module including those with other functions could have served as a scaffold for functional modules. The HTH module was also used as a scaffold for other proteins. The existence of the HTH module in a wide range of proteins suggests that the module was used as a scaffold from the early stage of protein evolution. If the HTH module was not used as a scaffold from the early stage, the module would not found in a variety of proteins.

One of the conceivable shuffling evolutions of the HTH module is depicted in Fig. 5. The original HTH module is considered to bind DNA. A module that had no obvious functions had no apparent advantage in existing by itself. The DNA-binding HTH module was, then, used in two different ways during protein evolution; one was the DNA-binding HTH module as in the HTH motif of repressors, and the other was as a scaffold module as in repressors and other proteins.

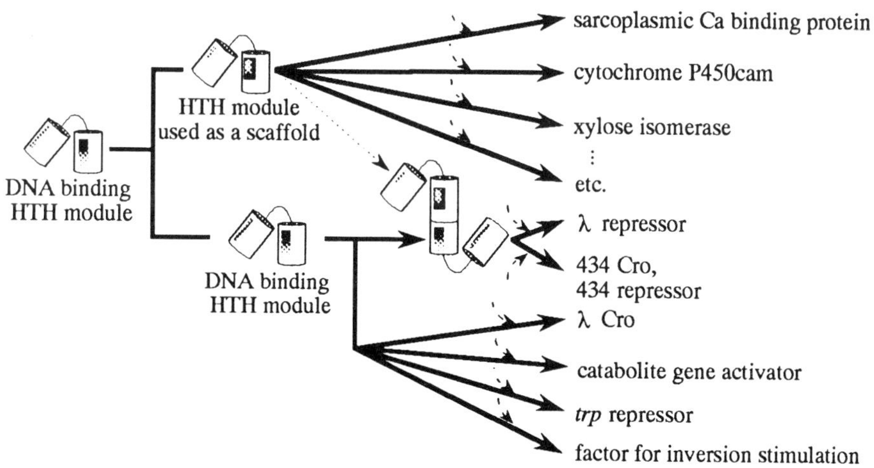

Fig. 5. A conceivable evolutionary pathway of proteins with the HTH module. A solid line indicates a hypothetical branching of HTH modules. A dashed line indicates recruitment of other modules into a protein. An HTH module could be considered to begin as a DNA-binding module. The same module could also be used as a scaffold in a protein. Duplication of the DNA-binding HTH module might create an HTH module as a scaffold for a DNA-binding HTH module. Addition of other modules resulted in λ and 434 repressors and 434 Cro. Other repressors evolved without duplication of HTH module. HTH module as a scaffold is found in many proteins. The proteins were evolved by a combination of HTH module as a scaffold and other modules. These proteins have no evidence of DNA/RNA interaction.

Acknowledgements

Part of this work was supported by Grants-in-Aid from the Ministry of Education, Science and Culture of Japan to K.Y. and M.G.

References

1. Brennan RG. DNA recognition by the helix-turn-helix motif. Curr Opin Struct Biol 1992;2:100–108.
2. Wright PE. POU domains and homeodomains. Curr Opin Struct Biol 1994;4:22–27.
3. Ramakrishnan V. Histone structure. Curr Opin Struct Biol 1994;4:44–50.
4. Ohlendorf DH, Anderson WF, Matthews BW. Many gene-regulatory proteins appear to have a similar α-helical fold that binds DNA and evolved from a common precursor. J Molec Evol 1983;19:109–114.
5. Yura K, Tomoda S, Gō M. Repeat of a helix-turn-helix module in DNA-binding proteins. Protein Eng 1993;6:621–628.
6. Gō M. Correlation of DNA exonic regions with protein structural units in haemoglobin. Nature 1981;291:90–92.
7. Gō M. Modular structural units, exons, and function in chicken lysozyme. Proc Natl Acad Sci USA 1983;80:1964–1968.
8. Gō M. Protein structures and split genes. Adv Biophys 1985;19:91–131.
9. Gilbert W, Marchionni M, McKnight G. On the antiquity of introns. Cell 1986;46:151–154.
10. Gō M, Nosaka M. Protein architecture and the origin of introns. Cold Spring Harbor Symp Quant Biol 1987;52:915–924.
11. Tittiger C, Whyard S, Walker VK. A novel intron site in the triosephosphate isomerase gene from the mosquito *Culex tarsalis*. Nature 1993;361:470–472.
12. Gilbert W, Glynias M. On the ancient nature of introns. Gene 1993;135:137–144.
13. Gilbert W. Why genes in pieces? Nature 1978;271:501.
14. Mondragon A, Harisson SC. The phage 434 Cro/O_R1 complex at 2.5 Å resolution. J Molec Biol 1991;219:321–334.
15. Mondragon A, Subbiah S, Alamo SC, Drottar M, Harrison SC. Structure of the amino-terminal domain of phage 434 repressor at 2.0 Å resolution. J Molec Biol 1989;205:189–200.
16. Jordan SR, Pabo CO. Structure of the lambda complex at 2.5 angstroms resolution. Details of the repressor-operator interactions. Science 1988;242:893–899.
17. Bernstein FC, Koetzle TF, Williams GJB, Meyer EF, Brice MD, Rodgers JR, Kennard O, Shimanouchi T, Tasumi M. The protein data bank: A computer-based archival file for macromolecular structures. J Molec Biol 1977;112:535–542.
18. Noguti T, Sakakibara H, Gō M. Localization of hydrogen bonds within modules in barnase. Proteins 1993;16:357–363.
19. McLachlan AD. Gene duplications in the structural evolution of chymotrypsin. J Molec Biol 1979;128:49–79.
20. Remington SJ, Matthews BW. A systematic approach to the comparison of protein structures. J Molec Biol 1980;140:77–99.
21. Vijay-Kumar S, Cook WJ. Structure of a sarcoplasmic calcium-binding protein from *Nereis diversicolor* refined at 2.0 Å resolution. J Molec Biol 1992;224:413–426.
22. Poulos TL, Finzel BC, Howard AJ. High-resolution crystal structure of cytochrome P450cam. J Molec Biol 1987;195:687–700.
23. Dauter Z, Terry H, Witzel H, Wilson KS. Refinement of glucose isomerase from *Streptomyces albus* at 1.65 Å with data from an imaging plate. Acta Cryst (Sect B) 1990;46:833–841.
24. Yanagawa H, Yoshida K, Torigoe C, Park J-S, Sato T, Shirai T, Gō M. Protein anatomy: Functional roles of barnase module. J Biol Chem 1993;268:5861–5865.

25. Aggarwal AK, Rodgers DW, Drottar M, Ptashne M, Harrison SC. Recognition of a DNA operator by the repressor of phage 434: A view at high resolution. Science 1988;242:899–907.
26. Anderson WF, Ohlendorf DH, Takeda Y, Matthews BW. Structure of the cro repressor from bacteriophage λ and its interaction with DNA. Nature 1981;290:754–758.
27. Schultz SC, Shields GC, Steitz TA. Crystal structure of a CAP-DNA complex: The DNA is bent by 90°. Science 1991;253:1001–1007.
28. Otwinowski Z, Schevitz RW, Zhang R-G, Lawson CL, Joachimiak A, Marmorstein RQ, Luisi BF, Sigler PB. Crystal structure of *trp* repressor/operator complex at atomic resolution. Nature 1988;355:321–329.
29. Yuan HS, Finkel SE, Feng J-A, Kaczor-Grzeskowiak M, Johnson RC, Dickerson RE. The molecular structure of wild-type and a mutant Fis protein: Relationship between mutational changes and recombinational enhancer function or DNA binding. Proc Natl Acad Sci USA 1991;88:9558–9562.
30. Gō M. Evolution of proteins viewed on their three-dimensional structures. Protein Nucl Acid Enzyme (in Japanese) 1994;39:2448–2456.

Tracing Biological Evolution in Protein and Gene Structures.
M. Gō and P. Schimmel, editors.

Module replacement converted coenzyme specificity of isocitrate dehydrogenase

Tairo Oshima[1], Takuro Yaoi[1] and Mitiko Gō[2]
[1]*Department of Molecular Biology, Tokyo University of Pharmacy and Life Science, Hachioji, Tokyo; and* [2]*Department of Biology, Faculty of Science, Nagoya University, Chikusa, Nagoya, Japan*

Abstract. The coenzyme specificity of isocitrate dehydrogenase was successfully converted from NADP to NAD specificity by a module replacement, but not by residue replacements. The module-replaced mutant enzyme was as stable as the original NADP-specific enzyme. These results support the idea that a module is a structural and functional unit of a protein.

Key words: exon-shuffling hypothesis, isopropylmalate dehydrogenase, molecular recognition.

Isocitrate and isopropylmalate dehydrogenases

Isocitrate dehydrogenase in the TCA cycle catalyzes dehydrogenation of isocitrate and at the same time decarboxylation at the C-3 carboxylic group of the substrate. The enzyme belongs to an enzyme family "decarboxylating dehydrogenases". Isopropylmalate dehydrogenase also belongs to the same enzyme family. These two enzymes are speculated to have evolved from a common ancestral enzyme, since they act on structurally similar substrate and essentially catalyze the same chemical reaction (Fig. 1). The primary and three-dimensional structures of these enzymes resemble each other [1,2]. These two enzymes differ in many respects from other well-known dehydrogenases such as alcohol dehydrogenase, lactate dehydrogenase, etc.; for example, the so-called Rossmann fold [3] is missing in the NAD(P)-binding regions of isocitrate and isopropylmalate dehydrogenases.

Though isocitrate and isopropylmalate dehydrogenases are similar to each other in many respects, their coenzyme specificity is different; usually bacterial isocitrate dehydrogenase transfers the hydrogen atom attached to the C-2 position to NADP, whereas isopropylmalate dehydrogenase prefers NAD as the hydrogen acceptor. The genes for these enzymes have been cloned and sequenced from an extreme thermophile, *Thermus thermophilus* (*T. thermophilus*) [4–6]. The three-dimensional structure of the thermophile isopropylmalate dehydrogenase has been solved [2]. The crystal structure of the thermophile isocitrate dehydrogenase is under investigation in cooperation with Prof N. Tanaka of the Tokyo Institute of Technology. However, the

---

Address for correspondence: Tairo Oshima, Department of Molecular Biology, Tokyo University of Pharmacy and Life Science, 1432 Horinouchi, Hachioji, Tokyo 192-03, Japan. Tel.: +81-426-76-7134. Fax: +81-426-76-7145.

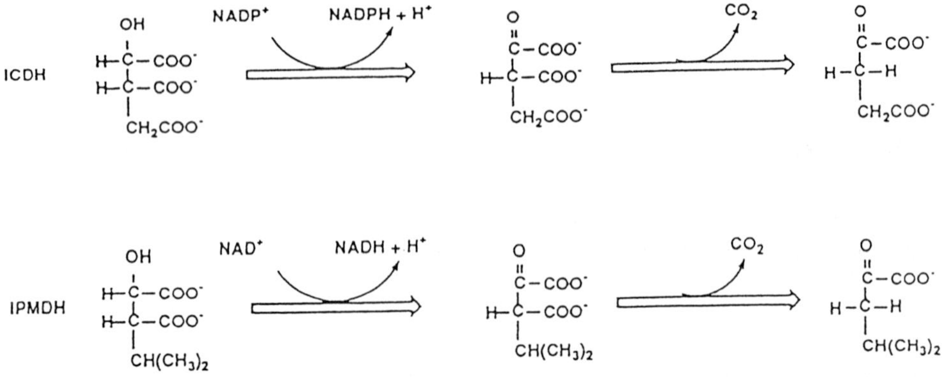

Fig. 1. Comparison of chemical reactions catalyzed by isocitrate dehydrogenase (ICDH, upper) and 3-isopropylmalate dehydrogenase (IPMDH, lower).

X-ray crystallographic structure of *Escherichia coli* (*E. coli*) isocitrate dehydrogenase has been reported and the NADP-binding site has been analyzed by the crystal structure of the enzyme-NADP complex [7]. The structure-function relationship of the thermophile isocitrate dehydrogenase can be speculated based on knowledge obtained from the analysis of the mesophile enzyme-NADP complex. Thus the thermophile isocitrate and isopropylmalate dehydrogenases seem to be good experimental materials for studying conversion of enzyme specificity using protein engineering techniques.

We have tried to convert their substrate specificity [8] using the thermophile enzymes. The thermophile enzymes are unusually resistant not only to heat, but also to chemical denaturants, and are easy to handle. For example, they can be easily purified by heating an *E. coli* extract after the gene coding for the thermophile enzyme is expressed in the mesophile host cells. Recently we succeeded in converting the coenzyme specificity of isocitrate dehydrogenase by module replacement.

Recognition of NADP by isocitrate dehydrogenase

Based on the crystallographic analysis on *E. coli* isocitrate dehydrogenase-NADP complex, Hurley and others proposed that four residues, Tyr345, Tyr391, Arg395, and Arg292′ (′ indicates the residue(s) from the second subunit, both isocitrate and isopropylmalate dehydrogenases are homodimers of identical subunits), interact with the 2′-phosphate group of NADP. However, two of them, Tyr391 and Arg395, are substituted with other residues in the thermophile isocitrate dehydrogenase, suggesting that these two residues are not essential for the NADP recognition. Only Tyr345 and Arg292 are conserved in the thermophile NADP-specific enzyme. In the thermophile enzyme these residues correspond to Tyr284 and Arg231. These two residues are conserved in NADP-specific isocitrate dehydrogenases from a variety of organisms, but substituted with other residues in NAD-specific isopropylmalate dehydrogenases suggesting that these two residues are important in the discrimination of NADP and NAD.

In addition, our sequence comparison revealed that Lys283 (numbering according to the thermophile isocitrate dehydrogenase), the adjacent residue to Tyr284, is conserved through NADP-dependent isocitrate dehydrogenases and substituted with Asp in NAD-dependent isopropylmalate dehydrogenases. Involvement of the corresponding Asp residue in NAD recognition in the active site of the thermophile isopropylmalate dehydrogenase is also suggested by the recent X-ray analysis of the enzyme-NAD complex [9]. Based on these results, it can be speculated that interactions between 2′-phosphate of NADP and these three amino acid residues discriminates between NADP and NAD in isocitrate and isopropylmalate dehydrogenases, as shown in Fig. 2. Our mutational analyses supported the hypothesis shown in Fig. 2 [10–12].

Residue substitutions

When these residues were substituted with the corresponding ones of NAD-specific isopropylmalate dehydrogenase, the affinity to NADP decreased, but that to NAD was not increased. For example, the catalytic efficiency defined as *k*cat/(*K*m for NADP) x (*K*m for isocitrate) of Arg231Ala or Tyr284Phe increased 2- to 3-fold, but those for NAD of these mutant enzymes were not improved significantly. Moreover, some mutant enzymes seemed to lose the conformational stability and were not expressed in *E. coli*. For example, the enzymatic activity of a double mutant Lys283Asp/-Tyr284Ile was not able to be detected in the host *E. coli* cell transformed with the mutant gene. The results suggest that the mutant enzyme cannot form the active conformation even at the growth temperature of the mesophile.

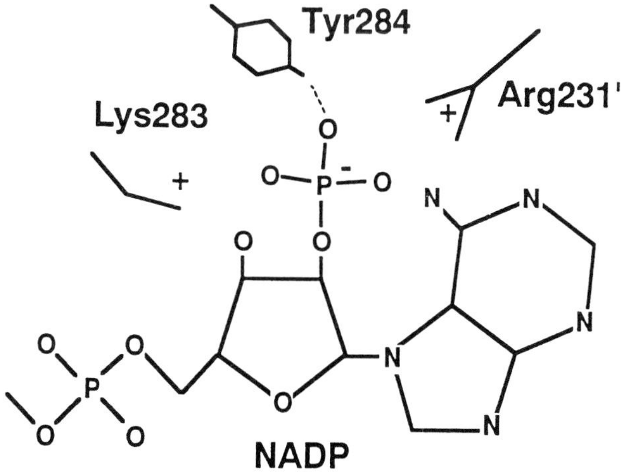

Fig. 2. Speculated interactions between *T. thermophilus* isocitrate dehydrogenase and the 2′-phosphate group of NADP.

Module replacement

Module structures of *E. coli* isocitrate dehydrogenase and *T. thermophilus* isopropylmalate dehydrogenase were analyzed by the centripetal profile method [13—15]. As expected from their high sequence homology, the module structures of these two enzymes resemble each other. The corresponding residues to Lys283 and Tyr284 in Fig. 2 are located in module-19 (numbering according to the module structure of the thermophile isopropylmalate dehydrogenase, module-19 of the thermophile 3-isopropylmalate dehydrogenase corresponds to module-26 in the mesophile isocitrate dehydrogenase). This module forms a loop connecting β-strand D and α-helix *i*.

Module-19 consists of 18 residues. Most of the residues are located around NAD(P). However, some do not interact with the coenzyme as shown in Fig. 3. Among these 18 residues, five, including Lys283 and Tyr284, are different between the thermophile isocitrate and isopropylmalate dehydrogenases.

However, when this module of the thermophile NADP-specific isocitrate dehydrogenase was replaced with the corresponding module of NAD-specific isopropylmalate dehydrogenase, the coenzyme specificity was dramatically altered without significant loss of the enzymatic activity as summarized in Table 1. As shown in the table, *K*m for NADP increased 180-fold by the module replacement. The ratio of catalytic efficiency for NAD and NADP (ratio of *k*cat/(*K*m for NAD) x (*K*m for isocitrate in the presence of NAD) and *k*cat/(*K*m for NADP) x (*K*m for isocitrate in the presence of NADP)) was improved 10^5-fold. The addition of an

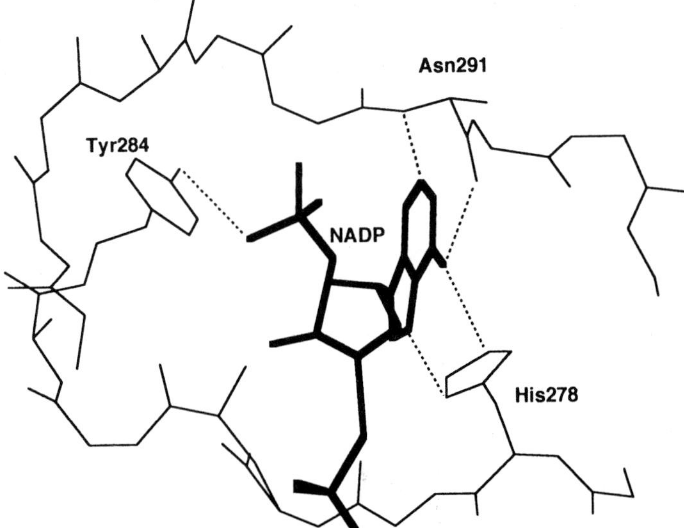

Fig. 3. Speculated interactions between module-19 (numbering according to the module structure of *T. thermophilus* isopropylmalate dehydrogenase) of the thermophile isocitrate dehydrogenase and NADP. This picture was drawn by superimposing side chains of the thermophile dehydrogenase on the three-dimensional structure of an *E. coli* isocitrate dehydrogenase-NADP complex.

Table 1. Kinetic parameters of the wild-type and module-19-replaced isocitrate dehydrogenase from *T. thermophilus*.

Enzyme	Coenzyme	*K*m for coenzyme (μM)	*K*m for isocitrate (μM)	*k*cat(s^{-1})
Wild-type	NADP	7.2	8.9	71
	NAD	3100	81	41
Module-19	NADP	1300	82	24
Replaced	NAD	220	22	41

Arg231Ala mutation to the module-replaced mutant enzyme further improved the affinity for NAD by about 6 times.

In contrast to some residue(s)-substituted mutations, the module-replaced mutants are as stable as the original wild-type thermophile enzyme. No significant loss of the activity was observed after the thermophile isocitrate dehydrogenase or the module-19-replaced mutant enzyme was heated at 80°C for 10 min. This observation suggests that a module is a structural unit of a protein.

Modules as functional and structural units

The Michaelis constant for NAD of the module-replaced mutant enzyme (220 mM) is only 30 times larger than the *K*m of the native enzyme (7 mM). The addition of a Arg231Ala mutation improved the affinity for NAD by 6 times. Thus the *K*m for NAD of the module-replaced/Arg231Ala mutant enzyme is in the "standard" value range of an enzyme affinity for coenzymes. It should be stressed that the alteration of the coenzyme specificity was achieved only by module replacement, but not residue(s) substitutions.

Often mutant enzymes lose their conformational stability, and it is difficult to analyze their functional properties. In this context, thermophile enzymes are good experimental materials in protein engineering, since they are unusually stable without any exception, and their mutants can be analyzed at the moderate temperatures even if the mutation affected the structural stability. However, in the present study, the module-replaced mutant is as stable as the thermophile wild-type enzyme. In the present study, the enzyme lost its ability to maintain the active conformation by replacing one to three residues located in module-19.

Gilbert [16] and Blake [17] proposed the exon-shuffling hypothesis for gene evolution, in which an exon is thought to act as a genetic unit in evolving a new gene, and encodes a structural and functional unit of a protein. Gō proposed the module structure and found a good correlation between the module structure of a protein and exon-intron structure of its gene [13,18]. Such a good correlation between exons and modules implies that an exon encodes a structural and functional unit of a protein and proteins have been evolved by exon shuffling, though there is a paper which claims that no significant relationship exists between module and exon structures in some proteins [19].

From the experimental side, the exon-shuffling hypothesis has been strongly supported by functional conversion between α-lactalbumin and lysozyme by exon exchange [20]. The present study also provides other supportive data for the exon-shuffling or module-shuffling theory on the enzyme evolution. Coenzyme specificity of the decarboxylating dehydrogenases can be converted by exon-exchange-encoding module-19, if the exon-intron structure of the genes coding for these enzymes corresponds to the module structure of the enzymes.

Coenzyme specificity of NADP-dependent isocitrate and NAD-dependent isopropylmalate dehydrogenases might have been diverged by a changing amino acid sequence of module-19 of the common ancestral enzyme which accepts both NAD and NADP. Module exchange could be a promising way to alter the function and physicochemical properties of a protein.

Acknowledgements

The study was supported by a Grant-in-Aid Scientific Research (02403029) from the Ministry of Education, Science, and Culture, the Japanese Government, and a fund from the Mitsubishi Foundation.

References

1. Hurley JH, Thorsness PE, Ramalingam V, Helmers NH, Koshland DE Jr. Structure of a bacterial enzyme regulated by phosphorylation, isocitrate dehydrogenase. Proc Natl Acad Sci USA 1989;86:8635–8639.
2. Imada K, Sato Y, Matsuura Y, Katsube Y, Oshima T. Three-dimensional structure of a highly thermostable enzyme, 3-isopropylmalate dehydrogenase of *Thermus thermophilus*. J Molec Biol 1991;222:725–738.
3. Rossmann MG, Lilgas A, Branden CI, Banaszak LJ. Evolutionary and structural relationships among dehydrogenases. Enzymes 1975;11:61–102.
4. Miyazaki K, Eguchi H, Yamagishi A, Wakagi T, Oshima T. Molecular cloning of the isocitrate dehydrogenase gene of an extreme thermophile, *Thermus thermophilus* HB8. Appl Environ Microbiol 1992;58:93–98.
5. Tanaka T, Kawano N, Oshima T. Cloning and expression of the leucine gene from *Thermus thermophilus*. J Biochem 1981;89:677–682.
6. Kirino H, Aoki M, Aoshima M, Hayashi Y, Ohba M, Yamagishi A, Wakagi T, Oshima T. Hydrophobic interaction at the subunit interface contributes to the thermostability of 3-isopropylmalate dehydrogenase from an extreme thermophile, *Thermus thermophilus*. Eur J Biochem 1994;220:275–281.
7. Hurley JH, Dean AM, Koshland DE Jr, Stroud RM. Catalytic mechanism of $NADP^+$-dependent isocitrate dehydrogenase: implications from the structures of magnesium-isocitrate and $NADP^+$ complexes. Biochemistry 1991;30:8671–8678.
8. Miyazaki K, Yaoi T, Oshima T. Expression, purification, and substrate specificity of isocitrate dehydrogenase from *Thermus thermophilus* HB8. Eur J Biochem 1994;221:899–903.
9. Hurley JH, Dean AM. Structure of 3-isopropylmalate dehydrogenase in complex with NAD^+: ligand-induced loop closing and mechanism for cofactor specificity. Structure 1994;2:1007–1016.
10. Yaoi T, Miyazaki K, Oshima T. Roles of Arg^{231} and Tyr^{284} of *Thermus thermophilus* isocitrate

dehydrogenase in the coenzyme specificity. FEBS Letters 1994;355:171–172.

11. Miyazaki K, Oshima T. Co-enzyme specificity of 3-isopropylmalate dehydrogenase from *Thermus thermophilus* HB8. Protein Eng 1993;7:401–403.
12. Yaoi T, Miyazaki K, Oshima T. His273 of 3-isopropylmalate dehydrogenase from *Thermus thermophilus* HB8 is involved in the coenzyme binding. Biochem Biophys Res Commun 1995;210:733–737.
13. Gō M. Correlation of DNA exonic regions with protein structural units in haemoglobin. Nature 1981;291:90–92.
14. Gō M. Modular structural units, exons, and function in chicken lysozyme. Proc Natl Acad Sci USA 1983;80:1964–1968.
15. Gō M, Nosaka M. Protein architecture and origin of introns. Cold Spring Harbor Symp Quant Biol 1987;52:915–924.
16. Gilbert W. Why genes in pieces? Nature 1978;271:501.
17. Blake CCF. Exons encode protein functional units. Nature 1979;277:598.
18. Gō M. Protein structures and split genes. Adv Biophys 1985;19:91–131.
19. Stoltzfus A, Spencer DF, Zuker M, Logsdon JM Jr, Doolittle WF. Testing the exon theory of genes: the evidence from protein structure. Science 1994;265:202–207.
20. Kumagai I, Takeda S, Miura K. Functional conversion of the homologous proteins α-lactalbumin and lysozyme by exon exchange. Proc Natl Acad Sci USA 1992;89:5887–5891.

Tracing Biological Evolution in Protein and Gene Structures.
M. Gō and P. Schimmel, editors.

The chimeric nature of nuclear genomes and the antiquity of introns as demonstrated by the GAPDH gene system*

Rüdiger Cerff
Institut für Genetik, Technische Universität Braunschweig, Braunschweig, Germany

Abstract. Genes for chloroplast and cytosolic glyceraldehyde-3-phosphate dehydrogenases of eukaryotes (class I GAPDH) are descendants of an ancient gene family that probably originated in the progenote. During eukaryotic evolution both genes were transferred to the nucleus from the progenitors of present-day chloroplasts and mitochondria, respectively, thereby replacing pre-existing GAPDH genes of the host cell. Chloroplast GAPDH genes contain introns at 10 different positions, five of which coincide precisely with intron positions found in glycolytic GAPDH genes from plants and animals. These findings, together with observations that introns tend to correlate with GAPDH protein structure, strongly support the exon theory of genes, suggesting that introns may be as old as the genes in which they reside. The data further demonstrate a lineage-specific loss of introns, in addition to "slippage" and "sliding" of introns in different GAPDH gene regions. We characterized a number of nuclear protein-coding genes of red algae, most of which contain a single short spliceosomal intron at their 5′ end. Intron splice junctions resemble those of higher plants, in agreement with phylogenetic trees, suggesting a sister relationship between red algae and green plants. The paucity and asymmetric localization of introns in red algae is similar to the situation found in yeast genes and may be interpreted in terms of differential intron loss due to conversion of genes by homologous recombination with cDNAs corresponding to reverse transcribed mRNAs or modified (incorrectly spliced) pre-mRNAs, respectively.

Key words: aldehyde dehydrogenase, endosymbiotic gene transfer, exon-shuffling, gene-protein correlations, glyceraldehyde-3-phosphate dehydrogenase, primary life forms.

Introduction

Ever since the discovery of spliceosomal introns 17 years ago there has been an ongoing debate about their origin (for reviews see [1–3]). The "introns early hypothesis" and its alias the "exon theory of genes" [4,5] propose that introns were present in the most primitive genes, where they accelerated the evolution of new functions by enhancing the recombination of proto-exons encoding small stably folding polypeptides to yield mosaic (sub-)genes encoding functional domains. These domains could then be "reshuffled" to form genes encoding the basic enzymes of primary metabolism found in all cells [6–10]. Prior to the discovery of introns by

---

Address for correspondence: Rüdiger Cerff, Institut für Genetik, Technische Universität Braunschweig, Spielmannstr. 7, D-38106 Braunschweig, Germany.
*Dedicated to the commemoration of the late Daniel I. Arnon, a pioneer of photosynthesis research in whose laboratory phosphorylating and nonphosphorylating GAPDH enzymes of higher plants were first characterised.

DNA sequencing techniques protein crystallographers had already recognized that structural domains, e.g., NAD(P)-binding domains of dehydrogenases, can be shared by otherwise unrelated proteins [11–13]. Opponents of the "introns early" view consider the complete absence of spliceosomal introns in present-day genes of bacteria and organelles as conclusive evidence for their eukaryotic origin. This "introns late" hypothesis suggests that introns invaded nuclear genomes after the separation of pro- and eukaryotes by insertion into contiguous genes via mechanisms possibly related to DNA transposition [14–16].

While "exon shuffling" [10] has been unambiguously demonstrated in the evolution of the genes for serine proteases [17] and low density lipoprotein receptor [18], only proteins belonging to the most ancient metabolic pathways, such as glycolysis or photosynthesis, are likely to be encoded by genes old enough to provide evidence for exon shuffling in the primordial assembly of structures conserved in the proteins of extant species [19]. Two glycolytic enzymes, triosephosphate isomerase (TPI) and glyceraldehyde-3-phosphate dehydrogenase (GAPDH), have played prominent roles in this "introns early versus introns late" debate. The maize gene encoding TPI is an impressive example of the precise conservation of intron positions across the plant/animal boundary [20,21]. Moreover, intron positions were shown to be related to TPI protein architecture by separating the polypeptide chain into compact Gō modules [22–25], although earlier reports exist on correlations of exons with protein structural domains for maize ADH [26] and chicken GAPDH [27,28]. The plant genes encoding chloroplast and cytosolic phosphorylating GAPDH, on the other hand, present a unique example for intron conservation in ancient duplicated genes of endosymbiotic, eubacterial origin.

Plant glyceraldehyde-3-phosphate dehydrogenases (classes I and III)

Chloroplast and cytosolic phosphorylating GAPDHs play key roles in the energy metabolism of plants. They are marker enzymes of two metabolic pathways, the photosynthetic Calvin cycle and glycolysis, sharing a number of reactions that are catalyzed by distinct (iso-)enzymes unique to each (Fig. 1). Chloroplast GAPDH (GapAB: EC 1.2.1.13) catalyzes the reductive step in the Calvin cycle by using NADPH generated in the photosynthetic light reaction. The enzyme has residual NAD activity in vitro [29,30] and is stimulated in vivo at the protein level via the thioredoxin system [31,32]. In angiosperms Calvin cycle GAPDH is composed of two isoenzymes, an A_2B_2 heterotetramer and an A_4 homotetramer [33–35]. Subunit B (GapB) differs from subunit A (GapA) by about 20% of its amino acid residues and has a slightly larger molecular weight owing to a flexible and highly charged C-terminal extension of about 30 additional amino acids conserved in all higher plants [36]. This C-terminal tail of unknown function is susceptible to proteolytic cleavage and is not essential for catalytic activity [37]. GapA and GapB are encoded in the nucleus, translated on free cytosolic ribosomes and imported into the chloroplast upon cleavage of an amino-terminal transit peptide [38]. *GapA* and *GapB* may be encoded

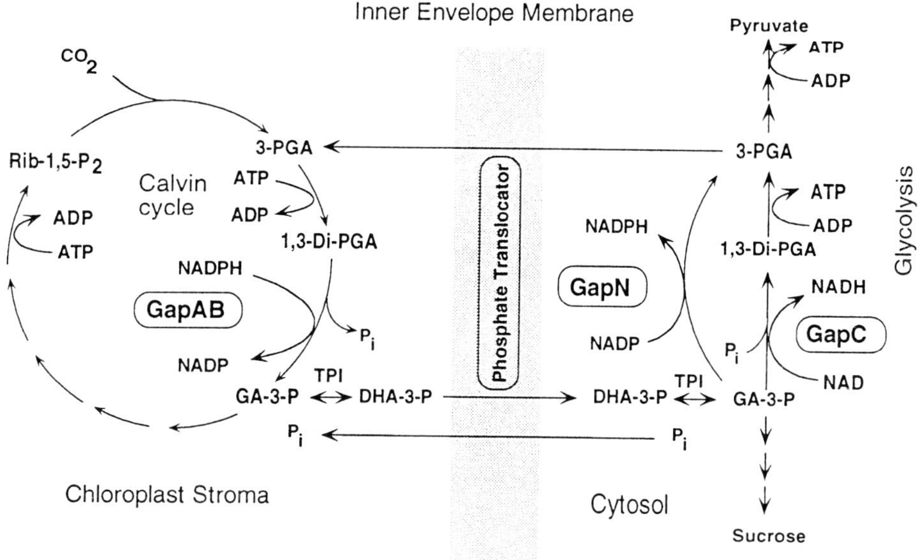

Fig. 1. Compartmentalisation and catalytic functions of plant glyceraldehyde-3-phosphate dehydrogenases, phosphorylating GAPDH (photosynthetic GapAB and glycolytic GapC) and nonphosphorylating GAPDH (cytosolic GapN), respectively. The reaction of GapN is irreversible. For phosphorylating GapAB and GapC, which catalyze the reversible oxidation of glyceraldehyde-3-phosphate (GA-3-P), only the directions dominating in vivo are indicated. Triose phosphates (GA-3-P and its isomer DHA-3-P generated by TPI) synthesized in the Calvin cycle or during starch degradation in plastids are exported to the cytosol in exchange for inorganic phosphate via the phosphate translocator [41,42]. Abbreviations: DHA-3-P — dihydroxyacetone phosphate; 3-PGA — 3-phosphoglycerate; 1,3-Di-PGA — 1,3-bisphosphglycerate; Rib-1,5-P2 — ribulose-1,5-bisphosphate; TPI — triosephosphate isomerase.

by multiple genes [39] and are subject to transcriptional activation by light [38,40].

Triose phosphates (D-glyceraldehyde-3-phosphate and its isomer dihydroxyacetone phosphate generated by TPI) synthesized in the Calvin cycle or during starch degradation in plastids are exported to the cytosol in exchange for inorganic phosphate via the phosphate translocator [41,42]. In the cytosol they are transformed into other substrates such as sucrose or reconverted into 3-phosphoglycerate, which is further metabolized via the glycolytic pathway or reimported into chloroplasts by the shuttle system of the phosphate translocator. Conversion of triose phosphates into 3-phosphoglycerate can be catalyzed by two different cytosolic GAPDH enzymes, the classic phosphorylating GAPDH of glycolysis (GapC: EC 1.2.1.12) and a non-phosphorylating NADP-specific GAPDH (GapN: EC 1.2.1.9), respectively (see Fig. 1). Glycolytic GAPDH (GapC) in angiosperms is a tetramer of identical or electrophoretically distinguishable subunits [34,43] encoded by multiple genes that are expressed constitutively or stimulated by environmental stress factors such as anaerobiosis, heat shock, and salinity [44–47]. The three-dimensional structure of the glycolytic GAPDH tetramer has been determined for both prokaryotic and eukaryotic sources [48–50], although no crystal structures for plant GAPDH enzymes are known.

To permit correlation between primary and tertiary structure codon numbering follows conventionally that of GAPDH from *Bacillus stearothermophilus* [51,52], the structure of which has been resolved at 1.8 Å [53].

All eukaryotic GapA and GapB proteins are chloroplast specific, while eukaryotic GapC enzymes usually function in the glycolytic pathway of the cytosol. Trypanosomes are exceptional in that they contain two largely different GapC enzymes, one located in the cytosol and the other in a microbody-like organelle, called the glycosome, harboring the major part of glycolysis from glucose to 3-phosphoglycerate in these organisms ([54], for evolutionary aspects see below and Figs. 4 and 5). Recently, a new *GapC* gene was discovered in Scots pine (*Pinus sylvestris*) encoding a GapC protein with a transit peptide that is targeted to chloroplasts [55]. This nuclear *GapCp* gene is highly expressed and has a wide distribution in gymnosperms and ferns. It encodes a NAD-specific GAPDH of unknown function which, however, may play a potential role in carbohydrate catabolism of the chloroplast (chlororespiration, see [56]). Also, entirely different functions may be envisaged since GAPDH has been implicated in a number of other cellular activities unrelated to glycolysis and photosynthesis, such as nuclear export of tRNAs [57], DNA repair [58], and protein phosphorylation [59]. *GapCp* is only marginally more closely related to *GapC* genes from chlorophytes than to the *GapC* gene of the red alga *Chondrus crispus*, suggesting that *GapCp* originated early in plant evolution, probably in green algae in striking parallel to the *GapA/GapB* gene duplication (see Fig. 2).

Nonphosphorylating GAPDH (GapN) was discovered 40 years ago by Arnon et al. [60]. It catalyzes an irreversible reaction (see Fig. 1) and seems to be restricted to the cytosol of photosynthetic eukaryotes including algae [61], where it generates NADPH for biosynthetic processes from triose phosphates exported from photosynthesizing chloroplasts [62]. In gluconeogenic tissues the enzyme activity may also be coupled to glycolysis. The native enzyme is a homotetramer with a subunit GapN of 53 kDa compared to only 36–42 kDa for GapA, GapB and GapC [38,63]. The recent cloning and characterization of GapN cDNAs [64] led to the surprising discovery that nonphosphorylating GAPDH is a member of the aldehyde dehydrogenase (ALDH) superfamily with no sequence homology to phosphorylating GAPDH. This ALDH

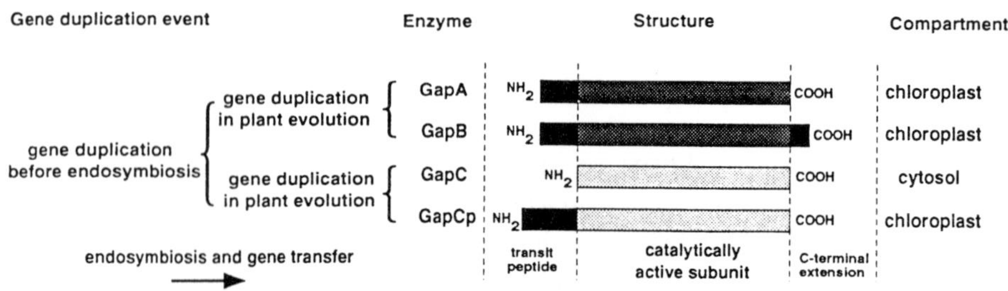

Fig. 2. Schematic structure of phosphorylating GAPDH proteins from plants. Transit peptide regions and the carboxyterminal extension of GapB are shown as solid boxes. Catalytic subunits are shaded for GapC-GapCp and GapA-GapB subunits, respectively, emphasizing their common ancestry through gene duplication during early plant evolution.

superfamily comprises multiple homologous enzymes from eubacteria and eukaryotes that are specific or nonspecific for particular aldehyde substrates. A phylogenetic analysis suggests that ALDH diversity reflects an early phase of intense metabolic radiation antedating the separation of pro- and eukaryotes [64].

These findings firmly establish that GAPDH originated at least twice independently in evolution, leading to three separate classes of enzymes as depicted in Fig. 3 (classes I, II and III). Although unrelated at the sequence level, phosphorylating and nonphosphorylating GAPDH resemble one another with respect to catalytic hydride transfer and covalent thioester formation in agreement with the concept of an early thioester world [65]. Whether or not phosphorylating and nonphosphorylating thioester mechanisms have a common evolutionary origin cannot be answered until three-dimensional structures of ALDH enzymes have become available. If they are homologous, as speculated in Fig. 3, then phosphorylating GAPDH originated after ALDH function and prior to the separation into two distinct GAPDH, classes I and II. The two classes share 10–15% amino acid identity and were thought to be restricted to eubacteria/eukaryotes (class I) and archaebacteria (class II), respectively [66]. The recent discovery of a class I GAPDH in the archaebacterium *Haloarcula vallismortis* [67] and the characterization of the corresponding operon (unpublished) suggest that also in the case of class I GAPDH the diversification of enzymes may have occurred prior to the separation of primary life forms. From these and other

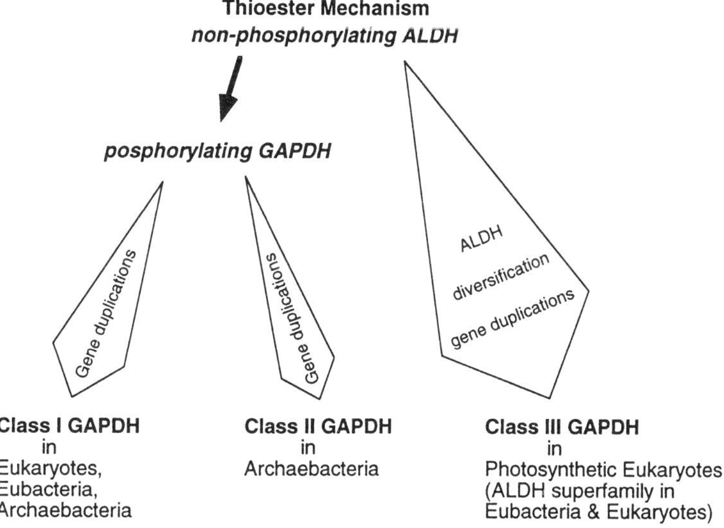

Fig. 3. Schematic outline of enzyme relationships and potential evolutionary origins of glyceraldehyde-3-phosphate dehydrogenases. Three different types of GAPDH can be distinguished, one of which, class III, is a nonphosphorylating aldehyde dehydrogenase (ALDH) with no sequence similarity to its phosphorylating counterparts, class I and class II. Substrate specificity of class III GAPDH for glyceraldehyde-3-phosphate probably originated independently by convergent evolution, while its basic catalytic mechanism (hydride transfer and thioester formation) may possibly be homologous to that of phosphorylating GAPDH (see text).

findings (see [68] and refs. therein) a rather complex picture of the last universal ancestor of pro- and eukaryotes emerges. Probably, this progenote was already an elaborated cell which not only had RNA, DNA, proteins, transcription, translation systems, and many metabolic pathways common to pro- and eukaryotes, but also multiple genes encoding the components of these pathways. As we will see below for class I GAPDH and as demonstrated previously for ALDH [64], these ancestral gene families in the progenote probably had many more members than the descendant families in present-day nuclear and eubacterial genomes, which seem to harbour just a few paralogous survivors of this ancient and diverse protein world.

Class I GAPDHs of eukaryotes are of eubacterial origin

Although plant GAPDH genes *GapA*, *GapB*, *GapC*, *GapCp*, and *GapN* are all encoded in the nucleus, it was suggested already some time ago that genes GapA/GapB encoding the Calvin cycle enzyme may be derived from the endosymbiotic genome of the chloroplast ancestor [69–71]. The characterization of the first GAPDH genes from cyanobacteria confirmed this prediction and, in addition, provided unexpected evidence for an endosymbiotic (eubacterial) origin of *GapC* [72]. In the meantime, more gene sequences from eubacteria and eukaryotes have become available, yielding a complex pattern of class I GAPDH gene evolution. As demonstrated in Fig. 4, various different types of GAPDH genes occur in eubacteria (grey lines) due to multiple early gene duplications. Up to three separate paralogous-GAPDH genes have been found in one and the same eubacterial genome, e.g., *gap1*, *gap2* and *gap3* in the cyanobacterium *Anabaena variabilis* [72]. Genes *gap1* and *gap2* are found in all cyanobacteria so far investigated including *Synechocystis* and *Synechococcus*, and an apparent orthologous gene *gap1* is also present in *E. coli* [73]. Surprisingly, *gap1* and *gap2* are closely related to eukaryotic genes *GapC* and *GapAB*, encoding glycolytic and photosynthetic GAPDH, respectively. To indicate this relationship genes *gap1* and *gap2* from cyanobacteria/*E. coli* will be called *gapC* and *gapA*, respectively.

The GAPDH tree clearly demonstrates that the *gapA/gapC* gene duplication occurred long before the separation of distinct organismal lineages. It probably antedates the divergence of eubacteria and archaebacteria as indicated by the recently characterized class I GAPDH gene of *Haloarcula vallismortis* (see above). All eukaryotic lineages on the GAPDH tree appear as terminal branches (black lines) grafted on two major eubacterial subtrees corresponding to *gapC* and *gapA*, respectively. This clearly suggests that eukaryotic genes *GapAB* and *GapC* are of eubacterial origin and became fixed in nuclear genomes probably via endosymbiotic gene translocations. All eukaryotic *GapAB* genes including those of red algae (*Chondrus* [74]) and *Euglena gracilis* [115] group on a common branch and emerge together from *gapA* of cyanobacteria, the free-living relatives of chloroplasts (see subtree *gapA* in Fig. 4).

The distribution of *GapC* genes in nuclear genomes is more complex than that of

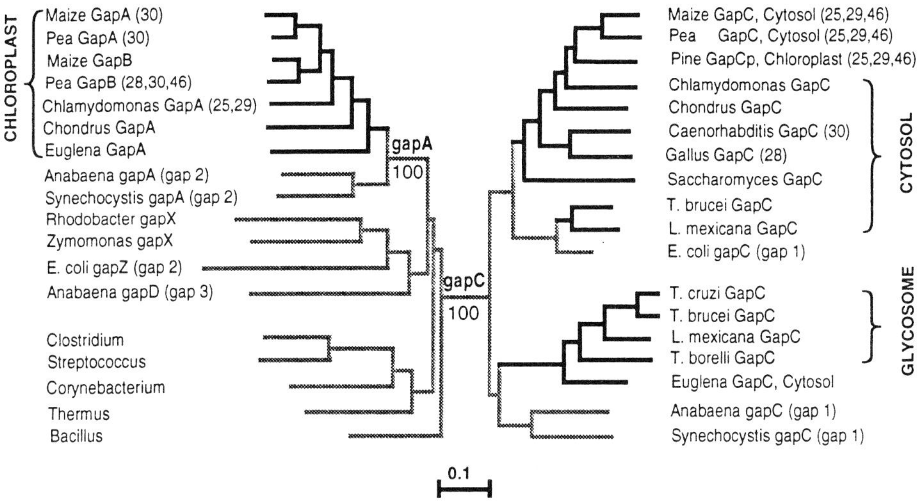

Fig. 4. Phylogenetic tree of class I GAPDH inferred by the neighbor-joining method [113] from a divergence matrix of non-synonymous substitutions [112]. There are multiple ancient eubacterial lineages (grey lines), two of which (*gapA* and *gapC*) gave rise to eukaryotic lineages bearing nuclear genes *GapA*, *GapB*, *GapC*, and *GapCp* (solid lines). All eukaryotic GapA and GapB proteins are chloroplast specific, while eukaryotic GapC enzymes are located in the cytosol, except for glycosomal GapC of trypanosomes and chloroplast GapCp of *Pinus*. Numbers in brackets after species names designate introns conserved precisely across the *GapAB/GapC* boundary (see Figs. 6 and 7). Numbers below the major branches *gapA* and *gapC* indicate that they were found in 100/100 parsimony replicates for the amino acid alignment (PAUP, version 3.0). Source of sequences: EMBL/GenBank database (see also [85,74,115]). The scale bar indicates 0.1 substitutions per site. Abbreviations: *T. brucei* — *Trypanosoma brucei*; *T. cruzi* — *Trypanosoma cruzi*; *T. borelli* — *Trypanoplasma borelli*; *L. mexicana* — *Leishmania mexicana*.

GapAB genes. Since *gapC* of the purple bacterium *E. coli* is more closely related to GapC genes from higher eukaryotes than to cyanobacterial *gapC* and, furthermore, purple bacteria probably gave rise to mitochondria [75,76], we have previously suggested that *GapC* of plants, animals, and fungi may be of mitochondrial origin [72,74]. In light of the new GAPDH data from trypanosomes [54] and *Euglena* [115] this hypothesis may now be integrated into a more complex scenario as depicted in Fig. 5, suggesting a biphyletic origin of the modern eukaryotic cell containing plastids and/or mitochondria. This model supposes that GapC of kinetoplastids/*Euglena* and probably also *Euglena GapA* were acquired independently of *GapC** (+mitochondria) and *GapA** (+chloroplasts) of higher eukaryotes. It does not require differential loss of chloroplasts and genes except for a single loss of *GapA* in the lineage leading to trypanosomes after their divergence from *Euglena*. This early gene transfer may have occurred in an endosymbiotic or nonendosymbiotic context. The kinetoplast and/or glycosome of trypanosomes and the mitochondrium of *Euglena* may or may not be remnants of such a potential early endosymbiont.

There are other conceivable scenarios such as a monophyletic origin model of two primary endosymbiotic transfers of paralogous pairs of genes, e.g., *gapC/gapC**

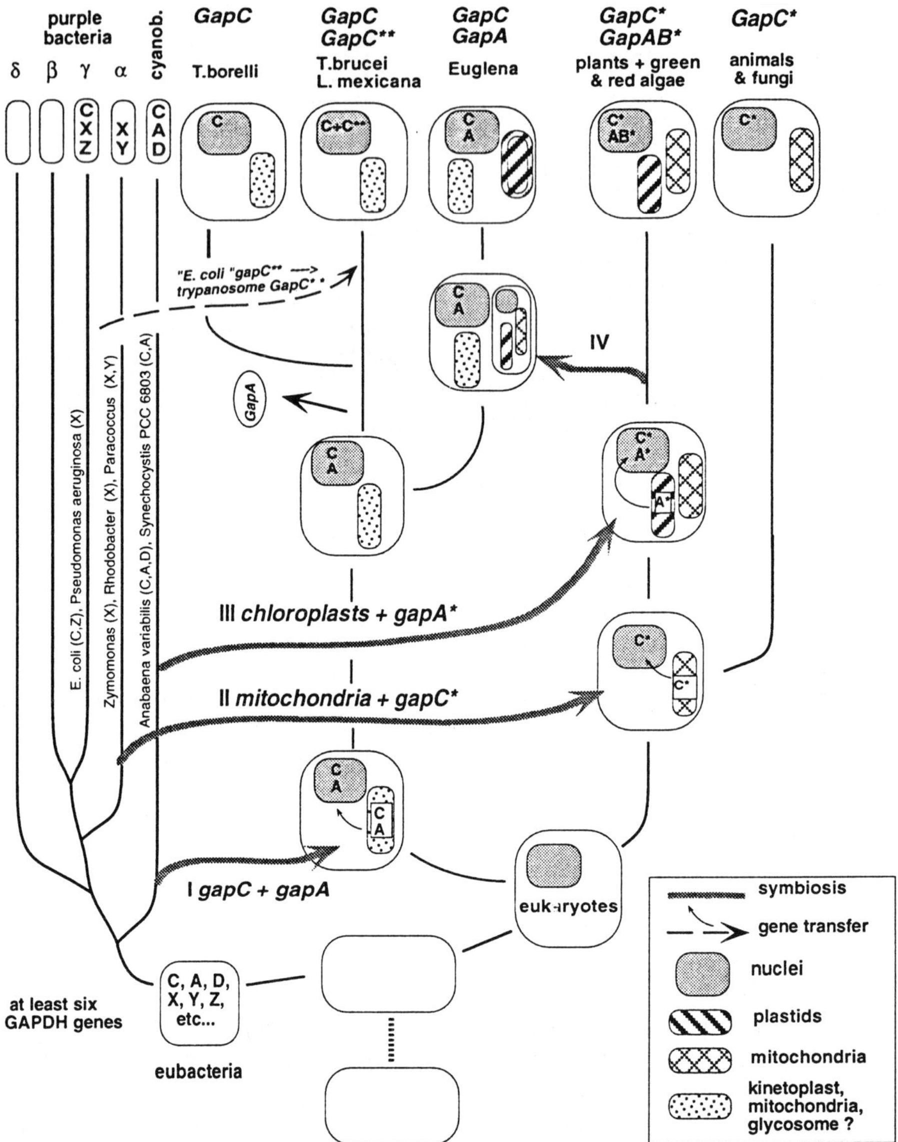

Fig. 5. Evolutionary scenario explaining the distribution of GAPDH genes in present-day eubacteria and eukaryotes in terms of three independent primary endosymbioses giving rise to two separate lineages leading to trypanosomes/*Euglena* (genes *GapC* and *GapA*) and higher eukaryotes (genes *GapC** and *GapA**), respectively. Gene *GapC*** was acquired recently by trypanosomes from an *E. coli*-like ancestor, probably in a nonendosmybiotic context. Endosymbiosis IV: Origin of the *Euglena* chloroplast via incorporation of a eukaryotic endosymbiont possibly related to green algae (see text). The schematic eubacterial tree on the left showing the branching order of purple bacteria (α, β, γ) and cyanobacteria is based on the 16S rRNA phylogeny of Woese [76]. There are six or more different paralogous GAPDH genes in present-day eubacteria showing between 50 and 60% amino acid sequence divergence in pairwise comparisons. Cyanobacteria contain types C, A and D and share type C with *E. coli*. Types X and Y occur in α-purple bacteria and types C, X and Z in γ-purple bacteria (see [74]).

(+mitochondria) and *gapA/gapA** (+chloroplasts), to a single host cell or variations on this theme with other combinations of paralogous gene pairs. However, these models all require multiple differential losses of genes and (proto-)chloroplasts. In the case of *Euglena*, which possibly received its chloroplast from a secondary endosymbiont related to green algae (symbiosis IV in Fig. 5 [77–80]), this scenario implies a relatively late transfer of the paralogous gene *GapA*. In all possible scenarios the strikingly close relationship between *E. coli gapC* and cytosolic *GapC*** of trypanosomes would be rationalized in terms of a relatively recent lateral gene transfer from an *E. coli*-like purple bacterium to an ancestor of *T. brucei* and *L. mexicana*. This interpretation seems inevitable, since a transfer in the opposite direction from trypanosomes to *E. coli* would require additional, highly speculative, eukaryote to eukaryote gene translocations to explain the origin of cytosolic *GapC*** in trypanosomes. This suggests that *GapC** and *GapC*** are orthologous genes that were acquired twice from purple bacteria by two successive gene transfers in an endosymbiotic (mitochondria) and nonendosymbiotic context, respectively. This interpretation predicts that gapC genes may eventually be found also in purple bacteria other than *E. coli*.

Intron conservation in class I GAPDH genes

Given their extremely ancient origin, probably antedating the separation of primary life forms, glycolytic and photosynthetic GAPDH genes are an interesting model system to test the "early versus late intron" problem. Genes *GapC*, *GapB* and *GapA* from higher plants have characteristic intron/exon patterns with high (8–11), moderate (6) and low (1 or 2) numbers of intervening sequences, respectively (Fig. 6), compatible with the idea that introns were lost progressively from these genes prior to their successive transfer to the nucleus [81]. Two introns strictly conserved across the *GapAB/GapC* boundary were reported in 1988 [82,83] and were regarded as strong evidence in favor of the "introns early" hypothesis [84]. But for critics of the "introns early" hypothesis, this evidence was not convincing and the identity of intron positions was dismissed as "parallel insertion of different introns" [15].

In Fig. 6 an updated version of the GAPDH intron matrix published previously [85] is shown, including the recently characterized gene *GapCp* encoding chloroplast *GapC* from *Pinus* [55]. There are 47 different intron positions in 28 separate genes including six genes encoding photosynthetic GAPDH (four *GapA*, two *GapB*) and 23 *GapC* genes encoding GapCp (gene 7) and cytosolic GAPDH from plants (genes 8–12), nonvertebrates (genes 13–16), vertebrates (genes 17 and 18), and fungi (genes 19–28). Out of 10 separate intron positions present in *GapAB* genes, five are precisely conserved in *GapC* genes (introns 25, 28, 29, 30 and 46, solid arrows at the bottom of Fig. 6) and three others (introns 14, 20, 32) have very close neighbors (introns 15, 21, 33) in *GapC* genes which are displaced 3, 6 and 8 nucleotides, respectively (see Fig. 7). The coding sequences flanking the conserved introns (exon/exon junctions) are shown in Table 1. They deviate considerably from the so-

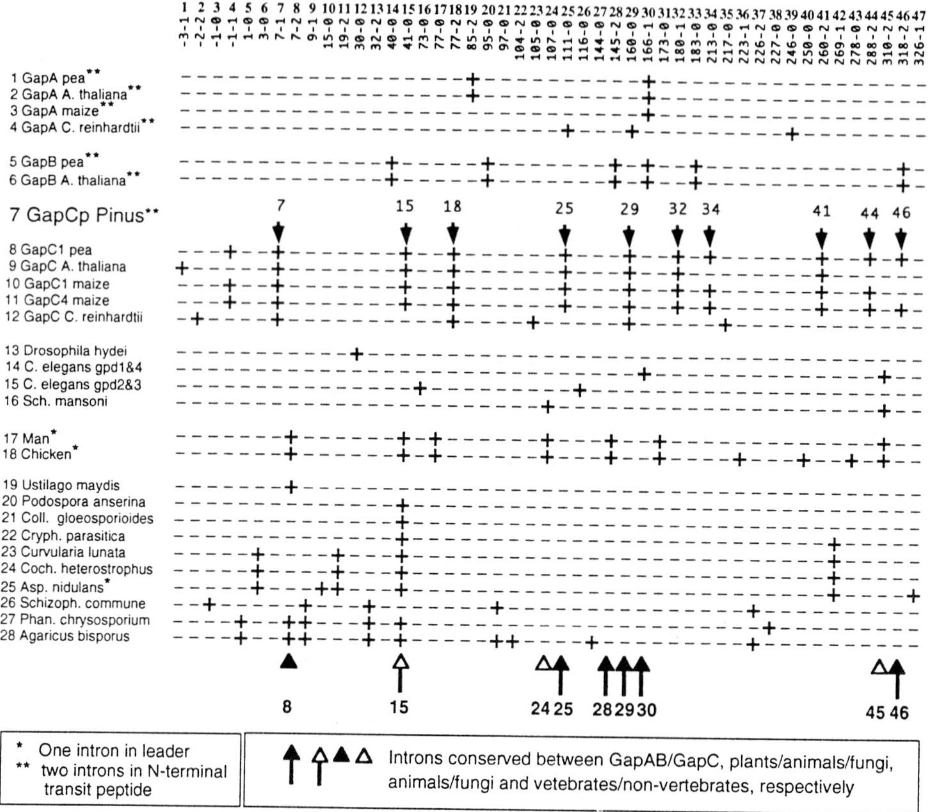

Fig. 6. The distribution of introns in present day GAPDH genes. The presence of an intron is indicated by "+", its absence by "–". There are 47 known intron positions, defined by codon number and phase, in 28 different genes. For introns separating triplets after the first or second base codon numbers are followed by –1 and –2, respectively. For introns falling between triplets the number of the codon located 3′ of the intron is given followed by –0. Codon positions are numbered with reference to GAPDH of *Bacillus stearothermophilus* [51,52].The 28 genes are shown from top to bottom as seven separate groups comprising genes encoding: lines 1–4, chloroplast GapA; lines 5 and 6, chloroplast GapB; line 7, chloroplast GapCp; lines 8–12, glycolytic GapC genes from plants; lines 13–16, nonvertebrates; lines 17 and 18, vertebrates; and lines 19–28, fungi. Arrows and arrow heads at the bottom of the figure indicate intron positions precisely conserved across *GapAB/GapC* genes (introns 25, 28, 29, 30, and 46) and across *GapC* genes of more or less distant eukaryotic organisms (introns 8, 15, 24, and 45), as indicated at the bottom of the figure. Arrows in line 7 indicate 10 introns in the recently discovered gene *GapCp* from *Pinus* [55] all of which are precisely conserved relative to cytosolic *GapC* genes of higher plants. Abbreviations: *A. thaliana — Arabidopsis thaliana; C. elegans — Caenorhabditis elegans; Sch. mansoni — Schistosoma mansoni; Coll. gloeosporioides — Colletotrichum gloeosporioides; Cryph. parasitica — Cryphonectria parasitica; Coch. heterostrophus — Cochliobolus heterostrophus; A. nidulans — Aspergillus nidulans; Schizoph. commune — Schizophyllum commune; Phan. chrysosporium — Phanerochaete chrysosporium.*

called "proto-splice" consensus C(A)AG/R thought to be the target of intron insertion in "introns late" scenarios [86].

In terms of the "introns early" hypothesis, these five intervening sequences are homologous and were present in the ancestral gene that gave rise to GapAB and

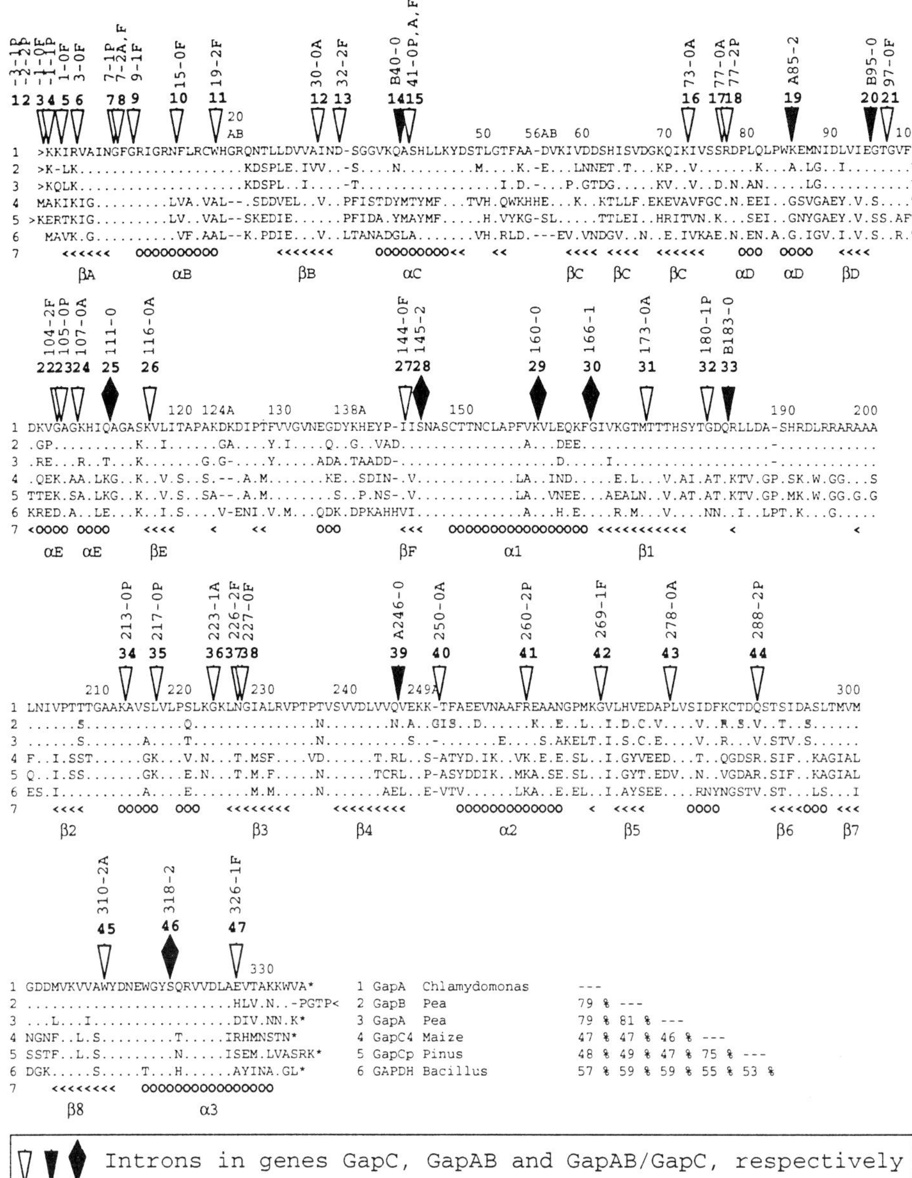

Fig. 7. Intron positions in class I GAPDH genes with respect to the primary and secondary structure of the protein. For definition of intron symbols see box at the bottom of the figure. Intron numbers and positions as in Fig. 6. Letters A and B preceding position designations (e.g., B40-0) stand for genes *GapA* and *GapB*, respectively. Letters P, A and F after intron position designations (e.g., 41-0P,A,F) specify *GapC* genes found in plants, animals and fungi, respectively. Symbols (oo) and (<<) in line 7 designate α-helices and β-sheets as determined for *Bacillus* GAPDH [48,53] and as specified in the SWISSPROT database. Arrow heads at the beginning (>) or end (<) of a sequence indicate the presence of amino-terminal and carboxy-terminal extensions, respectively (see Fig. 2). Percentage values specify sequence identities in pairwise comparisons. Abbreviations: *P. sativum* – *Pisum sativum; Z. mays* – *Zea mays.* Sources of sequence information: *Chlamydomonas* GapA and maize GapC4 [85]; GapB and GapA from pea [36]; *Pinus* GapCp [55]; *Bacillus* GAPDH [51].

Table 1. Exon/exon junctions of introns precisely conserved across genes *GapAB* and *GapC* encoding chloroplast and cytosolic GAPDH, respectively.

Intron no.	Position	Genes	Exon/exon junctions
25	111-0	*Chlamydomonas GapA*	CAG/G
		Maize *GapC4*	AAG/G
28	145-2	Pea *GapB*	AAG/C
		Chicken *GapC*	CAG/C
29	160-0	*Chlamydomonas GapA*	AAG/G
		Maize *GapC4*	AAG/G
30	166-1	Pea *GapB*	TCG/G
		Nematode *GapC*	TCG/G
46	318-2	Pea *GapB*	CAG/C
		Maize *GapC4*	CAG/C

GapC. As argued above, this gene duplication probably occurred in the progenote, the last common ancestor of all present-day organisms comprising archaebacteria, eubacteria, and eukaryotes. This seems, at first glance, strong evidence for the exon theory of genes [5] suggesting that introns are as old as the coding sequences in which they reside.

Viewed from the "introns late" standpoint, the five identical intron positions across genes *GapAB* and *GapC* are due to parallel insertions at common target sites rather than to common ancestry. Although a common "proto-splice" site is not apparent from the exon/exon junctions of present-day *GapAB/GapC* genes (see Table 1), its former existence in the ancestral genes would be essential, since the probability of finding these five identical intron positions as a result of random insertions is extremely low and has been estimated to be roughly 2×10^{-5} [85]. Since the coding sequences of *GapAB* and *GapC* are of endosymbiotic, eubacterial origin, this suggestion implies that introns in GAPDH genes were acquired subsequent to their transfer from the organelles to the nucleus. However, this scenario carries the compelling corollary that we would expect to find a larger number of coincident introns in *GapC* genes of common eukaryotic descent (e.g., *GapC* genes of plants, animals and fungi) than in comparisons across the *GapAB/GapC* boundary. Instead we find five introns precisely conserved across the highly divergent genes *GapAB* and *GapC*, and only one intron precisely conserved across *GapC* genes of plants, animals, and fungi (intron 15 at position 41-0, Fig. 6) which, curiously, has a very close neighbor in the ancient paralogous gene *GapB* (intron 14 at position 40-0) suggesting that also in this case we are dealing with a homologous *GapAB/GapC* intron of common ancestry. Phrased another way, why would introns transpose to distantly related "proto-splice sites" in the most divergent genes and avoid more closely related ones in genes of more recent common descent? Without invoking complex auxilliary assumptions, the "introns late" hypothesis cannot account for this observation in GAPDH genes, while retention of ancient introns and lineage-specific differential intron loss ("introns early" hypothesis) readily explain the data (see below and [87]).

Although the five introns conserved across present-day *GapAB* and *GapC* genes are removed with the aid of spliceosomes, the data do not indicate how their predecessors in the ancestral GAPDH gene were spliced. The discovery of group II introns in eubacteria [116] has given support to the notion that these elements may have been the precursors from which both spliceosomal introns and small nuclear RNAs in eukaryotic nuclei arose [14,88–90]. Introns in *GapAB* and *GapC* ancestors may have been mechanistically group II, and if so, may have evolved into contemporary spliceosomal introns in situ. This would imply that ancient group II introns were immobile and transformation into spliceosomal introns may have occurred in situ in eukaryotes after the emergence of mitochondria and chloroplasts. If, however, immobilization of group II introns is related to their evolutionary fragmentation into transacting snRNAs and the build-up of the spliceosomal machinery, then the present data suggest that bacteria had (proto-)spliceosomes. Indeed, independent support for this view has recently been obtained by Ushida and Muto [91], who discovered in *Mycoplasma capricolum* an abundant small RNA species that associates with endogenous proteins to form RNP particles and which is over 50% identical to eukaryotic U6 snRNAs known to play a critical role in pre-mRNA splicing [92]. However, definite conclusions about the evolutionary origin of this new U6-like RNA from *Mycoplasma* remain premature until more data on its phylogenetic distribution and function in eubacteria and, possibly, archaebacteria become available.

Gene-protein correlations, slippage and sliding of introns

In Fig. 7 all 47 intron positions in class I GAPDH genes are shown with respect to the primary and secondary structures of the protein. While fungal introns cluster in the 5′ part of the gene, nonfungal introns are spaced rather regularly with typical intervals measuring eight to 10 codons in the C-terminal region downstream from intron 40. As can be clearly seen from Fig. 7, introns in GAPDH genes avoid β-sheets and seem to prefer the border regions of secondary structure elements, especially in the 3′ half of the gene downstream from the conserved intron 28 separating the NAD-binding domain from the catalytic domain. These gene-protein correlations agree with previous observations by other authors [3,93] and seem difficult to reconcile with a model of late lineage-specific insertion of introns.

There are five "intron-free" regions corresponding to polypeptide segments of 20–30 residues between introns 15/16, 26/27, 33/34, 38/39, and 44/45, respectively, that harbor numerous structural elements (see Fig. 7) and most of the residues generating subunit-subunit contacts in the native GAPDH tetramer [49]. Interestingly, all five regions are symmetrical in that they are flanked by two introns of the same phase type, e.g., introns 0-0 for regions 1–4 and introns 2-2 for region 5 (see Fig. 7). Symmetrical modules with introns of the same phase type at both their 5′- and 3′-ends have been shown to be diagnostic for exon shuffling in the case of the evolution of plasma proteases and related proteins [94]. Since the successful insertion of symmetrical modules requires introns of the same phase class (e.g., 0-0 exons require phase 0 introns for recombination), exon shuffling leads to a nonrandom intron phase usage of the corresponding genes. Over 50% of the GAPDH introns (25 out of 47, see Fig. 6) occur in phase 0, suggesting that, indeed, exon shuffling via intronic recombination may have played a role in GAPDH gene evolution. In particular, the capacity of GAPDH monomers to form tetramers may have been acquired by insertion of 0-0 modules into phase 0 introns.

"Introns early" and "introns late" proponents not only disagree on the origin of spliceosomal introns but also on whether or not intron slippage can occur (the occasional displacement of an intron over a short distance of a conserved coding sequence). "Intron late" supporters, who believe in the transposition of spliceosomal introns, categorically reject the idea of intron slippage and on the basis of this implicit axiom "once inserted, always immobile" invoke "too tiny" ancestral GAPDH exons as evidence against the exon theory of genes [15]. There are numerous cases of quasi conservation of introns in GAPDH genes, where positions differ by only one to eight bases both across gene classes and major taxonomic groups, such as: introns 14/15 (B40-0/C41-0); introns 20/21 (B95-0/C97-0); introns 27/28 (C144-0/B145-2); introns 32/33 (C180-1/B183-0); introns 1 to 4 (C-3-1/C-2-2/C-1-0/C-1-1/C1-0: plants/fungi); introns 7/8 (C7-1/C7-2: plants/animals+fungi); introns 12/13 (C30-0/C32-2: *Drosophila*/fungi); introns 17/18 (C77-0/C77-2: vertebrates/plants); introns 22/23/24 (C104-2/C105-0/C107-0: fungi/*Chlamydomonas*/animals); and introns 37/38 (C226-2/C227-0: fungi/fungi); where B and C designate *GapB* and *GapC*, respectively. In terms of the "introns late" view, all these adjacent introns are non-homologous and have been inserted independently into the GAPDH coding sequence. We maintain that this interpretation is very unlikely, which is particularly evident for introns 37/38, which are separated by one nucleotide in *GapC* genes of three related basidiomycetes sharing several additional introns at identical positions (see Figs. 6 and 7).

We have previously argued [44,95] that both loss and displacement of introns over short distances can be explained in terms of occasionally (rarely) occurring splicing errors leading to modified pre-mRNAs which can then be reintroduced into the genome via reverse transcription and homologous recombination (gene conversion). This model is a modified version of that previously proposed by Fink [96] (see below). Splicing errors leading to pseudogenes have been described [97] and it cannot be excluded that they also cause structural modifications in functional genes. They may arise by premature release from spliceosomes of either partially spliced pre-mRNAs (intron loss) or of pre-mRNAs in which an intron has been reintroduced immediately after excision into a nearby site via reverse splicing (intron slippage). Reverse splicing has been described for group I and group II introns [98–100] but has been postulated to occur potentially also with pre-mRNA introns [101,102]. This "intron slippage – homologous recombination" scenario suggests that spliceosomal introns may eventually also be found in genes encoding transacting snRNAs that are implicated in the splicing process and interact with pre-mRNAs and introns within the spliceosome [92]. A striking confirmation of this prediction has been reported by Tani and Ohshima for the gene encoding U6 snRNA of *Schizosaccharomyces pombe*

[103].

RNA-dependent, spliceosome-mediated intron slippage should not be confused with DNA dependent "junction sliding", the sliding of a single intron-exon junction leading to insertions-deletions. This model has been proposed [104] to explain the observation that intron-exon junctions frequently coincide with internal length differences of the corresponding polypeptides as in the case of the serine protease and dihydrofolate reductase families. Junctional sliding may be caused by a mutation altering the intron-exon splice consensus, so that a functional mRNA can only be obtained by using a cryptic splice site of the same phase upstream or downstream from the mutated junction. Internal length differences are rare in GAPDH polypeptides, while the amino-termini may be variable. This amino-terminal polymorphism coincides with intron heterogeneity at the gene level (introns 1–6 in Figs. 6 and 7), suggesting the involvement of junction sliding in this region.

Lineage specific loss of introns

Intron patterns in class I GAPDH genes clearly demonstrate a lineage-specific loss of spliceosomal introns. For example, most fungal introns cluster in the 5′ part of the genes (Fig. 6), in agreement with the gene conversion model by Fink [96] suggesting a preferential loss of introns in the 3′ part of genes due to homologous recombination of genes with their reverse transcribed mRNAs. The paucity of introns in nonvertebrate GAPDH genes compared to their vertebrate counterparts is also more easily explained by intron loss than by intron gain. More specifically, *Arabidopsis GapC* lacks three introns (introns 34, 44, and 46 in Fig. 6), which are present in *Pinus GapCp*, pea *GapC1* and maize *GapC4*: three losses in *Arabidopsis* are clearly more likely than nine independent gains at three identical positions in pea and maize and *Pinus*. Moreover, *Chlamydomonas GapC* has six introns, only three of which coincide with GapC introns of higher plants, while *Pinus GapCp* which seems about as old or only marginally younger than *Chlamydomonas GapC* (see above and [55]) has the full complement of 10 introns present in higher plants. These observations clearly contradict the a priori notion of "introns late" proponents [15] that intron gain should be generally more plausible than intron loss.

Red algae represent a particularly impressive example for a massive lineage-specific loss of introns which occurred relatively recently. Up to now five different nuclear protein-coding genes have been characterized in rhodophytes: four genes of *Chondrus crispus* encoding β-tubulin (gene *TubB1* [105]), actin (gene *Act1*, [117]), chloroplast, cytosolic GAPDH (genes *GapA1* and *GapC1*, [106]) and one gene from *Gracilaria verrucosa* encoding chloroplast GAPDH (gene *GapA1* [107]). With the exception of the uninterrupted gene *GapC1* of *Chondrus crispus* [106] all genes contain at their 5′ end a single short spliceosomal intron (Fig. 8A), suggesting that this may be a general feature of nuclear protein-coding genes from red algae. The splice junctions of these rhodophyte introns (Fig. 8B,C) are relatively degenerate and show the highest similarity with the corresponding consensus sequences of green

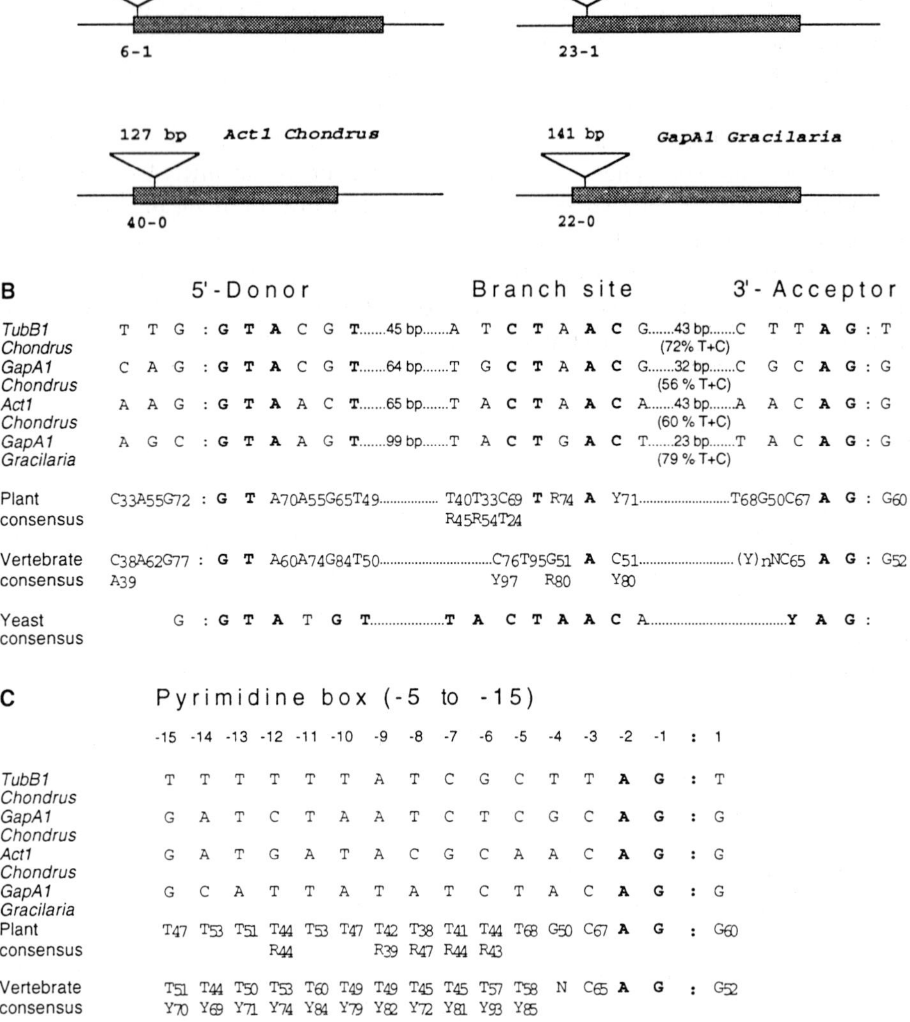

Fig. 8. Nuclear protein-coding genes of the red algae *Chondrus crispus* and *Gracilaria verrucosa* (A) contain 5′ introns resembling those of higher plants. Border junctions, putative branch sites (B) and pyrimidine boxes (C) of rhodophyte introns were compared to the corresponding consensus sequences from plants, animals and yeast [111,114]. Percentage values of nucleotide conservation are indicated as subscripts. Conserved nucleotides (~100%) are shown in bold. Letters R and Y in sequences correspond to purines (A + G) and pyrimidines (C + T), respectively. The T + C values (% T + C) given in (B) apply to the sequence between the branch site adenosine and the 3′-splice site AG. *TubB1* and *Act1* designate nuclear gene encoding β-tubulin and actin, respectively.

plants in agreement with recent suggestions based on the phylogenetic inference that red algae and green plants may be sister groups [74,107].

The paucity and asymmetric localization of spliceosomal introns in red algae is surprising, especially for the GAPDH genes which in higher eukaryotes (plants and vertebrates) and green algae (*Chlamydomonas reinhardtii*) contain many introns more or less evenly dispersed along the genes (Fig. 6). As in the case of fungal genes, these features may be explained by the Fink model [96] or its modified version (see above) rather than by a late lineage-specific acquisition of single introns after the rhodophyte/chlorophyte separation. While intron loss by cDNA-mediated gene conversion may explain the paucity and polar distribution of introns, it does not provide an answer to the question why 5′-introns should be retained at all in fungal and rhodophyte genes. For yeast it has been suggested that 5′-introns may be implicated in the control of pre-mRNA abundance via the nonsense-mediated mRNA decay pathway [108]. Another possibility may be the implication of 5′-introns in transcriptional control of gene expression for which evidence has been obtained for intron 1 of gene *GapA1* from maize [118].

Conclusions and perspectives

It is now widely accepted that ancient self-splicing introns related to present-day group II introns gave rise to their spliceosomal counterparts. When and in which lineages this conversion occurred are presently unknown. Concerning these questions three potential scenarios have recently been elaborated by Roger et al. [109]:

1. group II introns invaded the nucleus via endosymbiotic gene transfer from organelles and subsequently gave rise to spliceosomal introns in agreement with previous suggestions by Cavalier-Smith [14];
2. group II introns were present in the common ancestor of eukaryotes and eubacteria and gave rise to spliceosomal introns only in the former; or
3. group II introns gave rise to their spliceosomal counterparts before the separation of prokaryotes and eukaryotes.

The phylogenetic distribution and positional conservation of introns in class I GAPDH genes together with the observed gene-protein correlations support scenario 3, suggesting that a forerunner of the spliceosome existed in the common ancestor of pro- and eukaryotes. The data seem incompatible with scenario 1 and can only be explained by scenario 2 if one assumes that ancestral group II introns which gave rise to spliceosomal introns were immobile elements or elements of very limited mobility, and, in addition, were capable of promoting exon shuffling, which seems a rather unlikely possibility [94].

To obtain more information about the origin of spliceosomal introns, we recently characterized the genomic sequence of the *Euglena GapC* gene [115] representing the earliest eukaryotic lineage of the *GapC* subtree branching well below the eukaryotic crown taxa and *E. coli* and about as deeply as cyanobacterial *gapC* (Fig. 4). *Euglena GapC* contains four introns, at least three of which lack GT-AG consensus borders, and all of which fold into stable secondary structures that show no apparent "bulging A" motif (branch site) and no homology to self-splicing introns group I and group II. These data corroborate previous findings on the *Euglena rbcS* gene [110] and suggest that an unusual splicing mechanism exists in *Euglena*. All four introns in *Euglena GapC* are flanked by 2–3 bp repeats, so that their positions could not be determined to the nucleotide. In spite of this limitation, the three introns interrupting the coding region in *Euglena GapC* could be mapped to approximate positions very close or identical to GapB introns 14, 33 (positions 40-0, 183-0) and to GapC introns 37/38 (positions 226-2/227-0), all of which are border introns of the above described 0-0 modules, (see Fig. 6). The elucidation of the intron/exon organization of the *Euglena GapA* gene may be crucial for an adequate interpretation of these conspicuous findings.

Acknowledgements

The work from the author's laboratory has been supported by major financial contributions from the Deutsche Forschungsgemeinschaft (since 1975) and the Centre National de la Recherche Scientifique (1985 until 1989).

References

1. Doolittle WF. Understanding introns: origin and functions. In: Stone EM, Schwartz RJ (eds) Intervening Sequences in Evolution and Development. New York and Oxford: Oxford University Press, 1990;43–62.
2. Holland SK, Blake CCF. Proteins, exons and molecular evolution. In: Stone EM, Schwartz RJ (eds) Intervening Sequences in Evolution and Development. New York and Oxford: Oxford University Press, 1990;10–42.
3. Stone EM, Schwartz RJ. Intron-dependent evolution of progenotic enzymes. In: Stone EM, Schwartz RJ (eds) Intervening Sequences in Evolution and Development. New York and Oxford: Oxford University Press, 1990;63–91.
4. Doolittle WF. The origin and function of intervening sequences in DNA: a review. Am Nat 1987;130:915–928.
5. Gilbert W. The exon theory of genes. Cold Spring Harbor Symp Quant Biol 1987;52:901–905.
6. Blake CCF. Do genes-in-pieces imply proteins-in-pieces? Nature 1978;273:267–268.
7. Blake CCF. Exons present from the beginning? Nature 1983;306:535–537.
8. Darnell JE. Implications of RNA-RNA splicing in evolution of eukaryotic cells. Science 1978;202:1257–1260.
9. Doolittle WF. Genes in pieces: were they ever together? Nature 1978;272:581–582.
10. Gilbert W. Why genes in pieces? Nature 1978;271:501.
11. Rossman MG, Liljas A, Brändén C-I, Banaszak LJ. Evolutionary and structural relationships among dehydrogenases. In: Boyer PD (eds) The Enzymes. New York: Academic Press, 1975;62–102.
12. Rossmann MG. Introductory comments on the function of domains in protein structure. In: Stone EM, Schwartz RJ (eds) Intervening Sequences in Evolution and Development. New York and Oxford: Oxford University Press, 1990;3–9.
13. Rossmann MG, Moras D, Olsen KW. Chemical and biological evolution of a nucleotide binding domain. Nature 1974;250:194–199.
14. Cavalier-Smith T. Intron phylogeny: a new hypothesis. Trends Genet 1991;7:145–148.
15. Palmer JD, Logsdon JM. The recent origins of introns. Curr Opin Gen Devel 1991;1:470–477.
16. Rogers JH. How were introns inserted into nuclear genes? Trends Genet 1989;5:213–216.
17. Rogers J. Exon shuffling and intron insertion in serine protease genes. Nature 1985;315:458–459.

18. Südhof TC, Goldstein JL, Brown MS, Russell DW. The LDL receptor gene: a mosaic of exons shared with different proteins. Science 1985;228:815–822.
19. Fothergill-Gilmore LA, Michels PAM. Evolution of glycolysis. Prog Biophys Molec Biol 1993;59:105–235.
20. Gilbert W, Marchionni M, McKnight G. On the antiquity of introns. Cell 1986;46:151–154.
21. Marchionni M, Gilbert W. The triosephosphate isomerase gene from maize: introns antedate the plant-animal divergence. Cell 1986;46:133–141.
22. Doolittle WF, Stoltzfus A. Genes in pieces revisited. Nature 1993;361:403.
23. Gilbert W, Glynias M. On the ancient nature of introns. Gene 1993;135:137–144.
24. Gō M. Correlation of DNA exonic regions with protein structural units in haemoglobin. Nature 1981;291:90–93.
25. Tittiger C, Whyard S, Walker VK. A novel intron site in the triosephosphate isomerase gene from the mosquito *Culex tarsalis*. Nature 1993;361:470–472.
26. Brändén C-I, Eklund H, Cambillau C, Pryor AJ. Correlation of exons with structural domains in alcohol dehydrogenase. EMBO J 1984;3:1307–1310.
27. Cornish-Bowden A. Are introns structural elements or evolutionary debris. Nature 1985;313: 434–435.
28. Stone EM, Rothblum KN, Schwartz RJ. Intron-dependent evolution of chicken glyceraldehyde phosphate dehydrogenase gene. Nature 1985;313:498–500.
29. Cerff R. Glyceraldehyde-3-phosphate dehydrogenase (NADP) from *Sinapis alba*: steady state kinetics. Phytochemistry 1978;17:2061–2067.
30. Pupillo P, Giuliani-Piccari G. The reversible depolymerization of spinach chloroplast glyceraldehyde-phosphate dehydrogenase. Eur J Biochem 1975;51:475–482.
31. Li D, Stevens FJ, Schiffer M, Anderson LE. Mechanism of light modulation: identification of potential redox-sensitive cysteines distal to catalytic site in light-activated chloroplast enzymes. Biophys J 1994;67:29–35.
32. Scheibe R. Redox-modulation of chloroplast enzymes. A common principle for individual control. Plant Physiol 1991;96:1–3.
33. Cerff R. Quaternary structure of higher plant glyceraldehyde-3-phosphate dehydrogenases. Eur J Biochem 1979;94:243–247.
34. Cerff R. Separation and purification of NAD- and NADP-linked glyceraldehyde-3-phosphate dehydrogenases from higher plants. In: Edelman M, Hallick RB, Chua NH (eds) Methods in Chloroplast Molecular Biology. Amsterdam: Elsevier Biomedical Press, 1982;683–694.
35. Cerff R, Chambers SE. Subunit structure of higher plant glyceraldehyde-3-phosphate dehydrogenases (EC 1.2.1.12 and EC 1.2.1.13). J Biol Chem 1979;254:6094–6098.
36. Brinkmann H, Cerff R, Salomon M, Soll J. Cloning and sequence analysis of cDNAs encoding the cytosolic precursors of subunits GapA and GapB of chloroplast glyceraldehyde-3-phosphate dehydrogenase from pea and spinach. Plant Molec Biol 1989;13:81–94.
37. Zapponi MC, Iadarola P, Stoppini M, Ferri G. Limited proteolysis of chloroplast glyceraldehyde-3-phosphate dehydrogenase (NADP) from *Spinacia oleracea*. Biol Chem Hoppe-Seyler 1993;374: 395–402.
38. Cerff R, Kloppstech K. Structural diversity and differential light control of mRNAs coding for angiosperm glyceraldehyde-3-phosphate dehydrogenases. Proc Natl Acad Sci USA 1982;79: 7624–7628.
39. Quigley F, Brinkmann H, Martin WF, Cerff R. Strong functional GC pressure in light-regulated maize gene encoding subunit GapA of chloroplast glyceraldehyde-3-phosohate dehydrogenase: Implications for the evolution of GapA pseudogenes. J Molec Evol 1989;29:412–421.
40. Dewdney J, Conley TR, Shih MC, Goodman HM. Effects of blue and red light on the expression of nuclear genes encoding chloroplast glyceraldehyde-3-phosphate dehydrogenase of *Arabidopsis thaliana*. Plant Physiol 1993;103:1115–1121.
41. Flügge UI, Heldt HW. The phosphate-triosephosphate-phosphoglycerate translocator of the chloroplast. Trends Biochem Sci 1984;9:530–533.

42. Stitt M. The flux of carbon between the chloroplast and the cytosol. In: Dennis DT, Turpin DH (eds) Plant Physiology, Biochemistry and Molecular Biology. Singapore: Longman, 1990;309–326.
43. Russell DA, Sachs MM. The maize cytosolic glyceraldehyde-3-phosphate dehydrogense gene family: organ-specific expression and genetic analysis. Molec Gen Genet 1991;229:219–228.
44. Martinez P, Martin W, Cerff R. Structure, evolution and anaerobic regulation of a nuclear gene encoding cytosolic glyceraldehyde-3-phosphate dehydrogenase from maize. J Molec Biol 1989;208:551–565.
45. Ostrem JA, Vernon DM, Bohnert HJ. Increased expression of a gene coding for NAD:glyceraldehyde-phosphate dehydrogenase during the transition from C3 photosynthesis to crassulacean acid metabolism in *Mesembryanthemum crystallinum.* J Biol Chem 1990;265:3497–3502.
46. Russell DA, Sachs MM. Differential expression and sequence analysis of the maize glyceraldehyde-3-phosphate dehydrogenase gene family. Plant Cell 1989;1:793–803.
47. Yang Y, Kwon H-B, Peng H-P, Shih M-C. Stress responses and metabolic regulation of glyceraldehyde-3-phosphate dehydrogenase genes in *Arabidopsis*. Plant Physiol 1993;101:209–216.
48. Biesecker G, Harris JI, Thierry JC, Walker JE, Wonacott AJ. Sequence and structure of D-glyceraldehyde 3-phosphate dehydrogenase from *Bacillus stearothermophilus*. Nature 1977;266: 328–333.
49. Buehner M, Ford GC, Moras D, Olsen KW, Rossmann MG. Three dimensional structure of D-glyceraldehyde-3-phosphate dehydrogenase. J Molec Biol 1974;90:25–49.
50. Harris JI, Waters M. Glyceraldehyde-3-phosphate dehydrogenase. In: Boyer PD (eds) The Enzymes. New York: Academic Press, 1976;1–49.
51. Branlant C, Oster T, Branlant G. Nucleotide sequence determination of the DNA region coding for *Bacillus stearothermophilus* glyceraldehyde-3-phosphate dehydrogenase and of the flanking DNA regions required for its expression in *Escherichia coli*. Gene 1989;75:145–155.
52. Walker JE, Carne AF, Runswick MJ, Bridgen J, Harris JI. D-glyceraldehyde-3-phosphate dehydrogenase. Complete amino acid sequence of the enzyme from *Bacillus stearothermophilus*. Eur J Biochem 1980;108:549–565.
53. Skarzynski T, Moody PCE, Wonacott AJ. Structure of holo-glyceraldehyde-3-phosphate dehydrogenase from *Bacillus stearothermophilus* at 1.8 Å resolution. J Molec Biol 1987;193:171–187.
54. Michels PA, Marchand M, Kohl L, Allert S, Wierenga RK, Opperdoes FR. The cytosolic and glycosomal isoenzymes of glyceraldehyde-3-phosphate dehydrogenase in *Trypanosoma brucei* have a distant evolutionary relationship. Eur J Biochem 1991;198:421–428.
55. Meyer-Gauen G, Schnarrenberger C, Cerff R, Martin W. Molecular characterization of a novel, nuclear-encoded, NAD-dependent glyceraldehyde-3-phosphate dehydrogenase in plastids of the gymnosperm *Pinus sylvestris* L. Plant Molec Biol 1994;26:1155–1166.
56. Vermiglio A, Ravenel J, Peltier G. Chlororespiration: a respiratory activity in the thylakoid membrane of microalgae and higher plants. In: Wiessner W, Robinson DG, Starr RC (eds) Cell Walls and Surfaces, Reproduction, Photosynthesis. Berlin: Springer Verlag, 1990;188–205.
57. Singh R, Green MR. Sequence-specific binding of transfer RNA by glyceraldehyde-3-phosphate dehydrogenase. Science 1993;259:365–368.
58. Meyer-Siegler K, Mauro DJ, Seal G, Wurzer J, DeRiel JK, Sirover MA. A human nuclear uracil DNA glycosylase is the 37-kDa subunit of glyceraldehyde-3-phsophate dehydrogenase. Proc Natl Acad Sci USA 1991;88:8460–8464.
59. Kawamoto RM, Caswell AH. Autophosphorylation of glyceraldehyde-3-phosphate dehydrogenase and phosphorylation of protein from skeletal muscle microsomes. Biochemistry 1986;25:656–661.
60. Arnon DI, Rosenberg LL, Whatley FR. A new glyceraldehyde phosphate dehydrogenase from photosynthetic tissues. Nature 1954;173:1132–1134.
61. Mateos MI, Serrano A. Occurence of phosphorylating and nonphosphorylating NADP-dependent glyceraldehyde-3-phosphate dehydrogenases in photosynthetic organisms. Plant Sci 1992;163–170.
62. Kelly GJ, Gibbs M. Nonreversible D-glyceraldehyde 3-phosphate of plant tissue. Plant Physiol 1973;52:111–118.

63. Pupillo P, Faggiani R. Subunit structure of three glyceraldehyde 3-phosphate dehydrogenases of some flowering plants. Arch Biochem Biophys 1979;194:581–592.
64. Habenicht A, Hellman U, Cerff R. Nonphosphorylating GAPDH of higher plants is a member of the aldehyde dehydrogenase superfamily with no sequence homology to phosphorylating GAPDH. J Molec Biol 1994;237:165–171.
65. DeDuve C. The thioester world. In: Tran-Than-Van J, Tran-Than-Van K, Mounolu JC, Schneider J, McKay C (eds) Frontiers of Life. Gif sur Yvette: Editions Frontières, 1992;1–21.
66. Hensel R, Zwickl P, Fabry S, Lang J, Palm P. Sequence comparison of glyceraldehyde-3-phosphate dehydrogenases from the three urkingdoms: evolutionary implication. Can J Microbiol 1989;35:81–85.
67. Prüß B, Meyer HE, Holldorf AW. Characterization of the glyceraldehyde 3-phosphate dehydrogenase from the extremely halophilic archaebacterium *Haloarcula vallismortis*. Arch Microbiol 1993;160:5–11.
68. Forterre P, Benachenhou-Lahfa N, Confalonieri F, Duguet M, Elie C, Labedan B. The nature of the last universal ancestor and the root of the tree of life, still open questions. BioSystems 1993;28:15–32.
69. Brinkmann H, Martinez P, Quigley F, Martin W, Cerff R. Endosymbiotic origin and codon bias of the nuclear gene for chloroplast Glyceraldehyde-3-phosphate dehydrogenase from maize. J Molec Evol 1987;26:320–328.
70. Martin W, Cerff R. Prokaryotic features of a nucleus-encoded enzyme. cDNA sequences for chloroplast and cytosolic glyceraldehyde-3-phosphate dehydrogenases from mustard (*Sinapis alba*). Eur J Biochem 1986;159:323–331.
71. Shih M-C, Lazar G, Goodman HM. Evidence in favor of the symbiontic origin of chloroplasts: Primary structure and evolution of tobacco glyceraldehyde-3-phosphate dehydrogenase. Cell 1986;47:73–80.
72. Martin W, Brinkmann H, Savona C, Cerff R. Evidence for a chimeric nature of nuclear genomes: Eubacterial origin of eukaryotic glyceraldehyde-3-phosphate dehydrogenase genes. Proc Natl Acad Sci USA 1993;90:8692–8696.
73. Branlant G, Branlant C. Nucleotide sequence of the *Escherichia coli* gap gene. Different evolutionary behaviour of the NAD+-binding domain and of the catalytic domain of D-glyceraldehyde-3-phosphate dehydrogenase. Eur J Biochem 1985;150:61–66.
74. Liaud M-F, Valentin C, Martin W, Bouget F-Y, Kloareg B, Cerff R. The evolutionary origin of red algae as deduced from the nuclear genes enconding cytosolic and chloroplast glyceraldehyde-3-phosphate dehydrogenases from *Chondrus crispus*. J Molec Evol 1994;38:319–327.
75. Gray MW. Origin and evolution of organelle genomes. Curr Opin Gen Devel 1993;3:884–890.
76. Woese CR. Bacterial evolution. Microbiol Rev 1987;51:221–271.
77. Gibbs S. The chloroplasts of Euglena may have evolved from symbiotic green algae. Can J Bot 1978;56:2883–2889.
78. Gray MW. The endosymbiont hypothesis revisited. Int Rev Cytol 1992;141:233–357.
79. Hallick RB, Hong L, Drager RG, Favreau MR, Montfort A, Orsat B, Spielmann A, Stutz E. Complete sequence of *Euglena gracilis* chloroplast DNA. Nucl Acids Res 1993;21:3537–3544.
80. Martin W, Somerville CC, de-Goër SL. Molecular phylogenies of plastid origins and algal evolution. J Molec Evol 1992;35:385–404.
81. Liaud M-F, Zhang DX, Cerff R. Differential intron loss and endosymbiontic transfer of chloroplast glyceraldehyde-3-phosphate dehydrogenase genes to the nucleus. Proc Natl Acad Sci USA 1990;87:8918–8922.
82. Quigley F, Martin WF, Cerff R. Intron conservation across the prokaryote-eukaryote boundary: Structure of the nuclear gene for chloroplast glyceraldehyde-3-phosphate dehydrogenase from maize. Proc Natl Acad Sci USA 1988;85:2672–2676.
83. Shih M-C, Heinrich P, Goodman HM. Intron existence predated the divergence of eukaryotes and prokaryotes. Science 1988;242:1164–1166.
84. Doolittle WF. Whatever happened to the progenote? In: Fernholm B, Bremer K, Jörnvall H (eds)

The Hierarchy of Life. Amsterdam: Elsevier, 1989;65–72.
85. Kersanach R, Brinkmann H, Liaud M-F, Zhang D-X, Martin WF, Cerff R. Five identical intron positions in ancient duplicated genes of eubacterial origin. Nature 1994;367:387–389.
86. Dibb NJ, Newman AJ. Evidence that introns arose at proto-splice sites. EMBO J 1989;8:2015–2021.
87. Cerff R, Martin W, Brinkmann H. Origin of introns — early or late? Nature 1994;369:527–528.
88. Jarrell KA, Dietrich RC, Perlman PS. Group II intron domain 5 facilitates a *trans*-splicing reaction. J Molec Biol 1988;8:2361–2366.
89. Koch JL, Boulanger SC, Dib-Hajj SD, Hebbar SK, Perlman PS. Group II intron deleted for multiple substructures retain self-splicing activity. Molec Cell Biol 1992;12:1950–1958.
90. Roger AJ, Doolittle WF. Why introns in pieces? Nature 1993;364:289–290.
91. Ushida C, Muto A. A small RNA of *Mycoplasma capricolum* that resembles eukaryotic U6 small nuclear RNA. Nucl Acids Res 1993;21:2649–2653.
92. Wittop-Koning TH, Schümperli D. RNAs and ribonucleoproteins in recognition and catalysis. Eur J Biochem 1994;219:25–42.
93. Brändén CI. In: Fletterick R, Zoller M (eds) Current Communications in Molecular Biology: Computer Graphics and Molecular Modeling. Cold Spring Harbor, NY: Cold Spring Harbor Laboratories, 1986;45.
94. Patthy L. Intron-dependent evolution: preferred types of exons and introns. FEBS Letters 1987;214:1–7.
95. Liaud M-F, Brinkmann H, Cerff R. The beta-tubulin gene family of pea: Primary structures, genomic organization and intron-dependent evolution of genes. Plant Molec Biol 1992;18:639–651.
96. Fink GR. Pseudogenes in yeast. Cell 1987;49:5–6.
97. Sato N. Nucleotide sequence of a pseudogene for pea phytochrome reminiscent of an incorrect splicing event. Nucl Acids Res 1990;18:3632.
98. Lambowitz AM. Infectious introns. Cell 1989;56:323–326.
99. Mueller MW, Allmaier M, Eskes R, Schweyen RJ. Transposition of group II intron al1 in yeast and invasion of mitochondrial genes at new locations. Nature 1993;366:174–176.
100. Woodson SA, Cech TR. Reverse self-splicing of the *Tetrahymena* group I intron: implications for the directionality of splicing and for intron transposition. Cell 1989;57:335–345.
101. Cech TR. The generality of self-splicing RNA: Relationship to nuclear mRNA splicing. Cell 1986;44:207–210.
102. Sharp PA. On the origin of RNA splicing and introns. Cell 1985;42:397–400.
103. Tani T, Ohshima Y. The gene for the U6 small nuclear RNA in fission yeast has an intron. Nature 1989;337:87–90.
104. Craik CS, Rutter WJ, Fletterick R. Splice junctions: Association with variation in protein structure. Science 1983;220:1125–1129.
105. Liaud M-F, Brandt U, Cerff R. The marine red algae *Chondrus crispus* has a single highly divergent β-tubulin gene with a short spliceosomal introns at its 5′ end: functional and evolutionary implications. Plant Molec Biol 1995;28:313–325.
106. Liaud M-F, Valentin C, Brandt U, Bouget F-Y, Kloareg B, Cerff R. The GAPDH gene system of the red alga *Chondrus crisus*: promotor structures, intron/exon organization, genomic complexity and differential expression of genes. Plant Molec Biol 1993;23:981–994.
107. Zhou Y-H, Ragan MA. Cloning and characterization of the nuclear gene encoding glyceraldehyde-3-phosphate dehydrogenase from the marine red alga *Gracilaria verrucosa*. Curr Genet 1994;26:79–86.
108. He F, Peltz SW, Donahue JL, Rosbash M, Jacobson A. Stabilization and ribosome association of unspliced pre-mRNAs in a yeast *upf1*-mutant. Proc Natl Acad Sci USA 1993;90:7034–7038.
109. Roger AJ, Keeling PJ, Doolittle WF. Introns, the broken transposons. In: (eds.) Molecular Evolution of Physiological Processes. New York: The Rockefeller University Press, 1994;27–37.
110. Teissier L-H, Chan RL, Keller M, Weil J-H, Imbault P. The *Euglena rbcS* gene contains introns with unusual borders. FEBS Lett 1992;304:252–255.
111. Krainer AR, Maniatis T. RNA splicing. In: Hames BD, Glover DM (eds) Transcription and Splicing.

Oxford: IRL Press, 1988;131–206.
112. Nei M, Gojobori T. Simple methods for estimating the numbers of synonymous and nonsynonymous substitutions. Molec Biol Evol 1986;3:418–426.
113. Saitou N, Nei M. The neighbor-joining method: a new method for reconstructing phylogenetic trees. Molec Biol Evol 1987;4:406–425.
114. Sinibaldi RM, Mettler IJ. Intron splicing and intron-mediated enhanced expression in monocots. Prog Nucl Acid Res Molec Biol 1992;42:229–257.
115. Henze K, Badr A, Wettern M, Cerff R, Martin W. A nuclear gene of eubacterial origin in *Euglena* reflects cryptic endosymbioses during protist evolution. Proc Natl Acad Sci USA 1995;(in press).
116. Ferat J-L, Michel F. Group II self-splicing introns in bacteria. Nature 1993;364:358–361.
117. Bouget F-Y, Kerbourch C, Liaud M-F, de Goër S, Quatrano RS, Cerff R, Kloareg B. Structural features and phylogeny of the actin gene of *Chondrus crispus* (Gigartinales, Rhodophyta). Curr Genet 1995;28:164–172.
118. Donath M, Mendel R, Cerff R, Martin W. Intron-dependent transient expression of the maize *GapA1* gene. Plant Molec Biol 1995;28:667–676.

Tracing Biological Evolution in Protein and Gene Structures.
M. Gō and P. Schimmel, editors.

Putative origin of introns deduced from protein anatomy

Mitiko Gō and Tosiyuki Noguti
Department of Biology, Faculty of Science, Nagoya University, Nagoya, Japan

Abstract. *Background.* The origin of introns is an important question in molecular and evolutionary biology. Whether or not introns existed in ancestral genes of prokaryotes or whether they were inserted later only into the genes of eukaryotes remains an unanswered question. To gain insight into this question, we made an analysis based on accumulating gene structures. A module, the structural unit of proteins, was seen to have a close correlation with intron positions in several proteins. The correlation cannot be explained by the random insertion of introns, rather it suggests that 1) modules were encoded by exons at an early stage of molecular evolution, and they fused into new functional proteins, and 2) many introns of eukaryotes and all in prokaryotes were lost during biological divergence.

Methods. To investigate intron-module correlations more extensively, we made a statistical analysis of the correlation of module boundaries of four proteins and positions of introns of their genes. We decomposed triosephosphate isomerase (TPI), alcohol dehydrogenase (ADH), hemoglobin (HEM) and pyruvate kinase (PK) into modules, using a computerized method and atomic coordinates of their three-dimensional structures. Introns of the genes encoding those proteins were collected and their positions were compared with the module boundaries determined.

Results. The position of the introns was seen to have a significant correlation with module boundaries. Thus, the likelihood that all the introns were inserted into the eukaryotic genes can be discounted.

Conclusions. Our evidence confirms the view that modules are vestiges of the original building blocks of proteins and shuffling units in protein evolution. The introns were apparently located at every module boundary in ancestral genes but a majority of introns have been lost during processes of evolution.

Key words: exon shuffling, modules, molecular evolution, three-dimensional structure.

Introduction

Since introns were discovered in eukaryotic genes in 1977, two opposite views have been proposed; one is "introns early" [1–3] and the other is "introns late" hypothesis [4]. Were genes split from the beginning (introns early) [5] or did introns spread into eukaryotic genes as selfish elements (introns late) [6]? If introns had a role in the primordial assembly of minigenes and could speed up evolution, the remnant could be observed in contemporary protein architecture and function. On the contrary, if introns came late and inserted as selfish elements, the introns had almost no role in protein structure, function and evolution.

The biological importance of introns as mediators of exon shuffling [1], if it existed, can be traced in three-dimensional structures and functions of the contem-

Address for correspondence: Mitiko Gō, Department of Biology, Faculty of Science, Nagoya University, Chikusa-ku, Furo-cho, Nagoya 464-01, Japan. Tel.: +81-52-789-2976. Fax: +81-52-789-2977.

porary proteins. When the exon-shuffling hypothesis was proposed, only a few introns had been identified in genes such as immunoglobulin and hemoglobin. One of the authors (M.G.) showed that 1) hemoglobin subunit β-chain is decomposed into four compact structural elements M1 through M4, called modules [7], and 2) module M1 corresponds to exon 1, joint module M2 + M3 to exon 2, and module M4 to exon 3 [8]. The correspondence of introns to module boundaries led to a prediction of the presence of one more intron in the gene at the boundaries of modules M2 and M3. Later the expected intron at the boundary of M2 and M3 was found in the leghemoglobin gene [9]. Leghemoglobin is produced by plants and is homologous with the animal hemoglobin subunits α and β. The one-to-one correspondence between exons and modules suggests that the ancestral gene of animal hemoglobin and plant leghemoglobin was split by three introns at the boundaries of the four modules. One intron at the boundary of M2 and M3 was lost during evolution, thereby leading to vertebrate hemoglobin.

During the last decade, the exon-intron organization of genes and the three-dimensional structures of hemoglobins from various species have been reported. Based on this information we carried out an extensive analysis of hemoglobin and its gene, and found that the module-exon correlation is universally observed among the six taxa examined (K. Fukami-Kobayashi et al., manuscript in preparation).

The intron-module correlation was also observed in lysozyme [7], TPI [10,11], cytochrome c, etc. [12]. Usually only a few introns are present in a gene which encodes a protein, the three-dimensional structures of which are known. Thus, introns from various genes encoding proteins have to be collected and the correlation between their positions and module boundaries of the proteins have to be statistically tested. The unclear correlation between exons and structural elements reported by Stoltzfus et al. [6] can be discussed in this setting.

Materials and Methods

Globular proteins in general can be decomposed into modules. These modules are small compact pieces of contiguous 10–40 amino acid residues [8]. We developed an automatic computerized method to identify modules, using atomic coordinates of proteins [13,14] (M. Gō et al., manuscript in preparation) and determined the module organization of TPI, ADH, HEM and PK.

Using our new method, chicken TPI consisting of 247 amino acid residues was decomposed into 16 modules (Fig. 1). The new modules are essentially the same as previously reported modules of TPI [11], except that the three previous modules M1, M3 and M12, of the 13 identified modules, were each divided into two. The reason for increase in the number of modules in TPI is the improved sensitivity in the detection of the stable local minimum of centripetal profiles; the minimum are candidates of the module boundaries [15].

We asked whether module boundaries of the four proteins, TPI, ADH, HEM and PK would correlate with positions of introns in the corresponding genes. Many

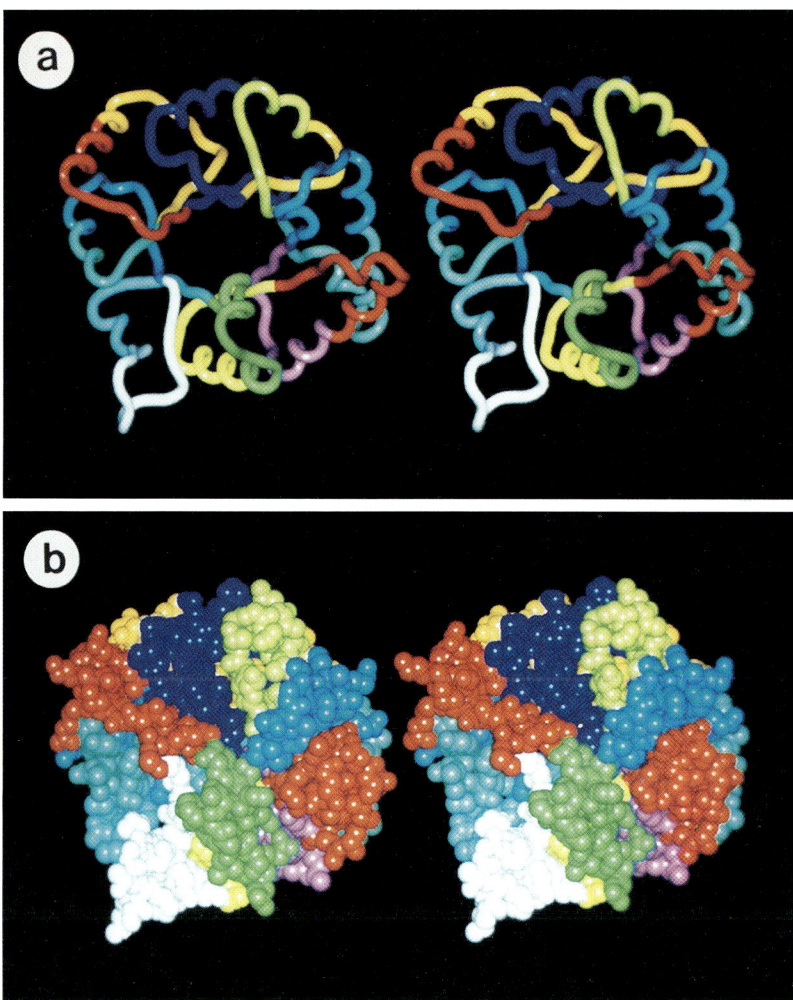

Fig. 1. Stereo diagrams of modules of triose phosphate isomerase (TPI) from cat muscle. TPI was decomposed into 16 modules. Each module is colored and shown in tube (a) and space-filling (b) models.

introns were located close to module boundaries. The introns are not always located at the exact module boundaries, however, and they are sometimes located within a few residues from the module boundaries. Thus, deviation of intron positions from the module boundaries has to be tested statistically to determine whether or not the deviation is the result of random insertion of introns.

We introduced a pseudomodule to carry out a control test for the intron-module correlation test. This pseudomodule is a module hypothetically defined as the segment starting at a center of a module and ending at the center of the next module (Fig. 2)

[15].

We used two different tests for the intron-module correlation. At first, we compared distribution of the deviation of intron positions from their nearest module boundaries and distribution from their nearest pseudomodule boundaries. Then, we tested the intron-module correlation by binomial distribution.

Given that the introns distributed randomly over the genes, the deviation of their positions from their nearest module boundaries should then be similar to that of the nearest pseudomodule boundaries. On the contrary, if the introns are located close to module boundaries, i.e., if the intron-module correlation model is correct, then deviation of their positions from their nearest module boundaries is distributed with a convex shape, as shown in Fig. 2, and deviation of the intron positions from their nearest pseudomodule boundaries is distributed with a concave shape, as shown in Fig. 2.

We further tested whether the intron-module correlation model is statistically significant by partitioning each intron into two areas, module territory or pseudo-module territory, depending on the nearest boundary of module or pseudomodule, of the intron position. Introns partitioned in the module territory and those in the pseudomodule territory have a binomial distribution. The number of introns in the module territory is expected to be larger than the number in the pseudomodule territory if the intron-module correlation is observed. The significance of these numbers is testable by binomial distribution.

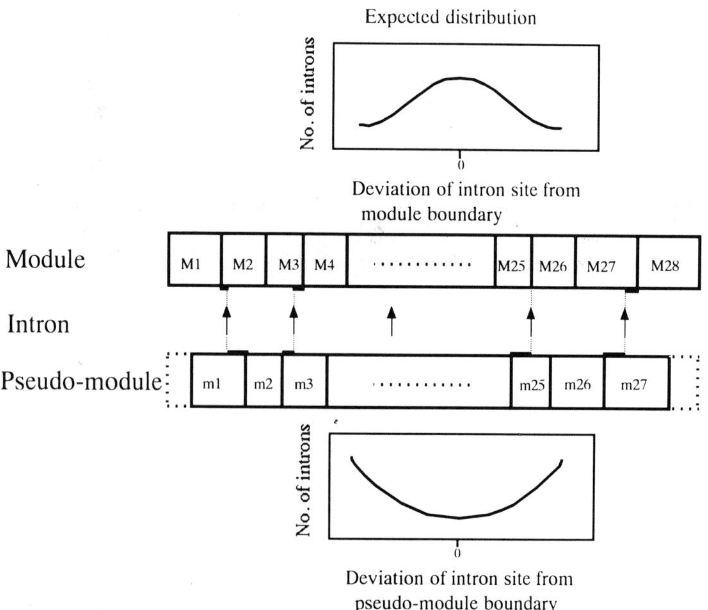

Fig. 2. A schematic illustration of protein's module and hypothetical pseudomodule boundaries, together with expected distribution of intron positions from the module and pseudomodule boundaries.

Results and Discussion

Distribution of introns

Figure 3 shows the distribution of the deviation of intron positions from their nearest module boundaries and from their nearest pseudomodule boundaries, for the four proteins, TPI, ADH, HEM and PK. We surveyed 81 intron positions. It is clear that each distribution has a shape similar to that expected by the intron-module correlation model. Thus, distributions of the deviations of intron positions from the module boundaries and pseudomodule boundaries clearly show that the intron positions are not random, but rather correlate with module boundaries.

The average of the intron deviation from the module boundaries was 2.9 amino acid residues. Close to module boundaries, 41 introns were located with a distance of less than 2.9 residues, but only 24 introns were located close to pseudomodule boundaries with less than the same distance. On the other hand, 40 introns deviated more than 2.9 residues from the module boundaries, but 57 deviated more than 2.9 residues from the pseudomodule boundaries. These differences in numbers show that the introns are not randomly distributed and that they tend to be located close to the module boundaries.

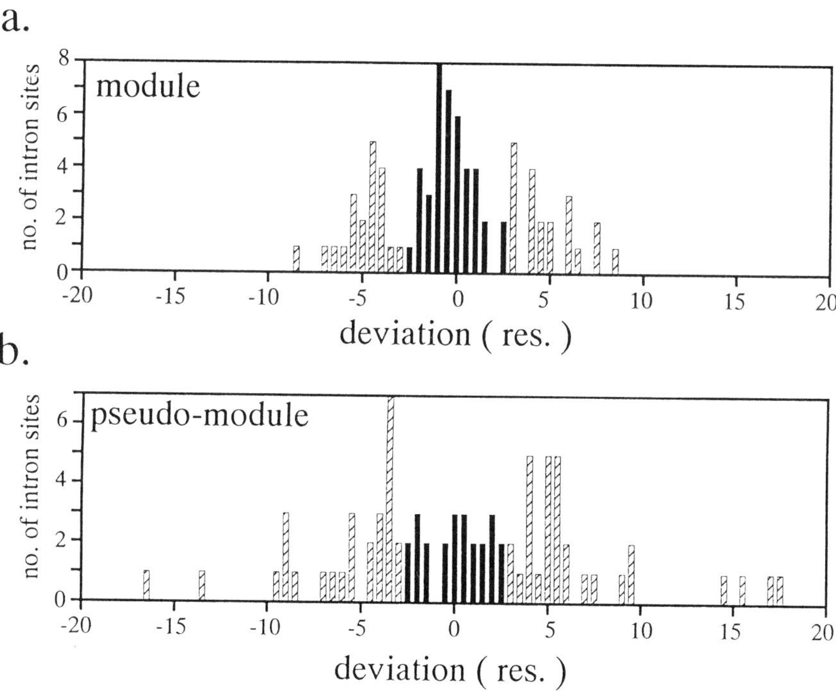

Fig. 3. Distribution of the deviation of the intron positions from module (a) and from pseudomodule (b) boundaries. The introns located within 2.9 residues from the module boundaries in (a) and the introns located within 2.9 residues from the pseudomodule boundaries in (b) are shown by black bars. The four protein families analyzed together are TPI, HEM, ADH and PK.

Statistical test of intron positions

With partition, 49 introns fall in the module territory and 32 in the pseudomodule territory. These values, by the binomial distribution test, support the intron-module correlation model with a statistical significance at the 5% level.

The correlation is statistically significant, although positions of some introns do deviate from the module boundaries. This deviation can be explained by at least three reasons; the first is the blurred boundaries of modules from the termini of the original building blocks, the second is intron sliding and the third is intron insertion. All would make correspondence of the intron positions to the module boundaries imperfect. However, we noted an intron-module correlation.

We recently tested the intron-module correlation for a number of proteins using the Akaike Information Criteria theory [16] and again found that the model was supported better than the intron-random model (M. Gō et al., manuscript in preparation).

Biological implications

Our statistical test shows that the observed intron positions correlate with module boundaries, thus the introns did not originate by random insertion of selfish elements. We then asked whether there were unique DNA sequences at terminal regions of modules which could be recognized as target signals by transposable elements. As no such unique sequences were evident, the possibility of targeted insertion of introns only at the module boundary regions can be excluded.

Our study on the relationship between protein modules and intron positions supports the "introns early" scheme and the postulation that modules are remnants of primitive miniproteins. Based on this scheme, eukaryotes have retained many of the original introns and prokaryotes have lost them. Exon-shuffling during biological evolution is equivalent to module-shuffling in proteins. We then made a module classification of proteins. Some modules are common to different proteins, thereby providing evidence of module-shuffling (K. Yura, manuscript in preparation). The result indicates the role of introns as device for exon-shuffling.

To elucidate the role of modules as miniproteins, we examined enzymatic functions of an isolated module of barnase, an RNase. Three of six modules of barnase had activity as RNase, thereby revealing that some modules are remnants of primitive minienzymes and that they and other modules were combined and became a stable globular domain with efficient catalytic activity [17].

While modules were originally introduced as structural units, protein engineering experiments showed that modules are also functional. The enzymatic function of lysozyme was converted by module replacement to α-lactalbumin [18]. Chimera hemoglobin of α- and β-subunits was designed and produced as a protein with the potential to form a tetramer with intact subunits [19]. Thus, modules are functional as well as structural units. Module-shuffling, functioning as a molecular mechanism of protein evolution, may become a technique which will facilitate acquisition of new, functional proteins.

Acknowledgements

We thank Dr H. Muirhead for providing the X-ray coordinates of PK and M. Ohara for critical comments on the manuscript. Part of this work was supported by Grants-in-Aid from the Ministry of Education, Science and Culture of Japan.

References

1. Gilbert W. Why genes in pieces? Nature 1978;271:501.
2. Doolittle WF. Genes in pieces: were they ever together? Nature 1978;272:581–582.
3. Blake CCF. Do genes-in-pieces imply proteins-in-pieces? Nature 1978;273:267.
4. Crick F. Split genes and RNA splicing. Science 1978;204:264–271.
5. Gilbert W, Glynias M. On the ancient nature of introns. Gene 1993;135:137–144.
6. Stoltzfus A, Spencer DF, Zuker M, Logsdon JM Jr, Doolittle WF. Testing the exon theory of genes: the evidence from protein structure. Science 1994;265:202–207.
7. Gō M. Modular structural units, exons, and function in chicken lysozyme. Proc Natl Acad Sci USA 1983;80:1964–1968.
8. Gō M. Correlation of DNA exonic regions with protein structural units in haemoglobin. Nature 1981;291:90–92.
9. Jensen EØ, Paludan K, Hyldig-Nielsen JJ, Jørgensen P and Marcker KA. The structure of a chromosomal leghaemoglobin gene from soybean. Nature 1981;291:677–679.
10. Straus D, Gilbert W. Genetic engineering in the Precambrian: Structure of the chicken triosephosphate isomerase gene. Molec Cell Biol 1985;5:3497–3506.
11. Gō M, Nosaka M. Protein architecture and the origin of introns. Cold Spring Harb Symp Quant Biol 1987;52:915–924.
12. Gō M. Protein structures and split genes. Adv Biophys 1985;19:91–131.
13. Bernstein FC, Koetzle TF, Williams GJB, Meyer EF Jr, Brice MD, Rodgers JR, Kennard O, Shimanouchi T, Tasumi M. The protein data bank: a computer-based archival file for macromolecular structures. J Molec Biol 1977;112;535–542.
14. Muirhead H, Clayden DA, Barford D, Lorimer CG, Fortergill-Gilmore LA, Schiltz E, Schmitt W. The structure of cat muscle pyruvate kinase. EMBO J 1986;5:475–481.
15. Noguti T, Sakakibara H, Gō M. Localization of hydrogen bonds within modules in barnase. Proteins 1993;16:357–363.
16. Akaike H. Information theory and an extension of the maximum likelihood principle. In: Petrov BN, Csáki F (eds) 2nd International Symposium on Information Theory. Budapest: Akadémiai Kaidó, 1973;267–281.
17. Yanagawa H, Yoshida K, Torigoe C, Park J-S, Sato K, Shirai T, Gō M. Protein anatomy: functional roles of barnase module. J Biol Chem 1993;268:5861–5865.
18. Kumagai I, Takeda S, Miura K-I. Functional conversion of homologous proteins α-lactalbumin and lysozyme by exon exchange. Proc Natl Acad Sci USA 1992;89:5887–5891.
19. Wakasugi K, Ishimori K, Imai K, Wada Y, Morishima I. "Module" substitution in hemoglobin subunits. J Biol Chem 1994;269:18750–18756.

Tracing Biological Evolution in Protein and Gene Structures.
M. Gō and P. Schimmel, editors.

Tests of the exon theory of genes

Walter Gilbert[1], Manyuan Long[1], Carl Rosenberg[2] and Manuel Glynias[3]
[1]The Biological Laboratories, Harvard University, Cambridge, Massachusetts, USA; [2]Whitehead Institute/MIT Center for Genome Research, Cambridge, USA; and [3]Research Institute, The Cleveland Clinic Foundation, Department of Cardiovascular Biology, Cleveland, Ohio, USA

Abstract. The exon theory of genes proposes that the first genes were assembled by recombination within introns that connected exons encoding short polypeptides that served as elements of structure or function. The primary paths of evolution were then exon shuffling and intron loss.

We discuss the relationship between exons and "modules," compact regions of polypeptide chain. Analysis of the triosephosphate isomerase genes argues that the set of "ancestral" intron positions is significantly related to the three-dimensional structure of the protein and suggests that the original gene was assembled from exons.

An analysis of a purged database of intron-containing genes shows that the distribution of intron phase within codons is nonrandom, 48% of the introns being in phase zero, and that there is a highly significant excess of symmetric exons (beginning and ending in the same phase), symmetric pairs, triples, etc., over expectation. This is evidence for exon shuffling in eukaryotic cells. By examining that part of the exon database that is homologous to prokaryotic proteins, we find again an excess of symmetric exons and hence evidence for exon shuffling in the original construction of genes.

Key words: Gō plot, modules, retroposition, triosephosphate isomerase.

The existence of the intron/exon structure of genes poses a problem: what is the origin of the introns? Can one deduce whether the introns were present from the beginning of evolution or were added during its course? One view is that the introns are old and were used to piece together the first complex genes. This view is summarized as "the exon theory of genes" [1]. The alternative view is that the introns are recent and were added to the structure of pre-existing genes by some process such as the transposition of a DNA or an RNA element to break up a continuous coding region.

The exon theory of genes expresses the idea that evolution began with short polypeptide chains, of the order of a single open reading frame, 15—20 amino acids long. That size estimate is based on arguments attributing introns to ancestral genes. That size also corresponds to the length of polypeptide chain that one would expect to have some structure in solution. This theory hypothesizes that these small exons are assembled into genes, that the mechanism of making novel genes is to shuffle exons, and that, over evolutionary time, there are processes that make complex exons

Address for correspondence: Dr W. Gilbert, The Biological Laboratories, Harvard University, 16 Divinity Ave., Cambridge, MA 02138, USA.

from simple ones. The dominant process for this last theory is retroposition, which takes out introns by the same process that makes pseudogenes. A reverse transcriptase copies a spliced message back into DNA that then reinserts into the chromosome. If the reinsertion of the intronless copy is into some nonexpressed area of the genome, that copy becomes a pseudogene. If that reinsertion is into an intron in some previously existing gene, the copy will become a complex 3' exon which will then proceed to evolve on its own.

A clear example of such retroposition to create a complex exon was worked out by Manyuan Long and Charles Langley [2] for the ADH gene in *Drosophila*, which normally has a complex structure involving four exons. In two *Drosophila* species that originated about 2.5 million years ago, *D. yakuba* and *D. teissieri*, there is a novel copy of an intronless ADH coding sequence attached to a 5' set of exons of some other gene, to make a new functional gene, the *jingwei* gene. Over evolutionary time, retroposition is a very important process leading both to the loss of single introns by recombination with cDNA copies and to the total loss of introns and the formation of complex exons.

The general hypothesis of this theory is that evolution is driven by recombination within introns, driven by the sliding and movement of introns to add new regions of polypeptide chain to the ends of exons and thus to produce new protein sequence at those boundaries, and driven by the loss of introns by rather common, but generally unselected, processes. On this picture the exons correspond to portions of protein sequence that can separately contribute to structure and function in an evolutionary sense. The simplest model would be that the exon products would be stable folding units, capable of independent existence as well as shuffling. More subtle models would permit exons to donate functions to their aggregates, such as a leader sequence, a hinge, a transmembrane element, or even a necessary amino acid.

The general prediction of the attitude contained in the exon theory of genes is that proteins will turn out to be made up of elements that ultimately correspond to exons and also correspond to “modules”. By “module” we mean a notion introduced by Mitiko Gō [3]: a region of polypeptide chain compactly folded in space and circumscribable by some maximum dimension, generally 28 Å.

There are three different models for intron evolution. On one extreme, gene evolution would correspond to a process in which the introns came into existence at the very beginning and have only been lost ever since. However, there are processes, as described by Tokio Tani (this symposium), of the addition of introns into previously existing genes, and a variant model in which introns play a dominant role in gene origination, but also are lost and gained afterwards which is also consistent with the exon theory of genes. A second type of model is that introns arose by addition to continuous genes at the origin of the eukaryotic cell or at the origin of the metazoan organisms. Simple variations on this model are that introns are only lost after that time or that they are both lost and gained. This second type of model is consistent with the use of exon shuffling to make vertebrate genes, but these models have no role for introns in prokaryotic proteins. The third type of model is that introns appeared only very late in evolution, even being added today, and there is no exon shuffling. These are three extreme views of the world. If the introns have been only lost, then, when we attribute all of the introns observed today in different homologous forms of a gene to an ancestral gene, all of those introns should match up to some aspect of the protein structure. However, if some introns have also been gained over evolutionary time, then the attributed "ancestral" gene will have both introns that are related to protein structure and some that are not. One of the deep problems in trying to construct tests of the exon theory is whether one can see signs of ancient introns against a background of loss and gain, or whether the loss and gain simply washes out any statistical test. Can we see past the present forms to the real biological phenomena?

Exons as compact structures

The argument that exons might represent compact structures in space has been used over the years to make several predictions as to the positions of novel introns. Mitiko Gō [3] first observed that the exons in globin could be described as compact structures, in the sense that the exon product folds back and forth in such a way that the polypeptide chain lies circumscribed by a sphere 28 Å in diameter, if there were three introns in globin rather than the two that had been observed in hemoglobin and myoglobin. This prediction was soon fulfilled by the discovery of a novel intron in leghaemoglobin [4], which was found to have a novel intron in the approximate position predicted. Her analysis used a two-dimensional plot (which we call a Gō plot), each point of which corresponds to a pair of amino acids in the protein structure. On this plot the distances between each pair of amino acids are indicated, usually by coloring the plot black for all amino acid distances greater than or equal to 28 Å. On such a plot, all pairs of amino acid residues in an exon are represented by a triangle along the diagonal of the plot. If that triangle includes no black regions or is just bordered by black regions, one concludes that all pairs of distances within the exon are less than 28 Å and that the exon is compact. If that triangle contains some black areas, one concludes that the exon is extended and could, by the addition of an intron, be broken into two portions, each of which would now correspond to smaller triangles which do not overlap the black regions. When we first examined such plots for the triosephosphate isomerase gene, we observed that three of the seven chicken exons contained black areas on the Gō plot, and we commented that two of these should probably be broken up by further introns [5]. A year later these introns were discovered, and we predicted in 1986 [6] that the last remaining exon was a further candidate for an intervening intron. In 1992 Claus Tittiger [7] discovered this intron in mosquito triosephosphate isomerase. Figure 1 shows the Gō plot for triosephosphate isomerase and the position of the novel intron in mosquito. This series of successful predictions makes us feel that there is some property in the world that is being encompassed in this way of looking at the exons as compact structures. Can we extend these ideas?

One can look at these predictions and say that they are examples of the matching

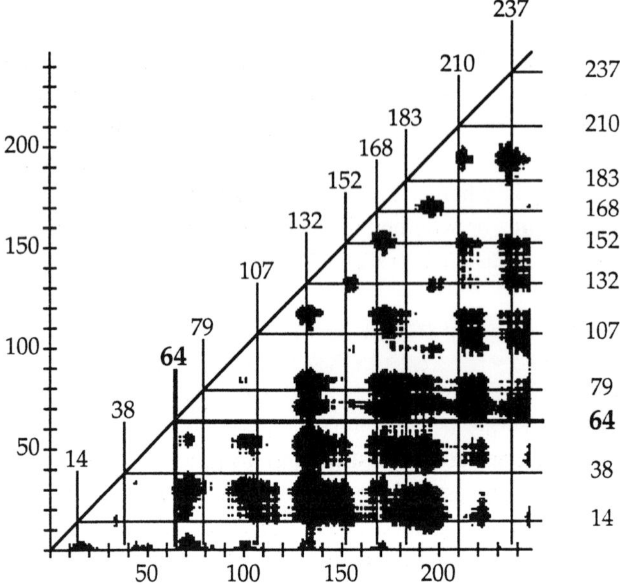

Fig. 1. A distance plot for triosephosphate isomerase. The black areas represent pairs of amino acid residues whose C^{α} carbons are more than 28 Å apart. The axes are the residues in the protein. The positions of the introns are shown, and the novel intron in mosquito is shown in bold.

of intron positions to a compactness definition of exons that reflects the three-dimensional structure of the protein. One could also look at the argument that has just been given and say: this fails to do anything significant because all that has been done is to break the protein up into small domains. If one breaks the plot up along the diagonal into small regions, those regions will eventually be compact, the corresponding "exons" will eventually be small in space. Can we find a method to show that the pattern of introns is very different from what one would find had one just broken the protein up into small pieces? The simplest approach is to break the protein up randomly and compare the distribution of hypothetical exons constructed by such a random pattern of "introns" with the real distribution. A measure of how good any breakup is in terms of compactness is found by counting the number of distances greater than 28 Å that are included in each exon (or in the case of a Gō plot, the dots contained in the triangle that corresponds to each exon). That procedure yields a numerical criterion that can be used to calculate what fraction of the alternative ways of breaking up the gene by random "introns" is better than the original breakup [8]. Table 1 shows that if we start with the observed gene, a triosephosphate isomerase (TPI) from chicken with five introns, or one from plants with eight introns, then many of the random alternatives have better scores, but as we add more and more introns to a hypothetical original gene, to get 10 introns (by combining corn and *Aspergillus*) or the 11 introns including one from mosquito, very few of the random alternatives are as good a fit to that distance plot as is the native set. The set of real positions are better than 99.9% of the random alternatives. This

Table 1.

Genes	Percentage of random alternatives with worse scores
TPI with 5 introns (*Aspergillus*)	46.4%
TPI with 6 introns (Chicken)	95.1%
TPI with 8 introns (Maize)	86.2%
TPI with 10 introns (Maize and *Aspergillus*)	99.3%
TPI with 11 introns (including *Culex*)	99.88%

is adequate to reject the null hypothesis that the introns are randomly placed in the gene, as we would expect if they had been added to DNA sequence by a process that did not look at protein structure. However, one might object to this null hypothesis and argue that random intron positions do not resemble intron positions in proteins. Random intron positions would mean that there are a lot of small exons and some very large ones, while in real genes most exons are about the same size. To take this into account, we can do a far more conservative calculation. We can maintain exon size in a variety of ways, the simplest is simply to permute the positions of the exons, to shuffle the exons, and ask how well does that alternative fit the distance plot. Table 2 shows that triosephosphate isomerase with 10 introns is better than 87% of the shuffled alternatives, while that of 11 introns is better than 96% of the shuffled alternatives. Once one adds the mosquito intron to the structure, one has a distribution of introns that is unusual, before that we cannot show significance by this very conservative test.

We generalized this idea by devising three more tests of exon structure [8]. A second test is compactness: are most of the close contacts between amino acid residues, those whose C^{α} positions are less than 12 Å apart, within the exons as opposed to between them. The third test counts and maximizes how many hydrogen bonds along the backbone are included within the exons, and the last test measures how much buried surface along the backbone is included within the exons. All four of these tests are related roughly to a notion of compactness along the protein chain, but they are not the same criterion. Each of these tests can be turned into a numerical score and tested against the original set of intron positions versus the random sets. These four tests do not correlate with each other on the randomized sets; they are not

Table 2.

Genes	Percentage of shuffled alternatives with worse scores
TPI with 5 introns (*Aspergillus*)	15.1%
TPI with 6 introns (Chicken)	29.8%
TPI with 8 introns (Maize)	15.8%
TPI with 10 introns (Maize and *Aspergillus*)	87.4%
TPI with 11 introns (including *Culex*)	96.1%

looking at the same parameter. We combine the four tests by ranking 10,000 alternative shuffled sets for each of the tests, averaging the ranks, and asking what rank is that average in that overall set of shuffles. This is a way of combining four tests in such a way as to show their independence and to get their distribution to approximate a Gaussian. The four tests show that the 10-intron structure for TPI is very significantly related to the protein structure, at the 95% level, and that relationship goes up above 99.4% as we add the last intron from mosquito.

These tests are reasonably robust. Logsdon and Palmer (unpublished) have found other introns in triosephosphate isomerase and they have used the positions of those other introns to argue against our first compactness test score, because introns at these extra positions deteriorate the score. However, the four tests are resistant to those added introns, and the score is still above 99%. That means that this set of introns including the added ones is still unusually related to the protein structure. Nonetheless, a true correlation between intron placement and protein structure can be obscured either by the loss of introns, so that one has too few introns to work with, or by the gain of a few adventitious introns, in which case one has too many.

We have been trying, but haven't yet succeeded to turn this way of looking at the gene and protein into a way to predict where introns might have been in the protein structure, but connections between looking at the protein structure statistically and predicting where the best intron positions might hypothetically be still do not work too well.

Last year Stoltzfus et al. [9] discussed a variety of tests of whether the exons corresponded to three-dimensional structures in proteins. In general, their efforts to find tests of the correlation between exons and three-dimensional structure were negative, and they took their inability to find such correlations as a disproof of the exon theory of genes. For the most part, they did not test the more usual interpretations of the exon theory of genes but examined only their own interpretations. The failure of their calculations reflects a failure of their interpretations, rather than the exon theory. They first discussed and disproved the concept that the exons would correspond to a secondary structure element such as an α-helix or a β-sheet, since they failed to find any correlation between introns and the boundaries of such secondary elements. They ignored the fact that, as early as 1979, it was pointed out that the introns in globin broke up the α-helices and that the exons were likely to represent more complex elements such as turns in the protein structure [10]. Stoltzfus et al. attacked at great length the idea of "centrality", the notion that introns lie close to the center of the proteins. For this notion they referenced one of our papers [5]. However, they failed to notice that this reference's only comment about "centrality" is that "centrality" does not work for triosephosphate isomerase. Lastly, they argued that calculations like the first test we discussed above fail when applied to a few other proteins. Their argument fails to take into account the problem posed by the biological reality. If introns existed only in the original protein gene and were only lost since, then one could establish an approximation of the original pattern by examining all of the currently existing forms of the gene. If the original pattern were one in which the introns were related to three-dimensional structure of the original gene, one would have some hope of finding some remnants of that relationship as one adds more and more introns into the structure. However, any ability to see a pattern would be defeated either if there was too great a loss or if there is any random process of intron addition, so that one has adventitious introns in addition to ancestral ones, or if there had been any alteration of the protein structure since its first assembly by the processes of evolution. Intron loss, and the possibility of intron gain make it extremely difficult to predict in advance which analyses of which gene structures will be revealing. Our own attitude is that the best approach is to identify a well-preserved fossil gene, not too ravaged by evolution, and to analyze the relationships of its structure to that of the protein. Only time will tell if enough clear examples will appear.

Intron positions in a database of exons

We have constructed a database of all genes that have an intron/exon structure and peptide sequence that can be identified automatically from their feature tables using GenBank. We capture about 95% of the genes that are there, about 5% have feature tables that cannot be read automatically. This process yielded, from GenBank 84, 9,300 sequences and 45,000 exons. These, of course, include very many repeats of things such as the globin sequences, the immunoglobulin sequences, and many fragmentary genes. We then purge that database automatically by various similarity measurements. In general, we purge the database by using FastA to construct a match between gene products, dividing the number of matching amino acids by the length of the shorter protein to provide a percentage match score, and for each match, above some cutoff, keeping only the entry with more exons. To identify automatically a set of reasonably independent proteins, we purged down to a similarity score of 20%. This yielded a set of 1,925 proteins and 13,042 exons. After dropping the first and last exons, we looked at the order of 1,600 genes with about 9,200 exons. Figure 2 shows the final exon distribution. As is always the case, this is a rather narrow distribution, peaking at an exon length of 35. This is not the 15 to 20 exon length we mentioned before, and one interpretation of this in the "exon theory of genes" is that these modern exons are already fusions of two to three primitive structures. (There are even in the database a few exons of length one and length two.)

Given all these exons, what can we say? The first thing we examined was the phase distribution of the introns. (The phase of an intron is its position in a codon: phase zero, one, or two corresponding to the intron lying before the codon, after the first base, or after the second base respectively.) If introns are added by any mechanism that looks at DNA sequence, one expects them to be added randomly with respect to the coding ability of the gene. One would expect them to be equally represented in all three phases. However, there is a great excess of introns in phase zero: 48% are in phase zero, 30% are in phase one, and 22% are in phase two. Rather than the 3,706 introns one expects in phase zero, there are 5,263, an excess of 1,557. These numbers are so large that this is wildly statistically significant. It is a surprise

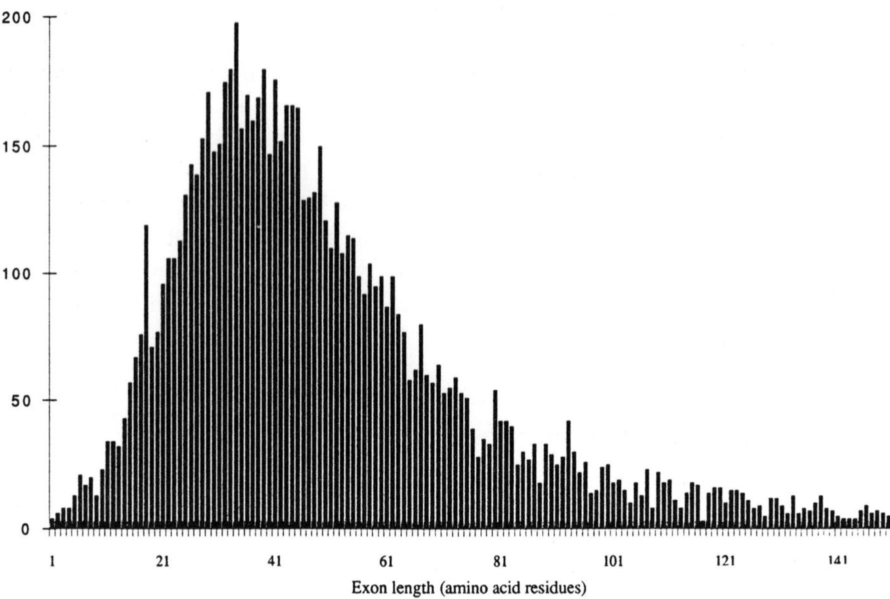

Fig. 2. The distribution of exon lengths in the final database.

to find an excess of intron positions relating to an aspect of protein coding, and the straightforward interpretation is that exon shuffling has produced an excess of phase zero introns, because exons ending in zero phase are more easily reused.

However, one might argue that perhaps some rule constraining the addition of introns accounts for this excess. If the introns entered into some sequence or sequences that were long enough to be themselves related to protein structure, this phase relationship would be only a consequence of the addition of introns. This argument is that the hypothetical target for introns (as transposons or as reverse splicing) is not short sequences, which one would expect to be random with respect to coding phase, but longer sequences which might be specific in respect to protein structure [11]. So here we simply note this unequal distribution in phase, and go on to examine another property that is not at all what we expect if introns were added. We examine symmetric exons, exons that begin and end in the same phase. Consider one process of exon shuffling in which exons add into the intron of some previously existing gene. That pre-existing intron is in some phase, let's say phase one, with respect to the surrounding exons. In order to add an exon to fit into this intron, that exon must have phase one at both ends; it must be symmetric. This is the observation that exon shuffling will work easier if the exons are symmetric, either having phase zero at both ends, phase one at both ends, or phase two at both ends. Thus, we asked is there an excess of symmetric exons or nonsymmetric exons? Furthermore, the criterion for inserting a pair of exons into an intron would be the same phase on the outside of the pair, but any phase in between. Thus we ask is there an excess of symmetric exons, symmetric pairs, symmetric triples, and so forth. In the reduced database there are 3,929 symmetric exons: 2,325 (0,0) exons, 1,085 (1,1) exons and 519 (2,2) exons. What do we expect? Since there is an excess of phase zero introns, we compare the observed number of symmetric exons to the expectation given the frequency of introns in each phase. (For example, we compare $P_1(j,k)$ with P_jP_k, where $P_1(j,k)$ is the probability of an exon being bounded by phase j, k, and P_j is the probability of an intron being in phase j). Then, even above the bias of intron phases, we observe a 12% excess of symmetric (0,0) exons, a 30% excess of symmetric (1,1) exons, and a 12% excess of symmetric (2,2) exons. Again, these numbers are large enough, the 12% excess difference is 250, the 30% excess is about 253, so that these are very, very significant differences. Table 3 shows that this excess is true both for symmetric exons, for pairs of exons asking if the outside is symmetric, for triples of exons, for quadruples, for quintuples. The excesses are consistent, but the absolute numbers fall as one looks at longer sets, so that eventually the excesses stop being significant. (We compare the pairs of exons with the prior probability based on the actual frequencies of all single exons, etc.). This is a very strong argument for exon shuffling, just seen by looking at the structure of the database. One might hypothesize that if there is a special rule that says there are phase zero introns in excess, maybe there should also be a rule that says there are phase (0,0) exons in excess. The striking thing, however, is that we observe a much greater, 30%, excess of (1,1) exons. This excess is direct evidence for repeated symmetric exons in the database, and that of course is a restatement of the exon shuffling hypothesis.

To restate the argument most bluntly: the great excess in (0,0) exons over the number that would be expected if the introns were random with respect to phase (2,325 versus 1,021), the null hypothesis, is directly an argument for exon shuffling. On this basis the excess of phase zero introns is a consequence of the excess of (0,0) exons. However, even if we accept the weaker argument, that perhaps a large part of this excess of (0,0) exons is due to the excess of phase zero introns due in turn to a (hypothetical) law of addition of introns so that they are more in phase zero, the fact then that the excess over this prior expectation is greatest for the (1,1) exons serves as independent evidence for exon shuffling (or, equivalently, for exons being related

Table 3. Symmetric exon patterns.

Exons	(0,0)	(1,1)	(2,2)	χ^2
1	2325/2075 (12%)	1085/832 (30%)	519/461 (13%)	200.37
2	1935/1829 (6%)	930/662 (40%)	421/373 (13%)	177.16
3	1593/1492 (7%)	715/518 (38%)	330/305 (8%)	132.76
4	1357/1230 (10%)	602/424 (42%)	311/252 (23%)	150.45
5	1060/1015 (4%)	461/350 (32%)	209/208 (0%)	53.64
6	899/842 (7%)	390/290 (34%)	183/172 (6%)	60.35
7	736/696 (6%)	318/240 (32%)	133/142 (–6%)	42.55

Note: Symmetric exons and exon sets in the purged database. Observed/expected (percentage difference) is shown. The expected values are calculated based on the prior expectation of frequencies for the composite exon sets. The χ square values are for the full set of nine intron combinations. The critical χ square value is 15.51 (df = 8) for the 0.05 level.

to the details of protein structure, since the phases, which are related to amino acid coding, are correlated at a distance.)

These numbers are independent of the purging. At every level of purging, from 60% down to 20% similarity, there is the same percentage excess of (1,1), (0,0), and (2,2) exons, and the same phase bias. If the distribution of the symmetric property had been what one expects for a random distribution of the ends of the exons, then duplications in the database, having an extra copy of a gene with a random distribution of exons, will not perturb the underlying randomness of the distribution. The fact that the distribution is nonrandom, an excess of symmetric exons, is not a property that could be created by having extra copies of genes in the database.

A considerable fraction of all exons are involved. Out of about 9,200 internal exons, if one asks how many exons in exon sets are involved, about 1,733 are involved in the excess. So at least 19% of the exons in the database were involved in exon shuffling.

Overall, the statistics of the distribution of symmetric exons provides evidence for exon shuffling. Our database is one of eukaryotic proteins, because those are the proteins that have an intron/exon structure. Thus the argument so far serves to prove exon shuffling for eukaryotic genes. However, we can ask how many of those proteins are homologous to prokaryotic proteins? We defined a set of those "ancient" proteins that are homologous to prokaryotic proteins by comparing our intron-containing database to the prokaryotic database (GenBank 85) using BLAST. We identify genes in our database that match prokaryotic proteins with BLAST scores above 75 (probabilities below 0.005 for the size of the prokaryotic database). These are genes like triosephosphate isomerase, or the "mitochondrial" genes that have introns in nuclear-encoded proteins, or other genes of ancient ancestry. We aligned each of the prokaryotic matches to the eukaryotic gene using the Smith-Waterman alignment, and defined a new database that contains only those introns that lie in the region of the eukaryotic protein that matches the prokaryotic one. This represents a class of genes (regions) that have introns but which arose as prokaryotic genes. These genes are not examples of exon shuffling in the eukaryotic world because these protein sequences came into existence in the progenote, and were present in the last common ancestor. We found 431 genes (and 1,587 internal exons) in the database to be homologous to prokaryotic genes, 22% of our database. The story for these genes is essentially the same. The intron phases are not random: 54% are in phase zero, 25%, in phase one, and 21%, in phase two. Again we can look at the symmetric exons in these bacterial-related genes. Again there is a dramatic excess of (0,0) exons over random expectation, and if we look at the excess over the biased expectation, Table 4 shows that there is a 31% excess of (1,N,1) exons (pairs of exons). However, there is lower significance, because the numbers begin to be small.

Now this is direct evidence that there has been exon shuffling in the ancestry of the genes with introns that are homologs of the prokaryotic genes. The implication of this is that ancient introns were used to assemble the prokaryotic genes. The advantage of this calculation is that it looks at a signal in the database that is derived from the act of exon shuffling, the *excess* of symmetric exons. If there is a

Table 4. Symmetric exon patterns for ancient genes.

	Nonpooled (nine exon types)				Pooled (two exon types)	
Exons	(0,0)	(1,1)	(2,2)	χ^2	(0,0)+(1,1)+(2,2)	χ^2
1	495/461 (12%)	109/96 (14%)	78/73 (7%)	10.01	682/630	6.96
2	404/359 (12%)	81/62 (31%)	50/52 (–4%)	20.96	535/473	13.45
3	315/270 (17%)	52/46 (13%)	34/39 (–13%)	16.96	401/355	10.07
4	243/206 (18%)	36/35 (3%)	26/23 (13%)	12.22	305/264	10.57

Note: Symmetric exons and exon sets for the regions corresponding to ancient genes. Observed/expected (percentage difference) is shown. The expected values are calculated based on the prior expectation of frequencies for the composite exon sets. The critical χ square value for the full set of nine intron combinations is 15.51 (df = 8) for the 0.05 level. For the two exon type calculation, the critical χ square value is 3.84 (df = 1) for the 0.05 level.

background of introns added to the database, it will not cancel the effect.

The excess of symmetric exons in this "ancient" part of the database has a simple interpretation in terms of original introns connecting exons that were shuffled at the origin of evolution and have simply survived. The excess has no interpretation in terms of an intron-added model.

Acknowledgements

This work was supported in part by the National Institutes of Health.

References

1. Gilbert W. The exon theory of genes. Cold Spring Harbor Symp Quant Biol 1987;52:901–905.
2. Long M, Langley C. Natural selection and the origin of *jingwei*, a chimeric processed functional gene in *Drosophila*. Science 1993;260:91–95.
3. Gō M. Correlation of DNA exonic regions with protein structural units in haemoglobin. Nature 1981;291:90–93.
4. Jensen EO, Paludan K, Hyldig-Nielsen JJ, Jorgensen P, Marcker KA. The structure of a chromosomal leghaemoglobin gene from soybean. Nature 1981;291:677–679.
5. Straus D, Gilbert W. Genetic engineering in the Precambrian: Structure of the chicken triosephosphate isomerase gene. Molec Cell Biol 1985;5(12):3497–3506.
6. Marchionni M, McKnight G, Gilbert W. On the antiquity of introns. Cell 1986;46:151–154.
7. Tittiger C, Whyard S, Walker VK. A novel intron site in the triosephosphate isomerase gene from the mosquito *Culex tarsalis*. Nature 1993;361:470–472.
8. Glynias M, Gilbert W. On the ancient nature of introns. Gene 1993;135:137–144.
9. Stoltzfus A, Spencer DF, Zuker M, Logsdon JM, Doolittle WF. Testing the exon theory of genes: the evidence from protein structure. Science 1994;265:202–207.
10. Gilbert W. Introns and exons: playgrounds of evolution. In: Axel R, Maniatis T, Fox CF (eds) Eucaryotic Gene Regulation. New York: Academic Press, 1979;1–12.
11. Fedorov A, Suboch G, Bujakov M, Fedorova L. Analysis of nonuniformity in intron phase distribution. Nucl Acid Res 1992;20:2553–2557.

Tracing Biological Evolution in Protein and Gene Structures.
M. Gō and P. Schimmel, editors.

Domain evolution of serine protease and its inhibitor genes

Takashi Gojobori[1], Toshinori Endo[1,2] and Kazuho Ikeo[1]
[1]*Center for Information Biology, National Institute of Genetics, Mishima; and* [2]*Department of Genetics, School of Life Science, Graduate University for Advanced Studies, Mishima, Japan*

Abstract. *Background.* The evolution of serine protease and its inhibitor is discussed with special reference to domain evolution. It is now known that most proteins are composed of more than one functional domain. Because serine proteases such as urokinase and plasminogen and some of their protease inhibitors are made of various functional domains, these proteins are typical examples of the so-called mosaic proteins.

Methods and Results. When kringle domains in serine proteases and a Kunitz-type protease inhibitor domain in the amyloid β precursor protein in Alzheimer's disease patients were examined by the molecular evolutionary analysis, the phylogenetic trees constructed showed that these functional domains had undergone dynamic changes in the evolutionary process. In particular, these domains are evolutionarily movable.

Conclusion. It is concluded that these functional domains evolved independently of each other and that they have been shuffled to create the existent mosaic proteins. This conclusion leads us to the reasonable speculation that these functional domains must have been minigenes, possibly at the time of primordial life or the origin of life. We may call these minigenes "ancestral minigenes".

Key words: domain shuffling, kringle, minikringle, molecular evolution, sequence motif.

Introduction

It is known that a group of serine proteases are involved in the systems of blood coagulation and fibrinolysis [1]. Each of these serine proteinases is separated into two chains. For example, uPA (urinary plasminogen activator) can be separated into two chains, A-chain and B-chain. The B-chain corresponds to a protease domain having serine at its active site, whereas the A-chain contains various functional domains such as an EGF (epidermal growth factor)-like domain and the so-called kringle domain [2]. This kind of protein can be called a "mosaic protein" because it has more than one different domain.

The kringle domain is a characteristic secondary structure of approximately 80 amino acids which is made of three pairs of cysteine-cysteine bonds. The number of the kringle domains in a serine protease depends upon the kind of proteases (Fig. 1). For example, as shown in Fig. 1, the uPA and factor XII have single-kringle domains in their A-chains [3,4]. On the other hand, the tPA (tissue-type plasminogen activator) and prothrombin each have two [5]. Furthermore, the plasminogen has five kringle

Address for correspondence: Dr Takashi Gojobori, Center for Information Biology, National Institute of Genetics, Mishima 411, Japan.

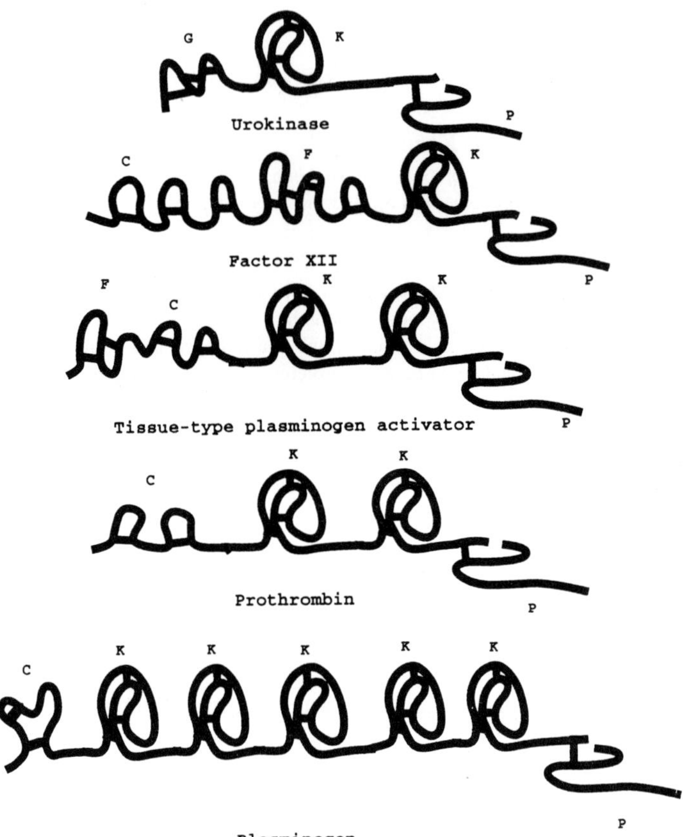

Fig. 1. Domain structures of serine proteases involved in the blood coagulation and fibrinolysis system. K — kringle domain; G — EGF-like growth factor domain; C — calcium-binding domain; F — finger-type domain; and P — protease domain.

domains [6].

Interestingly, it has recently become known that apolipoprotein(a) contains 38 kringle domains together with a serine protease-like domain. We have to note that apolipoprotein(a) has the function of transferring lipids in blood (Fig. 2). Moreover, the hepatocyte growth factor (HGF) was found to have four kringle domains as well as serine protease-like domain (Fig. 3) [7,8]. Recently, a kringle domain was found in a receptor-type protein called Ror [9]. Because the Ror is possibly a signal transduction-related protein, the receptor-type kringle appears irrelevant with serine protease.

A serine protease inhibitor is generally much smaller in size than serine proteases, and it has usually a single domain which functions as a protease inhibitor. However, it has been recently found that a serine protease inhibitor domain called the Kunitz-type inhibitor has existed in a β-amyloid precursor protein (βAPP) in the brain of an

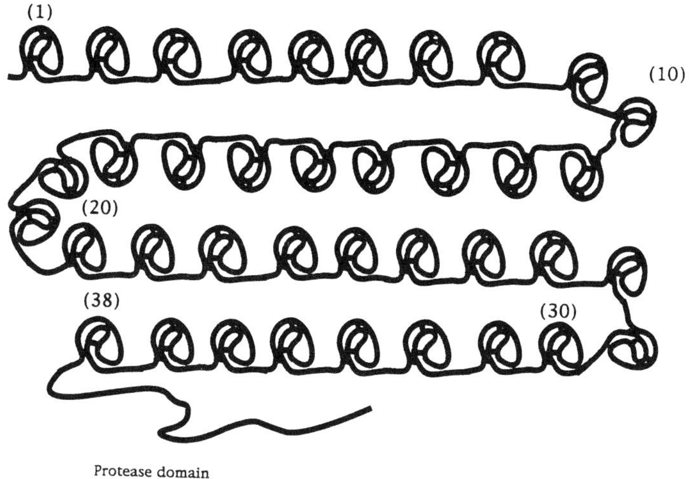

Fig. 2. Domain structures of apolipoprotein(a) that is composed of 38 kringle domains and a serine protease-like domain.

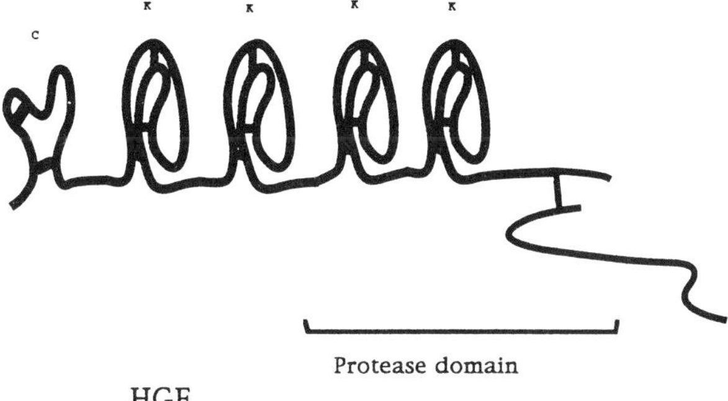

Fig. 3. Domain structures of HGF. C — a calcium-binding domain; K — kringle domains, and a protease domain.

Alzheimer's disease patient [10]. It seems that the protease inhibitor has become one of the functional domains in βAPP during evolution. It is well known that the brain of an Alzheimer's disease patient contains big cavities and accumulates βAPP. There are two kinds of βAPP with different molecular weights, APP751 and APP695 (Fig. 4). APP751 has a molecular weight larger than APP695. It has been said that a patient tends to express APP751 at a larger quantity than APP695 [11]. APP751 contains an additional domain, compared with APP695 [12–14]. This additional domain is a Kunitz-type protease-inhibitor domain. In fact, the amino acid sequence of this domain can inhibit a serine protease and has a characteristic sequence pattern that is called the "minikringle". This is because the secondary structure of the "minikringle" is very similar to that of the kringle, although the former is much

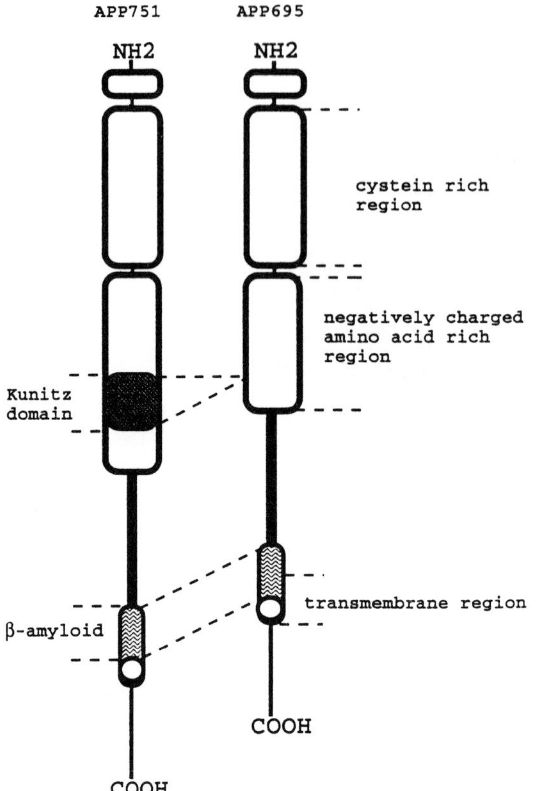

Fig. 4. The minikringle (namely, the Kunitz-type inhibitor domain) inserted into βAPP.

shorter in length than the latter. As shown in Fig. 5, the same type of structure can be seen in bovine pancreatic trypsin inhibitor (BPTI), snake toxins, inter-α trypsin inhibitor (IαTI), lipoprotein associated coagulation inhibitor (LACI), and other various proteins [15].

It is of particular importance to elucidate the evolutionary mechanisms of such mosaic proteins and the evolutionary origin of the inserted protease-inhibitor domain. In the present paper, we shall pay our attention to the evolutionary process of serine proteases and its protease inhibitors by exemplifying domain evolution. In particular, we shall review our studies on molecular evolution of kringle domains and discuss the evolutionary relationship between the inserted protease-inhibitor domain in βAPP and other protease inhibitors. We would then show our observations that these functional domains must have undergone a remarkably dynamic evolution. Moreover, we point out that different functional domains are evolving independently of each other.

It is quite possible that these functional domains may have been smaller genes in the evolutionary past. Therefore, we may call those ancestral genes "ancestral minigenes". In the world of primordial life, there may have existed at least a minimum set of ancestral minigenes that were necessary for sustaining life forms.

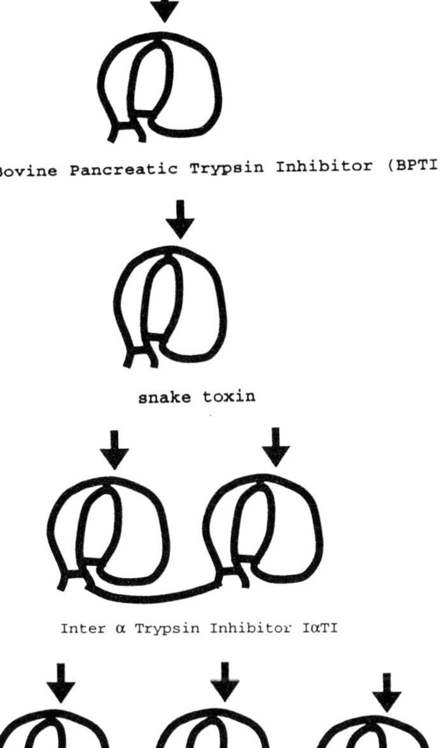

Fig. 5. The minikringle structures of serine proteases. BPTI — bovine pancreatic trypsin inhibitor; snake — snake toxin; IαTI — inter-α trypsin inhibitor; LACI — lipoprotein-associated coagulation inhibitor. Arrow heads represent active sites.

During evolution, these minigenes must have been duplicated repeatedly and must have been combined with other minigenes, so that many mosaic proteins are found in the existent proteins [16]. Along with accumulation of mutations such as nucleotide substitutions, the mosaic proteins must have survived until the present time.

Materials and Methods

Kringle and protease domains

The nucleotide and amino acid sequence data for human HGF [7,8], human tPA [5], human urokinase [3], factor XII [4], human prothrombin [17], bovine plasminogen [6], human apolipoprotein(a) [18], and Ror [9] were respectively collected from the DDBJ/EMBL/GenBank DNA database and literature. The nucleotide sequences were translated into amino acid sequences according to the universal genetic codes.

Minikringle domains

We collected about 50 amino acid sequence data of the Kunitz-type protease inhibitors (namely, minikringles) from the protein database and the DNA sequence database. The data collected includes BPTI, snake toxins, IαTI, and LACI as well as βAPP.

Multiple alignment

We separated all the amino acid sequence data into the kringle, and protease, minikringle domains. The amino acid sequences for each domain were aligned with each other by our newly developed algorithm of multiple alignment in which Needleman and Wunsch's algorithm [19] and the repeated procedures of phylogenetic tree construction are simultaneously employed. However, the multiple alignments obtained here will remain almost unchanged even if they are made by the other available algorithms such as TreeAlign [20].

Construction of phylogenetic trees

For kringle, protease, and minikringle domains, we estimated the number of amino acid substitutions per site from comparisons between all possible pairs of amino acid sequences. For the correction of possible multiple substitutions at a site, we used formula $d_a = -\log_e(1 - p - 0.5p^2)$, where d_a and p are the number of amino acid substitutions per site and the proportion of different amino acids, respectively [21].

Phylogenetic trees for the amino acid sequences of the kringle domains and kinase domains were constructed by the unweighted pair-grouping (UPG) method [22] and the neighbor-joining (NJ) method [23], using the d_a values. The UPG method requires a constant rate of amino acid or nucleotide substitutions over time, whereas the NJ method does not. We also constructed phylogenetic trees using nucleotide sequences. The number of nucleotide substitutions was estimated by the two- and six-parameter methods [24]. However, the phylogenetic tree obtained for the nucleotide sequences was virtually the same as that for the amino acid sequences.

Results and Discussion

Evolution of kringle domains

The phylogenetic tree of kringle domains constructed by the NJ method suggests that the plasminogen-type ancestor protein having five kringles appeared first (Fig. 6). Then, the dynamic evolutionary process regarding the number of kringles took place. Figure 6 shows a summary of evolution of kringle domains. The single kringle domain duplicated itself about 500 million years ago. It is interesting to note that the receptor-type kringle appeared about 400 million years ago. Then, approximately 300 million years ago, the ancestral gene of plasminogen having five kringles emerged, and the entire gene was duplicated about 80 million years ago. One of them became the present plasminogen as it is, but the other underwent very dynamic evolution. In particular, the first three kringle domains must have been deleted from the plasminogen-type ancestor protein. Then, the fourth kringle of the remaining two kringles, the fourth and fifth ones, repeated itself many times to become apolipoprotein(a). This repetition seems to have taken place within the last 5 million years [25].

The kringle domain appears to be evolving as a unit because neither a half nor a quarter of the kringle has been found yet. If each kringle domain evolutionarily behaves as a unit of evolution, how can we understand it at the genome level [26]? Gilbert [27] proposed the exon shuffling theory that exons are units of evolution at the genome level. It is known that a kringle domain consists of several exons in a genomic sequence of serine protease. If the exon can act evolutionarily as a unit of

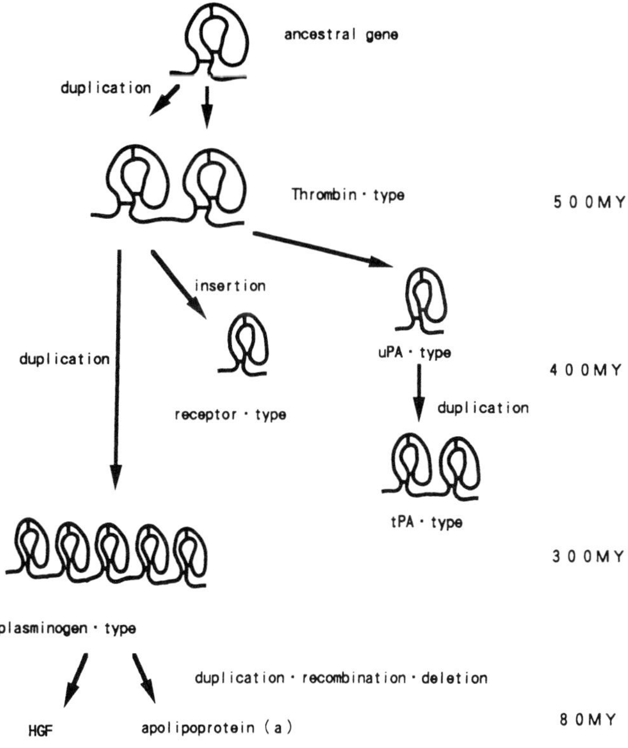

Fig. 6. The evolutionary process of kringle domains. For detailed explanation, please see the explanation in the text.

evolution, it will be possible that incomplete kringle domains such as a half of a kringle domain and a quarter of the domain appear during evolution. Although these incomplete kringle domains are not observed yet, we may be able to find them in the future. Alternatively, it can be explained by the possibility that the incomplete kringle domains were selected out quickly because of functional disadvantages. Anyway, further and extensive studies of domain evolution will give clearer insight into the relationship between the mode of domain evolution and the exon shuffling theory.

Different evolutionary histories of kringle and protease domains

We constructed phylogenetic trees for kringle structures and protease domains separately with the use of the same data set. In Fig. 7, topologies of the phylogenetic trees can be compared between kringle and protease domains. In the case of kringle domains, 38 kringle domains in apolipoprotein(a) and five kringle domains in plasminogen are represented by apo and plgen, respectively. Similarly, one (kringle A) of two kringle domains (kringles A and S) of prothrombin, a single kringle domain of urokinase, two kringle domains of tissue-type plasminogen activator, and four kringle domains of HGF are represented by pro, uPA, tPA, and HGF, respectively. Interestingly enough, two topologies for the kringles and protease domains are different from each other (Fig. 7). This strongly suggests that the kringles and protease domains are evolving independently of each other [28].

As mentioned before, the kringle structure only has been found in the possibly receptor gene for signal transduction in connection with the nervous system (called *ror*). This gene does not have a protease domain at all and its gene product is apparently a transmembrane-bound protein. This gene is not a protease gene. Thus, the kringle structure seems to have been inserted into this gene in the evolutionary

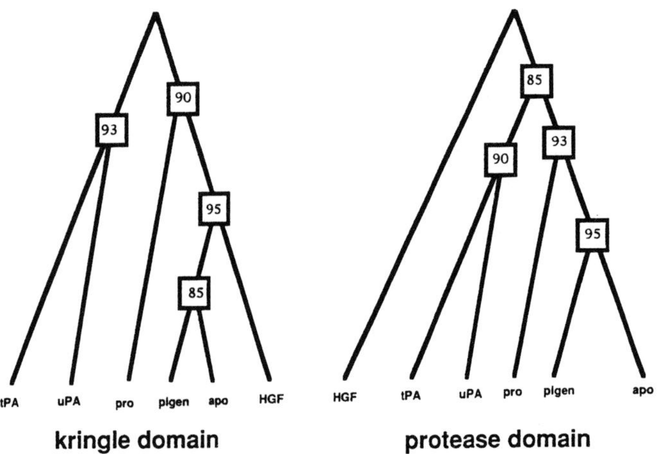

Fig. 7. Different topologies in phylogenetic trees which were constructed with the use of the same data set. tPA — tissue-type plasminogen activator; uPA — urinary plasminogen activator; pro — prothrombin; plgen — plasminogen; apo — kringle structures; HGF — hepatocyte growth factor.

history after the appearance of the ancestral gene of *ror*. From these evolutionary observations, we may conclude that the domains are evolving dynamically. Recently, Doolittle [29] also proposed a similar hypothesis, that the domain is evolutionarily movable. Our study supports this idea very strongly.

Evolution of minikringles

Let us pose the following questions to ourselves. What is the evolutionary origin of the minikringle domain? When was the domain of APP751 inserted into the ancestral form of APP695 or deleted from the ancestral form of APP751? The phylogenetic tree obtained shows that the minikringle domain of APP751 has the closest common ancestor with trypstatin and IαTI. Although a biological function of the minikringle

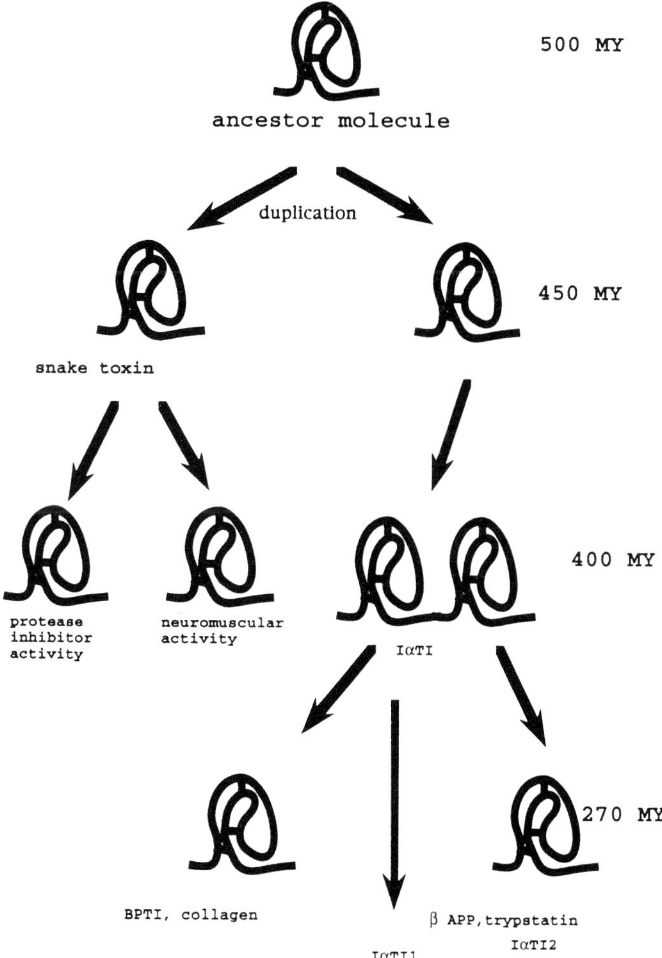

Fig. 8. The evolutionary processes of minikringle domains. MY – million years.

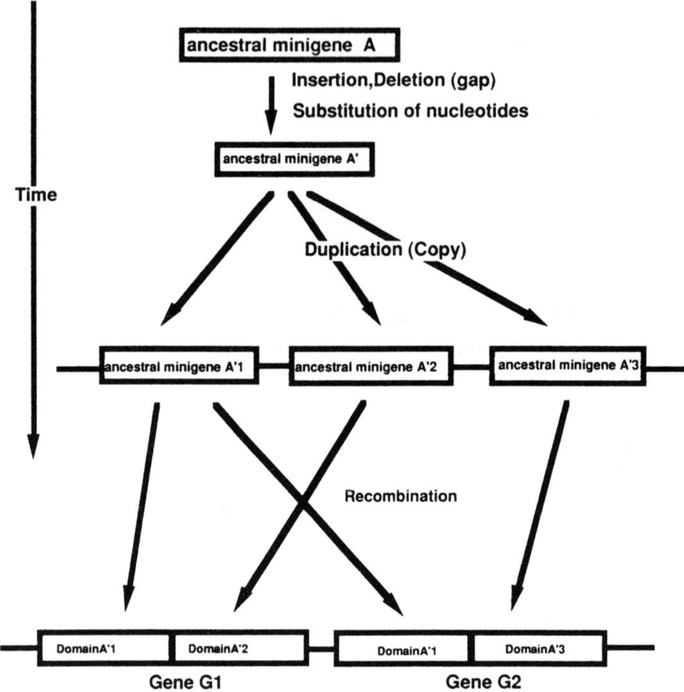

Fig. 9. Domain evolution based on the ancestral minigenes.

domain of APP751 has not been found, the phylogenetic tree suggests that the domain may have a function related to the self-immune mechanisms or the signal transduction in the immune system [15]. This is because trypstatin is involved with those biological functions and it is expressed on the cell surface of T cells.

It is also shown that the minikringle domain of βAPP has a long history (Fig. 8). The ancestral domain seems to have diverged from its common ancestor about 450 million years ago. Then, domain duplication must have taken place and one of the duplicated domains was inserted into βAPP about 270 million years ago. Because 270 million years is so old that it is corresponding to the divergence time between human and chicken, it follows that human βAPP must have retained this domain for a long time.

IαTI ancestral minigenes

As we have seen so far, it seems that a functional domain is a unit of evolution and is evolutionarily movable. It is quite possible that these functional domains were ancestral minigenes.

Our observation leads us to the following chart (Fig. 9). In the past, each domain could have been a minigene in charge of a particular function. These minigenes must have a common ancestor called "ancestral minigene". By duplication, recombination, and possibly other mechanisms, the present gene was created by a mosaic gene. Then, a very big question might arise from here. At the time of origin of life, what is the minimum set of ancestral minigenes that are necessary for maintaining the life phenomena? How many genes are there? Of course, we cannot answer this question now. In the future, however, we should be able to answer that question by analyzing the vastly accumulating DNA sequence data.

Acknowledgements

We would like to thank Dr Mitiko Gō for giving us such a wonderful opportunity to write the present paper in this volume of book. The present study has been supported in part by the Ministry of Education, Science and Culture, Japan.

References

1. Furie B, Furie CB. The molecular basis of blood coagulation. Cell 1988;53:505–518.
2. Pathy L. Evolution of the proteases of blood coagulation and fibrinolysis by assembly from modules. Cell 1985;41:657–663.
3. Verde P, Stoppelli MP, Galeffi P, Nocera PD, Blasi F. Identification and primary sequence of an unspliced human urokinase poly(A) + RNA. Proc Natl Acad Sci USA 1984;81:4727–4731.
4. McMullen BA, Fujikawa K. Amino acid sequence of the heavy chain of human a-Factor XIIa (activated Hageman factor). J Biol Chem 1985;260:5328–5341.
5. Pennica D, Holmes WE, Kohr WJ, Harkins RN, Vehar GA, Bennet WM, Yelventon E, Seeburg PH, Heyneker L, Goeddel DV. Cloning and expression of human tissue-type plasminogen activator cDNA in *E. coli.* Nature 1983;301:214–221.
6. Schaller J, Moser W, Dannegger-Muller GAK, Roselt SJ, Kampfer V, Rikli EE. Eur J Biochem 1985;149:267–278.
7. Nakamura T, Nishizawa T, Hagiya M, Seki T, Shimonishi M, Sugimura A, Tashiro K, Shimizu S. Molecular cloning and expression of human hepatocyte growth factor. Nature 1989;342:440–443.
8. Miyazawa K, Tsubouchi H, Naka D, Takahashi K, Okigaki M, Arakaki N, Nakayama H, Hirono S, Sakiyama O, Takahashi K, Gohda E, Daikuhara Y, Kitamura N. Molecular cloning and sequence analysis of cDNA for human hepatocyte growth factor. Biochem Biophys Res Commun 1989;163: 967–973.
9. Mastakowski P, Carroll RD. A novel family of cell surface receptors with tyrosine kinase-like domain. J Biol Chem 1992;267:26181–26190.
10. Ponte P, Gonzalez-DeWhitt P, Schilling J, Miller J, Hsu D, Greenberg B, Davis K, Wallace W, Lieberburg I, Fuller F, Cordell B. A new A4 amyloid mRNA contains a domain homologous to serine inhibitors. Nature 1988;333:525–527.
11. Yamada T, Sasaki H, Dohura K, Goto I, Sakaki Y. Structure and expression of alternatively spliced forms of mRNA for the mouse homolog of Alzheimer's disease amyloid β precursor. Biochem Biophys Res Commun 1989;15:8906–8912.
12. Tanzi RE, McClatchey AI, Lamperti ED, Villa-Komaroff L, Gusella JF, Neve RL. Protease inhibitor domain encoded by an amyloid protein precursor mRNA associated with Alzheimer's disease. Nature 1988;33:1528–1530.
13. Kang J, Mueller-Hill B. The sequence of the two extra exons in rat pre-A4. Nucl Acid Res 1989;17:2130.
14. Kitaguchi N, Takahashi Y, Oishi K, Shiojiri S, Tokushima Y, Utsunomiya T, Ito H. Enzyme

specificity of proteinase inhibitor region in amyloid precursor protein of Alzheimer's disease: different properties compared with protease nexin I. Biochem Biophys Acta 1990;1038:105–113.

15. Ikeo K, Takahashi K, Gojobori T. Evolutionary origin of Kunitz-type trypsin inhibitor domain inserted in the amyloid β precursor protein of Alzheimer's disease. J Molec Evol 1992;34:536–543.
16. Gojobori T, Ikeo K. Molecular evolution of serine protease and its inhibitor with special reference to domain evolution. Phil Trans R Soc 1994;344:411–415.
17. Sandra JFD, Ross T, MacGillivary A, Davie EW. Characterization of the complementary deoxyribonucleic acid and gene coding for human prothrombin. Biochemistry 1983;22:2087–2097.
18. McLean JW, Tomlison JE, Kuang W-J, Eaton DL, Ellson YC, Fless GM, Scanu AM, Lawn RM. cDNA sequence of human apolipoprotein(a) is homologous to plasminogen. Nature 1987;330:132–137.
19. Needleman SB, Wunsch CD. A general method applicable to the search for similarities in the amino acid sequence of two proteins. J Molec Biol 1970;48:443–453.
20. Hein J. A unified approach to alignments and phylogeny reconstruction. Meth Enzymol 1990;183: 626–645.
21. Kimura M. The Neutral Theory of Molecular Evolution. Cambridge: Cambridge University Press, 1983.
22. Nei M. Molecular Population genetics and Evolution. Amsterdam: North-Holland, 1975.
23. Saitou N, Nei M. The neighbor-joining method: a new method for reconstructing phylogenetic trees. Molec Biol Evol 1987;4:406–425.
24. Kimura M. The Neutral Theory of Molecular Evolution. Cambridge: Cambridge University Press, 1983.
25. Ikeo K, Takahashi K, Gojobori T. Evolutionary origin of numerous kringles in human and simian apolipoprotein(a). FEBS Lett 1991;287:146–148.
26. Holland SK, Blake CCF. In: Stone EM, Schwartz RJ (eds) Proteins, Exons, and Molecular Evolution in Intervening Sequences in Evolution and Development. New York: Oxford University Press, 1990;10–42.
27. Gilbert W. Why genes in pieces? Nature 1978;271:501.
28. Ikeo K, Takahashi K, Gojobori T. Different evolutionary histories of kringle and protease domains in serine proteases — a typical example of domain evolution. J Molec Evol 1995;40:331–336.
29. Doolittle RF, Bork P. Evolutionary mobile modules in proteins. Sci Am 1993;Oct:50–56.

Tracing Biological Evolution in Protein and Gene Structures.
M. Gō and P. Schimmel, editors.

Sketch of landscapes in the protein sequence space around catalase I from *Bacillus stearothermophilus*

Tetsuya Yomo, Savitr Trakulnaleamsai and Itaru Urabe
Department of Biotechnology, Faculty of Engineering, Osaka University, Suita, Osaka, Japan

Abstract. A landscape in protein sequence space shows the relationship between the primary structure and the level of a property of each protein. We developed methods for observing local landscapes experimentally through catalase I from *Bacillus stearothermophilus* (*B. stearothermophilus*) with respect to its catalatic activity, peroxidatic activity, and thermostability. For these three properties, the wild-type enzyme is located near the top of the hill, indicating that the present enzyme is fairly optimized for these properties. A positive correlation was found between the altitudes of the catalatic and the peroxidatic activities, indicating that the location of the hills and valleys in the landscapes of the two activities roughly correspond with each other. In contrast, the thermostability landscape appeared quite differently. Thus, the results show that even from a rough sketch of the landscapes based on the experimental data, the characteristic features of catalase I can be elucidated. The sketch of a landscape, therefore, provides a new view in understanding enzymes.

Introduction

The properties of an enzyme are thought to have been improved and optimized based on the need of the living cell during evolution. The course of the evolution of an enzyme, i.e., the course of the change in its amino acid sequence, can be expressed as a locus on the sequence space [1,2], where each point corresponds to each amino acid sequence, and the newest end is the position of the present enzyme. This concept of "protein sequence space" was introduced by Smith for explaining the evolution of proteins by natural selection [3]. Each protein sequence has various properties, such as catalytic activity, stability, and so forth. For any such property, we can ask about the distribution of its values across the sequence space. The distribution of values of a property across the space is defined as the "landscape" with respect to the property [2,4]. Hence, the process of enzyme evolution is thought to be hill climbing on the landscape by adaptive walks using mutation and selection [1–5].

As the landscape shows the direct relationship between the primary structure and the level of a function or a property of all the proteins in the sequence space, then it can also be useful for the understanding of proteins as a whole. One of the main purposes of enzymologists has been to understand the relationships among the primary structure, the tertiary structure, and the function of proteins. In short, understanding the relationship between the primary structure and function is actually

Address for correspondence: Dr T. Yomo, Department of Biotechnology, Faculty of Engineering, Osaka University, 2-1 Yamadaoka, Suita, Osaka, Japan.

drawing the landscape itself. As no method has been proposed yet for observing the landscapes through experimental data, it is worthwhile to develop one.

Recently, the mechanism of evolution has been applied for improving enzymes and macromolecules [6—12]. For this new type of protein engineering by artificial evolution, it is important to know the landscape around a target enzyme, because the landscape provides a guide map for directed walking. Random mutagenesis is a good method for preparing random samples for looking at the local landscape around an enzyme. In addition, it is important to develop analytical methods for measuring many samples and extracting information about the landscape from the data. As for a model enzyme, we used catalase I of *Bacillus stearothermophilus* (*B. stearothermophilus*), a member of the bacterial catalases with broad-spectrum peroxidatic activity [13—16]. In this work we examined the local landscapes around catalase I with respect to catalatic activity, peroxidatic activity and heat stability. The importance of the information about the landscapes on the protein engineering by artificial evolution is discussed.

Materials and Methods

Bacterial strains and plasmids

Escherichia coli (*E. coli*) UM228, a catalase HPI-deficient mutant [17], was kindly provided by Dr P.C. Loewen, University of Manitoba, Winnipeg, Canada. The pOD64, a derivative of pUC19 containing the gene of catalase I from *B. stearothermophilus*, was prepared previously [18].

Chemical mutagenesis of catalase gene

The pOD64 was digested with *Eco*RI, *Pst*I, and *Dra*I, and the 2.7-kb DNA fragment containing the catalase gene was inserted into M13mp18. Random mutagenesis was carried out by the method of Myers et al. [19], where the single-stranded DNA of the hybrid plasmid was first treated with sodium nitrite at room temperature for 1 h before converting it to the duplex form. The duplex DNA was digested with *Bst*EII and *Pst*I, and the resulting 2.5-kb fragment was ligated to pOD64 that had been digested with *Bst*EII and *Pst*I to remove its wild-type catalase gene. The *E. coli* UM228 was then transformed with the hybrid plasmids in which the 5′-upstream region and most of the coding region of the catalase gene were subjected to the random mutagenesis.

Colony assay of heat-resistant catalatic and peroxidatic activities

For catalatic activity, the filter paper containing the colonies was placed into an empty plate where it was moistened with 0.1 M potassium phosphate (pH 7.0). The plate was then heated at 70°C for 10 min to reduce the background activity. The catalatic activity was tested by applying a drop of 30% H_2O_2 on each colony, as described by Loewen [20]. For peroxidatic activity, the colony assay method was as described by Loprasert et al. [21] with slight modification. The cells transferred to filter paper were heat treated as described above after the lysozyme treatment [21] and the peroxidatic activity was tested in 0.1 M potassium phosphate (pH 7.0) instead of 20 mM sodium/potassium phosphate (pH 7.0).

Enzyme characterization

For enzyme characterization, cell lysates of the *E. coli* before and after heat treatment were prepared. *E. coli* UM228 cells harboring plasmid containing the wild-type or mutant gene of catalase I were grown on a Luria-Bertani medium [22] containing 50 μg/ml ampicillin. The cell lysate before heat treatment was obtained from the resulting supernatant after centrifugation of the sonicated cells in 0.1 M potassium phosphate (pH 7.0). The lysate was heated at 70°C for 10 min and centrifuged, the supernatant was then used as the heat-treated lysate.

These lysates were analyzed by SDS-PAGE [23], and the concentration of the catalase protein in the lysates was determined through the intensity of the protein bands as visualized by Coomassie brilliant blue (details of the procedure will be published elsewhere). Heat stability of catalase was estimated from the ratio of the catalase protein concentration measured before and after the heat treatment described above. This is based on the fact that heat denaturation causes insolubilization and precipitation of this enzyme. Therefore, a lower value of the ratio corresponds to a lower heat stability.

Catalatic activity and peroxidatic activity of the lysate after heat treatment were measured as described previously [13]. Both values were divided by the protein concentration obtained above, and the specific activities were expressed as the number of catalytic cycle per second per catalase protein.

Results and Discussion

Random mutant population

The gene encoding *B. stearothermophilus* catalase I [13] was subjected to random mutation with sodium nitrite [19] and the mutated gene was introduced into *E. coli* UM228 cells [17], after which 2,648 transformants were obtained. The catalatic and peroxidatic activities of all the heat-treated transformants were examined by colony assay. The transformants were classified into S group with both activities at the same level as the wild type, M group where both or either one of the activities is weaker than the wild type, and B group where neither activities are detectable. The number of transformants which belonged to the S, M and B groups were 1,323, 191, and 1,134, respectively.

The productivity of mutant catalase proteins of the transformants as analyzed by

SDS-PAGE were of different levels or even less than the detectable level for a substantial number of the mutants. Of the transformants so far examined, all the arbitrarily chosen 20 S-group, 55 M-group, and none of the arbitrarily chosen 328 B-group transformants produced a detectable amount of the enzyme protein in their heat-treated lysates. Due to the possibility that some transformants in the S group may produce the wild-type enzyme, it is the 55 M-group transformants that was used as the source of the random mutant sample of catalase I for observing the landscape. To check the degree of gene damage caused by the random mutagenesis, seven mutant genes out of the 55 M-group mutants were sequenced. A total of 39 mutations were found on the nucleotide sequence of the 2.5-kb *Bst*EII-*Pst*I DNA fragment containing the catalase gene.

Landscapes around catalase I

The properties as to the heat stability (S) and to the two specific activities (CA and PA, corresponding the catalatic and peroxidatic activities, respectively) of the 55 mutant catalases were measured as described under Materials and Methods. From these results, the altitude of the position of the present enzyme in the three kinds of landscapes drawn for the properties were estimated using an index T for a property of an enzyme defined by the following equation:

$$T = \frac{(x - \mu)}{\sigma} \tag{1}$$

where x is the observed value of the property of the enzyme under study, μ is the mean value and σ is the standard deviation of the values of the property of the mutant population obtained after random mutagenesis. Namely, T serves as an index showing the extent of deviation of the property of the enzyme from the mean value of the mutant population. As the mutants are random samples taken from a local sequence space around the wild-type enzyme, T also serves as an index showing the altitude of an enzyme measured from the level of the mean value in the local landscape. In addition, as T value is a normalized altitude, T can provide the convenience for direct comparison of the different properties of an enzyme as well as of different enzymes. The normalization done in Eqn. 1 means that the ruler of the altitude is defined for each local landscape searched by random mutagenesis. Hence, T shows a relative altitude in each local landscape and can be compared in this meaning.

For the comparison between landscapes of different properties, the values of the properties must be of the same dimension. Therefore, before calculating the T values of catalase I for the three properties, the values of S, CA, and PA were converted into the forms of $-\ln(-\ln(S))$, $\ln(CA)$, and $\ln(PA)$, respectively. As $-\ln(S)$, CA, and PA correspond to the rate constants, these properties can be compared at the free energy level by the above conversion. Table 1 shows the estimated μ and σ values of each

Table 1. Altitude of catalase I in the landscapes of stability and activity.

	Heat stability $-\ln(-\ln(S))$	Catalactic activity $\ln(CA)$	Peroxidatic activity $\ln(PA)$)
Wild-type enzyme (x_o)	0.760	6.77	0.263
Mean value (μ)	–0.534	3.92	–1.42
Standard deviation (σ)	0.749	2.31	1.48
Altitutde of wild type (T_o)	1.73	1.24	1.14

The heat stability (S) and two specific activities (CA and PA) of the wild-type and the 55 mutant catalases were measured as described under Materials and Methods; μ was estimated to be the same as the average ($\bar{x}$) of the 55 mutants, and σ was estimated by the following calculation:

$$\left\{\frac{\sum_{i=1}^{55}(x_i-\bar{x})^2}{54}\right\}^{\frac{1}{2}},$$

where x_i is the observed value of each property of the *i*th mutant. Altitude of the wild-type enzyme (T_o) was calculated by Eqn. 1.

property of the mutant population together with the T values of the wild-type enzyme as calculated using Eqn. 1. These results show that the order of the altitude of catalase I in the landscapes for the three properties is stability > catalatic activity > peroxidatic activity.

The T values of the three properties of all the 55 M-group mutants were calculated using Eqn. 1, and are arranged in the order of the highest to the lowest T to form a cumulative frequency curve (Fig. 1). The cumulative frequency (upper abscissa) corresponds to the relative area to the whole space area searched by the random mutagenesis (lower abscissa), because each mutant enzyme or each point of the sequence space occupies the same area. On the other hand, the ordinate shows the T value corresponding to the altitude of the property. In this sense, these curves show a rearranged landscape of the sequence space around the wild-type enzyme.

The rearranged landscape shows that for these three properties, the wild-type enzyme is located near the top of the hill, indicating that the present enzyme is fairly optimized for these properties. However, it should be noted that there are some points at which altitudes are higher than that of the present enzyme.

Figure 2 shows the relationship between two properties using the T values. In Fig. 2, a tendency is observed that mutants with higher catalatic activity have higher peroxidatic activity and mutants with lower catalatic activity have lower peroxidatic activity. As a mutant with high (or low) activity is located at a hill (or valley) region

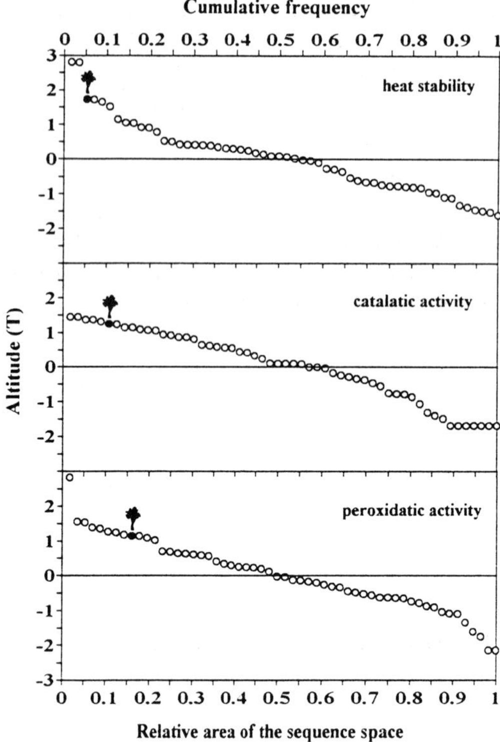

Fig. 1. Rearranged landscape drawn by cumulative frequency curve of T. T values for the 55 M-group mutants were calculated from Eqn. 1 using the μ and σ values given in Table 1, and arranged in the order of the highest to the lowest T to form a cumulative frequency curve of T. The cumulative frequency on the upper abscissa corresponds to the relative area of the sequence space on the lower abscissa. For the catalatic activity, CA values of less than 0.1 s^{-1} were regarded as 0.1 s^{-1}. The position of the wild-type enzyme is shown by the filled circle with the tree mark.

in the activity landscape, the tendency observed in Fig. 2 means that the location of the hills and valleys in the landscapes of the catalatic activity roughly corresponds with that of the peroxidatic activity. This may be due to the fact that the catalatic and the peroxidatic reactions are catalyzed at the same active site and share a common intermediate of compound I [24,25]. Although the positive correlation between the two activities means that most sequences in the local space have similar reaction specificity, the presence of the mutant indicated by the open arrow shows that some sequences have different reaction specificity which are peroxidase rather than catalase. At the positions of these sequences, a hill region for the peroxidatic activity coincides with a valley region for the catalatic activity.

For the other two pairs of the properties, on the other hand, clear correlation was not observed (Fig. 2). This means that the landscape of the heat stability is quite different from those of the two activities.

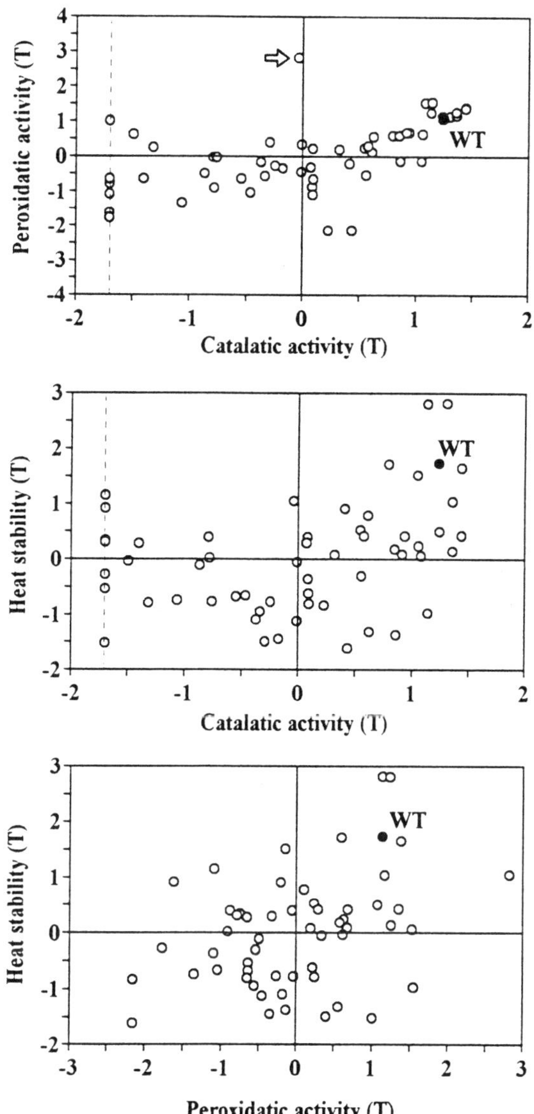

Fig. 2. Correlation between two properties. The open circles show the positions of the 55 M-group mutants having the coordinates expressed by the T values of the indicated properties. The mutant indicated by the open arrow has a different reaction specificity. The filled circle shows the position of the wild-type enzyme. The value of the lower limit for the detection of the catalatic activity is shown by the broken line.

Application of landscapes for artificial evolution

To know the landscapes in the sequence space around an enzyme is to understand the characteristics of the proteins in the searched local sequence space, and to understand the situation of the enzyme among the neighbors in the space. The landscapes obtained for catalase I show that there are many variant sequences having similar properties to those of catalase I. This means that there are many possible pathways of evolutional walks by mutation and selection, which will lead to a divergent population of an enzyme family. Similarly, this wide allowance in designing the sequence of an enzyme enables us to improve enzymes and macromolecules by artificial evolution through the process of random mutation and selection [6–12].

For this new type of protein engineering by artificial evolution, the landscape in the sequence space provides a guide map for directed walking, just as the three-dimensional structure of a protein provides a guide for the site-directed mutagenesis in the classic protein engineering. For example, the ease or difficulty in obtaining a mutant enzyme having a desired property can be estimated from the rearranged landscape as shown in Fig. 1. According to the results on catalase I, the position of the wild-type enzyme is nearer to the top of the stability hill than to the peroxidatic activity hill. Therefore, to increase the peroxidatic activity is easier than further stabilization. Thus, we can estimate the possibility of improvement of the present enzyme. In addition, we can walk further toward a more improved point by the second random mutagenesis starting from the best point in the landscape observed by the first mutagenesis. These successive walks correspond to the adaptive walks in evolution and sooner or later, will bring us to the most improved point practically attainable in the sequence space; then, we can decide to stop further walking. This strategy is generally applicable to any properties of any enzymes. Therefore, the adaptive walks with a map of the landscape is very powerful method of protein engineering.

References

1. Kauffman SA. Autocatalytic sets of proteins. J Theor Biol 1986;119:1–24.
2. Kauffman SA. Adaptation on rugged fitness landscapes. 618. In: Stein DL (ed) Studies in the Sciences of Complexity, Vol. 1. Redwood City: Addison-Wesley Publishing Company, 1989;527–618.
3. Smith JM. Natural selection and the concept of a protein space. Nature 1970;225:563–564.
4. Kauffman S, Levin S. Towards a general theory of adaptive walks on rugged landscapes. J Theor Biol 1987;128:11–45.
5. Macken CA, Perelson AS. Protein evolution on rugged landscapes. Proc Natl Acad Sci USA 1989;86:6191–6195.
6. Makino Y, Negoro S, Urabe I, Okada H. Stability-increasing mutant of glucose dehydrogenase from *Bacillus megaterium* IWG3. J Biol Chem 1989;264:6191–6195.
7. Arase A, Yomo T, Urabe I, Hata Y, Katsube Y, Okada H. Stabilization of xylanase by random mutagenesis. FEBS Lett 1993;316:123–127.
8. Ellington AD, Szostak JW. In vitro selection of RNA molecules that bind specific ligands. Nature 1990;346:818–822.
9. Tuerk C, Gold L. Systematic evolution of ligands by exponential enrichment: RNA ligands to bacteriophage T4 DNA polymerase. Science 1990;249:505–510.
10. Bock LC, Griffin LC, Latham JA, Vermaas EH, Toole JJ. Selection of single-stranded DNA

molecules that bind and inhibit human thrombin. Nature 1992;355:564–566.
11. Roberts BL, Markland W, Ley AC, Kent RB, White DW, Guterman SK, Ladner RC. Directed evolution of a protein: selection of potent neutrophil elastase inhibitors displayed on M13 fusion phage. Proc Natl Acad Sci USA 1992;89:2429–2433.
12. Beaudry AA, Joyce GF. Directed evolution of an RNA enzyme. Science 1992;257:635–641.
13. Trakulnaleamsai S, Aihara S, Miyai K, Suga Y, Sota M, Yomo T, Urabe I. Revised sequence and activity of *Bacillus stearothermophilus* catalase I (formerly peroxidase). J Ferment Bioeng 1992;74:234–237.
14. Claiborne A, Fridovich I. Purification of o-dianisidine peroxidase from *Escherichia coli* B. J Biol Chem 1979;254:4245–4252.
15. Loewen PC, Stauffer GV. Nucleotide sequence of katG of *Salmonella typhimurium* LT2 and characterization of its product, hydroperoxidase I. Molec Gen Genet 1990;224:147–151.
16. Welinder KC. Bacterial catalase-peroxidases are gene duplicated members of the plant peroxidase superfamily. Biochem Biophys Acta 1991;1080:215–220.
17. Triggs-Rain BL, Loewen PC. Physical characterization of *katG*-encoding catalase HPI of *Escherichia coli*. Gene 1987;52:121–128.
18. Loprasert S, Urabe I, Okada H. Overproduction and single-step purification of *Bacillus stearothermophilus* peroxidase in *Escherichia coli*. Appl Microbiol Biotechnol 1990;32:690–692.
19. Myers RM, Lerman LS, Maniatis T. A general method for saturation mutagenesis of cloned DNA fragments. Science 1985;229:242–247.
20. Loewen PC. Isolation of catalase-deficient *Escherichia coli* mutants and genetic mapping of *katE*, a locus that affects catalase activity. J Bacteriol 1984;157:622–626.
21. Loprasert S, Negoro S, Okada H. Cloning, nucleotide sequence, and expression in *Escherichia coli* of the *Bacillus stearothermophilus* peroxidase gene (*perA*). J Bacteriol 1989;171:4871–4875.
22. Maniatis T, Fritsch EF, Samgrook J. Molecular Cloning: A Laboratory Manual. New York: Cold Spring Harbor Laboratory, 1982.
23. Laemmli UK. Cleavage of structural proteins during the assembly of the head of bacteriophage T4. Nature 1970;227:680–685.
24. Schonbaum GR, Chance B. Catalase. In: Boyer PD (ed) The Enzymes, Vol. 13, 3rd Edn. New York: Academic Press, 1976;363–408.
25. Dawson JH. Probing structure-function relations in heme-containing oxygenases and peroxidases. Science 1988;240:433–439.

Tracing Biological Evolution in Protein and Gene Structures.
M. Gō and P. Schimmel, editors.

Correlation between module boundaries and intron positions in hemoglobins from various taxa

Kaoru Fukami-Kobayashi, Masateru Mizutani and Mitiko Gō
Department of Biology, Faculty of Science, Nagoya University, Nagoya, Japan

Abstract. *Background.* Recent reports on invertebrate hemoglobin genes revealed a variety of exon-intron structures. With accumulation of these data, one can examine correlations between module boundaries of hemoglobins and intron positions of their genes, based on wider evolutionary lineages than was feasible a decade ago.

Methods. Hemoglobins from nine different lineages were chosen for identification of modules by centripetal profile. Alignment of amino acid sequences was carried out for hemoglobins whose three-dimensional or gene structures are known. A phylogenetic tree of hemoglobins was constructed to trace the process of intron loss/gain.

Results. All hemoglobins were similarly decomposed into four modules. In alignment of hemoglobin amino acid sequences, a good correlation was observed between the intron positions and the module boundaries. A reconstructed process of intron loss/gain supports the ancient origin of hemoglobin introns.

Conclusions. The most probable explanation for the correlation is that the introns were located in the ancestral gene of hemoglobin, at its origin. This proposal is tenable if the ancestral hemoglobin derived from the combination of four modules, each encoded by one exon. This putative conclusion obtained from hemoglobin data can explain the existence of introns in the earliest stage of protein evolution, as mediators of module combination.

Key words: domain, intron-early, phylogeny, protein evolution.

Introduction

Correlation of DNA exonic regions with protein structural units was first proposed in 1981 [1]. Using mammalian hemoglobin β-chain, it was shown that conformations of segments encoded by exons are the least extended, and that one more cut can be introduced into the second segment dividing it into two least extended conformations. Soon after that, an intron corresponding to the position of the proposed cut was shown to exist in the leghemoglobin gene from soybean *Glycine max* [2]. This gene has three introns [3], two of which are located in homologous sites with the intron positions of vertebrate hemoglobins, and the remaining locates close to the predicted site. Such a structural unit as found in hemoglobin was designated "module" [4]. Modules were defined as compact substructures within globular domains and they consist of 10–40 amino acid residues [5].

Address for correspondence: Mitiko Gō, Department of Biology, Faculty of Science, Nagoya University, Furo-cho, Chikusa-ku, Nagoya 464-01, Japan. Tel.: +81-52-789-2976. Fax: +81-52-789-2977.

The one-to-one correspondence between modules and exons in hemoglobin implies that the primordial hemoglobin gene originated by fusion of four exons, each of which encodes a module, through the intron-mediated exon shuffling. Thus, the origin of hemoglobin introns seems ancient; introns already existed in the hemoglobin gene with its origin. The correlation may therefore occur in all hemoglobins from different evolutionary lineages which share the same ancestral gene.

In a wide sense, hemoglobin is defined as a heme-protein that reversibly binds the oxygen molecule. Such proteins are found in organisms that do not have blood, insects and plants for instance, and even in single-celled organisms such as protozoa, yeast and bacteria [6]. On the basis of similarities of their three-dimensional structures and amino acid sequences, the oxygen-binding heme-proteins seem to have three roots (Fig. 1). The first group includes hemoglobins from yeast, plants and invertebrates as well as vertebrate hemoglobins and myoglobins. The group also includes prokaryotic hemoglobin which was first reported in 1986 [7]. The discovery of the prokaryotic hemoglobin indicates that the origin of the hemoglobins belonging to the first group is older than eukaryotes. In this sense, we call this group the universal-type hemoglobin. Crystal structures of the hemoglobins in this group show the same protein fold, which strongly supports the common evolutionary origin of members of this group. The hemoglobins of protozoa [8,9] and cyanobacteria [10] belong to the second group, the cyanobacteria-type hemoglobin. *Paramecium* hemoglobin was discovered by Satoh and Tamiya in 1937 [11]. Hemoglobins in this group were also found in *Chlamydomonus* [12]. The amino acid sequences of this group are only distantly related to those of the first group, they show low sequence identities to the universal-type hemoglobins and the sequence length is about 20% (~30 amino acid residues) shorter than this type. Thus, one cannot relate these hemoglobins to the universal-type hemoglobins on the basis of their primary structures. The three-dimensional structures of the cyanobacteria-type hemoglobins will have to be determined to clarify evolutionary relationships with other types of hemoglobin. A crystal structure of the second group has not been documented. The third group, the IDO-type hemoglobin, consists of hemoglobin from *Sulculus*, abalone. The amino acid sequence is twice as large as that of the universal-type

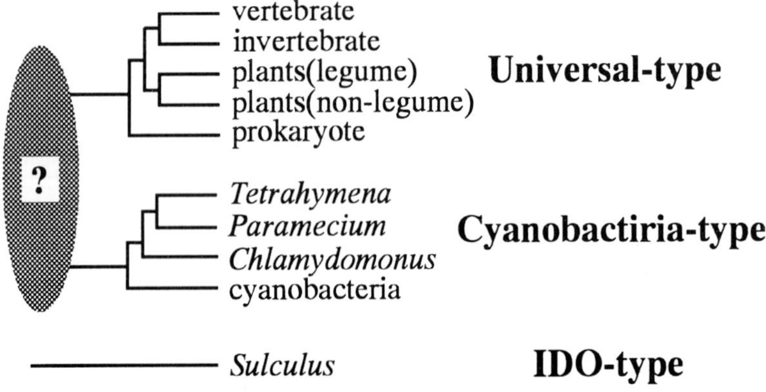

Fig. 1. Different roots of hemoglobins.

hemoglobins and has a 35% identity to that of indoleamine 2,3-dioxygenase (IDO) [13]. The evolutionary origin of the IDO-type hemoglobin may be unrelated with that of the universal-type hemoglobins. Therefore, at least for the present, the three groups will be regarded as having three different origins. The correlation between three-dimensional and gene structures of hemoglobins can be assessed only for the first group, vertebrate hemoglobins and their homologs. In the present report, "hemoglobin" will be taken to mean the universal-type hemoglobins.

Hemoglobins are sometimes found as a domain in a form fused with other functional domain(s). Flavohemoglobins from *E.coli* [14] and yeast [15] consist of two domains that have heme and flavin prosthetic groups, respectively. Amino acid sequences of the heme-binding domains of flavohemoglobins have low identity with those of the universal-type hemoglobins. However, the heme-binding domains are considered to be homologous with the universal-type hemoglobins because they have residues similar to those which are conserved among the universal-type hemoglobins at the important sites for function or conformational stability of hemoglobin. The existence of hemoglobin domains in enzymes suggests that these domains function not only as oxygen transporter protein but also as a molecular switch to control the function of the second domain associated with oxygen binding to the hemoglobin domain. Some hemoglobins have multiple heme-binding domains, for example, hemoglobins from *Barbatia* [16] and *Ascaris* [17] have two domains, each of which structurally corresponds to a subunit of vertebrate hemoglobin. Hemoglobin from brine shrimp consists of nine domains [18]. Thus, there is a definite evolutionary variety in the structure and function of hemoglobin.

To date, three-dimensional structures of hemoglobins from more than a dozen species have been determined and their atomic coordinates are kept at the Brookhaven Protein Data Bank [19]. Gene structures of hemoglobins in many taxa have also been characterized. Genes of invertebrate hemoglobins show exon-intron structures slightly different from those of vertebrate or legume hemoglobin genes [20]. The accumulation of data allows for examination of correlations between the modules of hemoglobins and the intron positions of their genes, based on wider evolutionary lineages.

Material and Methods

Hemoglobins

Up to the present, the three-dimensional structure of hemoglobin has been determined in five taxa, and intron positions have been identified in six, including the five taxa. Table 1 shows hemoglobin three-dimensional and gene structures used for the present study. This situation makes hemoglobin one of the most suitable protein families for investigating correlations between protein three-dimensional structure and exon-intron structure of its gene. We examined the correlation in both time scales extended over the six taxa and confined within each of the five taxa.

Identification of modules

Modules were originally determined using a distance map [1]. In the present analysis, however, identification of module boundaries was carried out using centripetal profiles as well as extension profiles [5]. By locating local minima of centripetal profiles, the centers of contiguous segments of the hemoglobins could be identified. Lengths of the segments used in this detection were 61, 71, 81, 91, 101, 111 and 121 residues. The residues for the minima of the centripetal profiles are candidates for module boundaries and they were screened with extension profiles. The identified module boundaries are shown in distance maps where the pairs of amino acid residues whose α-carbons are separated by >30 Å are shaded.

Alignment

To examine the correlation between intron positions and module boundaries, it is ideal to simultaneously be aware of intron positions of a gene and the three-dimensional structure of its product. Human hemoglobin α-chain [21,22], β-chain [22,23] and seal myoglobin [24,25] are such a case, but usually, one has either regarding each hemoglobin. Alignment of amino acid sequences was thus carried out among the hemoglobins whose three-dimensional structures and/or the intron positions in their genes are known. For the present purpose, the alignment must be

Table 1. Hemoglobins for the study of module-intron correlation.

Group	Taxon	3-D structure (species/protein)	PDB code	Refs.	Gene structure (species/protein)	Refs.
A	Vertebrates	human α	1THB (subunit A)	[22]	human α	[21]
		human β	1THB (subunit B)	[22]	human β	[23]
		sea lamprey	2LHB	[33]	seal Mb	[24]
		seal Mb	1MBS	[25]		
B	*Arthropoda*	*Chironomus* III	1ECA	[34]	*Chonomus* IIb	[35]
					brine shrimp T4	[32]
C	*Annelida*	*Urechis caupo*	1ITH (subunit A)	[36]	earthworm	[37]
D	*Nematoda*				*C.elegans*	[31]
					codworm	[38]
					Ascaris	[17]
E	*Mollusca*	*Scapharca*	2SDH (subunit A)	[39]	*Anadara*	[40]
		sea hare Mb	1MBA	[41]	*Barbitia*	[16]
F	Plants	yellow lupine	2LH1	[42]	swamp oak	[43]
					soybean	[3]

accurate, with an error of less than a few residues. In the present case of hemoglobins, we obtained an accurate alignment within each taxon. The amino acid sequences of the hemoglobins from species which belong to different taxa are distantly related, so we aligned them while directing attention to positions of some conservative residues and to locations of α-helices, to facilitate an accurate alignment.

Phylogenetic tree

To trace the process of intron loss/gain, a phylogenetic tree of hemoglobins was constructed. Gene structures of *Vitreoscilla* hemoglobin [7], hemoglobin domain in yeast flavohemoglobin [15] and hemoglobin domain in *Escherichia coli* (*E.coli*) dihydropteridine reductase [14], and *Chironomus* hemoglobin IV [26] were used, in addition to sequences of hemoglobins whose intron positions in their genes had been documented. The evolutionary distance between two amino acid sequences was computed taking into account the evolutionary similarity [27]. Data on the accepted point mutations and frequencies for the 20 amino acid residues [28] were used for computation. Using the distance matrix, a phylogenetic tree was constructed by the neighbor-joining method [29].

Results

Figure 2 shows the identified module boundaries in distance maps of six hemoglobins from the five different taxa. In all the hemoglobins, deep and stable minima were detected in the centripetal profiles at spatially equivalent residues that were determined as module boundaries with a distance map of horse hemoglobin β-chain. The three minima were identified as module boundaries. In the distance maps, a pair of horizontal and vertical lines from the diagonal point and corresponding to the module boundaries scarcely crossed the dark islands. Other than the module boundaries, there were some less stable minima on the profiles, because the method involving use of centripetal profiles is more sensitive for detecting the local centers of a protein. Although the minimum at residue 93 in numbering of leghemoglobin is not so deep and stable, and although the minimum at residue 139 is too close to the C-terminus to yield a compact C-terminal module, they could be additional module boundaries. For the present, however, only the three minima were used as module boundaries; all hemoglobins were similarly decomposed into at least four modules.

Since identities of amino acid sequences are high enough within each taxon, we could make an accurate alignment of each taxon. Both of the three-dimensional and gene structures of hemoglobins were reported in all taxa, except taxon D. We thus found a correlation between module boundaries and intron positions in the five taxa, while excluding almost completely any uncertainty of alignment. In each taxon, introns were localized within six residues from module boundaries.

The module boundaries were localized to three limited regions of a few residues each in the alignment of the hemoglobins from different taxa. All introns were located close to any of the three regions. Figure 3 summarizes results of the

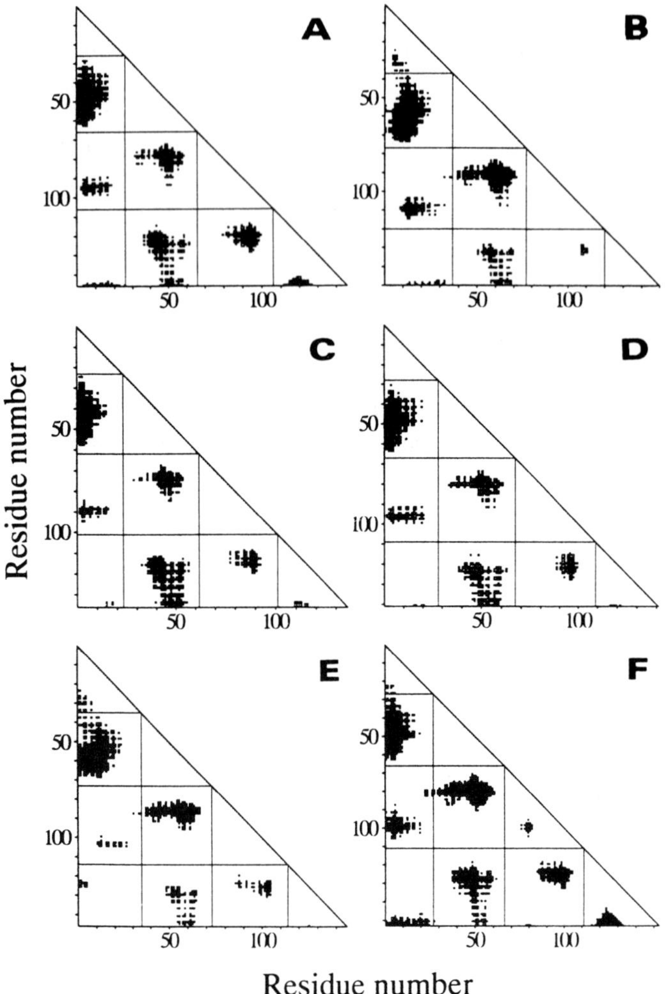

Fig. 2. Distance maps of (A) human hemoglobin β-chain, (B) sea lamprey hemoglobin, (C) *Chironomus* III hemoglobin, (D) *Urechis caupo* hemoglobin, (E) *Scapharca* hemoglobin and (F) yellow lupine leghemoglobin. For the codes in the Protein Data Bank at Brookhaven National Laboratory [19], see Table 1. The dark regions indicate the pairs of amino acid residues whose α-carbons are separated by >30 Å. A module boundary is shown by a pair of horizontal and vertical lines in the maps.

alignment. Introns fall into seven positions, which are localized to three regions. They are all within six residues from module boundaries, thereby indicating a nonrandom distribution and a close correlation with module boundaries.

Such a situation could be presumed if the hemoglobin gene was formed by combining four exons which encode one module each, through the intermediation of introns. Some introns might then be lost during and after the joining process. The intron-loss process could be traced in the contemporary exon-intron structure of hemoglobins. To gain support for this notion, a process of intron loss/gain was reconstructed for gene structures of hemoglobins, on the basis of their phylogenetic relationship.

The tree in Fig. 4 shows the evolutionary relationship of hemoglobins or hemoglobin domains whose gene structures are known. Gene structures of hemoglobins from various taxa are illustrated with boxes on the right of the tree. Starting from a gene with three introns at all three module boundaries, a process of intron loss/gain was traced. Surprisingly, one of the most parsimonious processes, which is shown in Fig. 4, did not include any intron gain.

Discussion

Introns were localized at seven sites around module boundaries in the alignment of

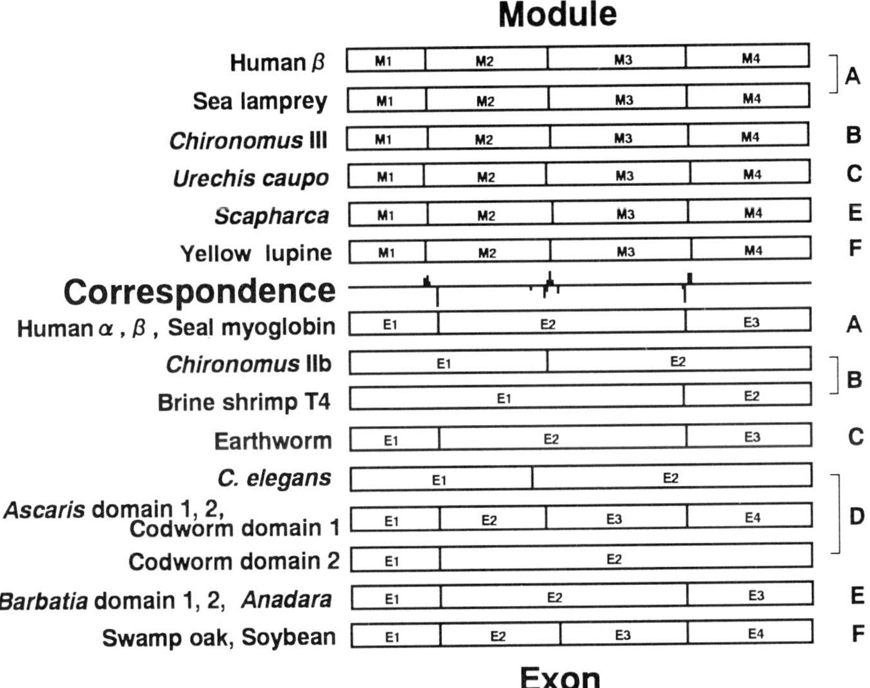

Fig. 3. A schematic diagram of module boundaries and the intron positions. The hemoglobin sequences from six taxa were aligned; (A) vertebrates, (B) *Arthropoda*, (C) *Annelida*, (D) *Nematoda*, (E) *Mollusca* and (F) higher plants, for which the three-dimensional structures or intron positions in their genes were determined. The intron positions were taken from GenBank/EMBL/DDBJ database accession nos. V00488 for human α-chain [21], V00499 for human β-chain [23], V00471, V00472, and V00473 for seal myoglobin [24], L03409 for *Chironomus* IIb [35], J05161 for earthworm [37], Z11681 for codworm [38], L20699 for *Anadara* [40], X53950 for swamp oak [43], and V00452, and J01300 for soybean [3], and were taken from literature for brine shrimp T4 [32], *C. elegans* [31], *Ascaris* [17], and *Barbatia* [16].

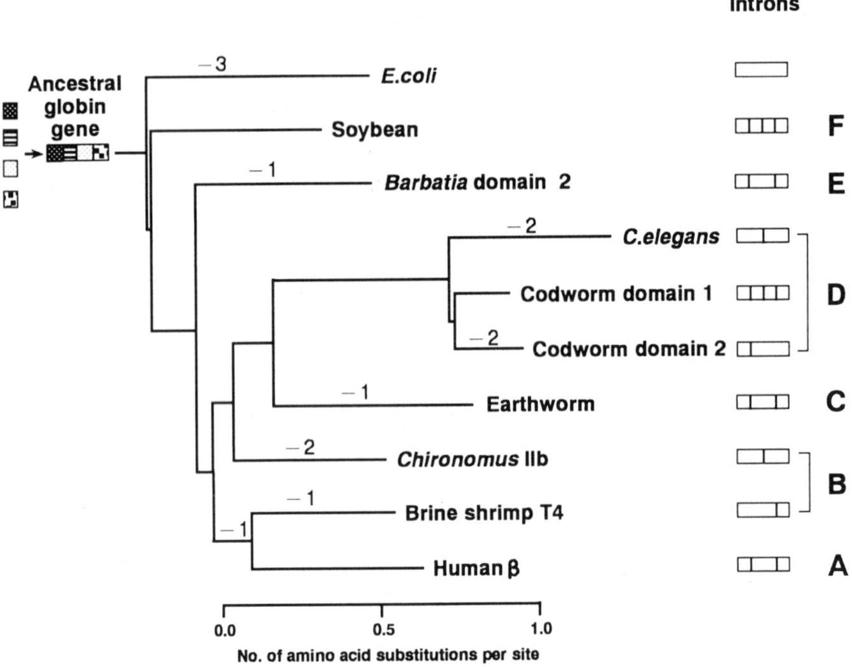

Fig. 4. Presumed intron loss in hemoglobin genes along the branches of a phylogenetic tree. For the six taxa A through F see Table 1.

hemoglobins. There are at least four possible explanations for the observation (Fig. 5). The former three postulate that introns were inserted into the hemoglobin gene after the ancestral intron-free gene was formed. The last explanation is based on the assumption that introns existed in the ancestral hemoglobin gene. The four are: 1) introns randomly inserted were localized to the specific three regions by chance; 2) a specific nucleotide sequence exists in the regions of the hemoglobin gene corresponding to the vicinity of module boundaries, and introns were inserted into the regions using the sequences as a target; 3) some selection functioned to the randomly inserted introns, and only introns in the regions corresponding to module boundaries survived; and 4) introns in the hemoglobin gene are remnants of the exon-joining process in the primordial evolution of hemoglobin.

If we assume that introns were inserted randomly, and that the localization occurred by chance (explanation one), the probability is extremely low. In the case of hemoglobin, the intron that is most distant from a module boundary is that of *C. elegans*: it is six residues away from the nearest module boundary in the hemoglobin alignment. The probability of finding seven introns within six residues from module boundaries as a result of random insertion is estimated as follows: Let hemoglobin length be 150 residues. Sum of the number of residues within six residues from any one of the three module boundaries is 39. Probability that all seven introns are inserted independently into one of the 39 residues out of 150 is $(39/150)^7 \fallingdotseq 8.0 \times$

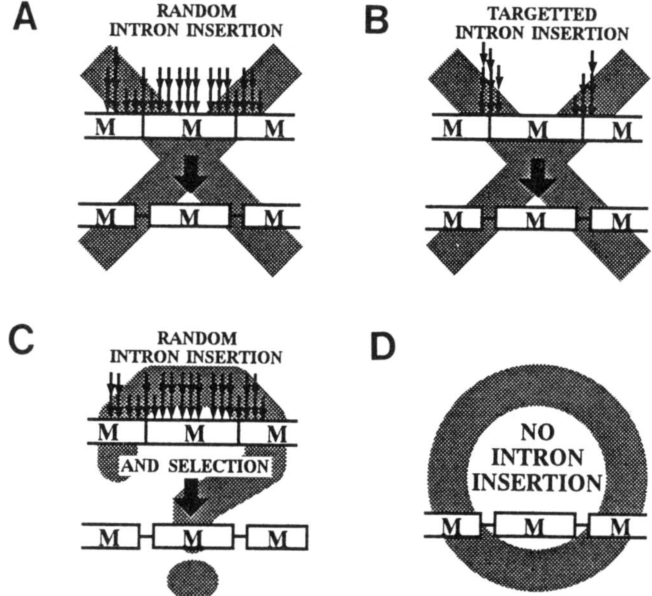

Fig. 5. Four possible explanations for the localization of introns around module boundaries.

10^{-5}. This would only occur in about one case out of 10,000.

Some deviation in frequencies of four bases is observed around the exon-intron boundaries [30]. The deviation shows that an intron is most frequently sandwiched between the sequences [CA] AG and [AG] ([CA] means C or A). The sequence [CA] AG [AG], called proto-splice site, is not enough to make up a target sequence for intron insertion. The sequence of the proto-splice site, whose length is only 4 nt, can be found everywhere in the genome, and is no more frequent in regions corresponding to the vicinity of module boundaries. For example, the frequency of appearance of proto-splice site in coding regions of 12 hemoglobins (those shown in Table 1 except brine shrimp hemoglobin) was 0.025, that at the exon-intron junctions was 0.135, and that in regions within six residues from the homologous residues corresponding to the module boundaries of human hemoglobin β-chain in each hemoglobin was 0.022. Even if introns are preferably inserted into the sequences, the probability of finding seven introns within six residues from module boundaries would hardly be increased. Therefore explanation two is not plausible.

It may be possible to invoke some selection pressure in favor of introns inserted into positions corresponding to module boundaries (explanation three). Exons that encode a peptide with a module structure would have a great advantage to survive in a process of exon rearrangement to make up a protein. Because the conformation of modules is essentially determined within self (K. Takahashi et al., submitted), a combination of modules has a high potential to produce new functional proteins without changing conformation. During formation of a protein, genomes with introns on positions corresponding to module boundaries in their hemoglobin gene would have an advantage over those whose hemoglobin introns were not located at positions corresponding to module boundaries, because exons that encode the hemoglobin module structure could serve for the process.

If introns were inserted randomly into various sites in the ancestral hemoglobin gene, the cost of selection in favor of introns inserted into positions corresponding to module boundaries would be too large for a genome to survive. As discussed for explanation one, the probability that randomly inserted introns fall into the limited regions around module boundaries by chance is extremely low. The selected introns would have derived from an enormous number. As discussed for explanation two, an efficient target sequence for localization of intron insertion was not observed, thus explanation three is not realistic. The most reasonable explanation for the correlation between module boundaries and intron positions is that the introns were present in the ancestral hemoglobin gene (explanation four).

In addition, the reconstructed process of intron loss/gain can explain the present gene structure using only intron losses. It also supports the proposal that introns existed in the ancestral gene of hemoglobins. Since the origin of hemoglobin must have preceded divergence of prokaryotes and eukaryotes, the origin of hemoglobin introns is accordingly very ancient.

Deviation of positions of introns corresponding to the same module boundary (Fig. 3) can be explained by 1) uncertainty of the alignment, 2) intron sliding and 3) intron slippage. In this analysis, the possibility of 1) was excluded, because we compared closely related sequences within each taxon. The intron of *C. elegans* falls under the second exception: The position of this intron is somewhat distant from module boundaries between M2 and M3, and it is different even from those of other nematoda. Kloek et al. [31] showed evidence for intron sliding of this *C. elegans* intron from the original position still occupied by codworm and *Ascaris* introns. Possibility three is that intron slippage could occur on the intron of brine shrimp hemoglobin domain T4, the position of which is one base deviated from those of others [32].

There are more introns in the hemoglobin alignment that were not made the object of examination of correlation. These introns are found between hemoglobin domains in multiple heme-binding domain hemoglobins, or between signal peptide and hemoglobin domains. Because we knew the three-dimensional structure of only a single hemoglobin domain, we searched for the module-intron correlation within hemoglobin domains. Introns on the boundaries of hemoglobin domains probably have more recent origins; they might have occurred when hemoglobin domains duplicated, joined with other kinds of domains, or acquired signal sequences. On the other hand, introns within hemoglobin domains likely originated when the hemoglobin domain itself emerged.

Acknowledgements

We thank M. Ohara for assistance with the manuscript. A part of this work was supported by Grants-in-Aid for General Scientific Research (M.G.) and for Encouragement of Young Scientists (K.F.) from the Ministry of Education, Science and Culture, Japan and by the Ogasawara Foundation.

References

1. Gō M. Correlation of DNA exonic regions with protein structural units in haemoglobin. Nature 1981;291:90–92.
2. Blake CCF. Exons and the structure, function and evolution of haemoglobin. Nature 1981;291:616.
3. Hyldig-Nielsen JJ, Jensen EØ, Paludan K, Wiborg O, Garret R, Jørgensen P, Marcker KA. The primary structures of two leghemoglobin genes from soybean. Nucl Acid Res 1982;10:689–701.
4. Gō M. Modular structural units, exons, and function in chicken lysozyme. Proc Natl Acad Sci 1983;80:1964–1968.
5. Gō M, Nosaka M. Protein architecture and the origin of introns. Cold Spring Harbor Symp Quant Biol 1987;52:915–924.
6. Takagi T. Hemoglobins from single-celled organisms. Curr Opin Struct Biol 1993;3:413–418.
7. Wakabayashi S, Matsubara H, Webster DA. Primary sequence of a dimeric bacterial haemoglobin from *Vitreoscilla*. Nature 1986;322:481–483.
8. Iwaasa H, Takagi T, Shikama K. Protozoan myoglobin from *Paramecium caudatum*. Its unusual amino acid sequence. J Molec Biol 1989;208:355–358.
9. Iwaasa H, Takagi T, Shikama K. Protozoan hemoglobin from *Tetrahymena pyriformis*. Isolation, characterization, and amino acid sequence. J Biol Chem 1990;265:8603–8609.
10. Potts M, Angeloni SV, Ebel RE, Bassam D. Myoglobin in a cyanobacterium. Science 1992;256: 1690–1692.
11. Satoh T, Tamiya H. Über die Atmungsfarbstoffe von *Paramecium*. Cytologia 1937;(Fujii Jubilee volume):1133–1138.
12. Couture M, Chamberland H, St.-Pierre B, Lafontaine J, Guertin M. Nuclear genes encoding chloroplast hemoglobins in the unicellular green alga *Chlamydomonus eugametos*. Molec Gen Genet 1994;243:185–197.
13. Suzuki T, Takagi T. A myoglobin evolved from indoleamine 2,3-dioxygenase. J Molec Biol 1992;228:698–700.
14. Vasudevan SG, Armarego WLF, Shaw DC, Lilley PE, Dixon NE, Poole RK. Isolation and nucleotide sequence of the *hmp* gene that encodes a haemoglobin-like protein in *Escherichia coli* K-12. Molec Gen Genet 1991;226:49–58.
15. Zhu H, Riggs AF. Yeast flavohemoglobin is an ancient protein related to globins and a reductase family. Proc Natl Acad Sci 1992;89:5015–5019.
16. Naito Y, Riggs CK, Vandergon TL, Riggs AF. Origin of a "bridge" intron in the gene for a two-domain globin. Proc Natl Acad Sci 1991;88:6672–6676.
17. Sherman DR, Kloek AP, Krishnan BR, Guinn B, Goldberg DE. *Ascaris* hemoglobin gene: plant-like structure reflects the ancestral globin gene. Proc Natl Acad Sci 1992;89:11696–11700.
18. Manning AM, Trotman CNA, Tate WP. Evolution of a polymeric globin in the brine shrimp *Artemia*. Nature 1990;348:653–656.
19. Bernstein FC, Koetzle TF, Williams GJB, Meyer EF Jr, Brice MD, Rodgers JR, Kennard O, Shimanouchi T, Tasumi M. The protein data bank: A computer-based archival file for macromolecular structures. J Molec Biol 1977;112:535–542.
20. Moens L, Vanfleteren J, De Baere I, Jellie AM, Tate W, Trotman CNA. Unexpected intron location in non-vertebrate globin genes. FEBS Lett 1992;312:105–109.
21. Liebhaber SA, Goossens MJ, Kan YW. Cloning and complete nucleotide sequence of human 5'-α-globin gene. Proc Natl Acad Sci 1980;77:7054–7058.
22. Waller DA, Liddington RC. Refinement of a partially oxygenated T state human haemoglobin at 1.5

Å resolution. Acta Cryst 1990;B46:409–418.
23. Lawn RM, Efstratiadis A, O'Connell C, Maniatis T. The nucleotide sequence of the human β-globin gene. Cell 1980;21:647–651.
24. Blanchetot A, Wilson V, Wood D, Jeffreys AJ. The seal myoglobin gene: an unusually long globin gene. Nature 1983;301:732–734.
25. Scouloudi H, Baker EN. X-ray crystallographic studies of seal myoglobin. The molecule at 2.5 Å resolution. J Molec Biol 1978;126:637–660.
26. Antoine M, Niessing J. Intron-less globin genes in the insect *Chironomus thummi thummi*. Nature 1984;310:795–798.
27. Fukami-Kobayashi K. Estimation of evolutionary distance between distantly related sequences of amino acids, taking account of patterns of amino acid replacement. Molec Biol Evol 1994;11:99–105.
28. Jones DT, Taylor WR, Thornton JM. The rapid generation of mutation data matrices from protein sequences. Comp Appl Biosci 1992;8:275–282.
29. Saitou N, Nei M. The neighbor-joining method: A new method for reconstructing phylogenetic trees. Molec Biol Evol 1987;4:406–425.
30. Dibb NJ, Newman AJ. Evidence that introns arose at proto-splice sites. EMBO J 1989;8:2015–2021.
31. Kloek AP, Sherman DR, Goldberg DE. Novel gene structure and evolutionary context of *Caenorhabditis elegans* globin. Gene 1993;129:215–221.
32. Moens L, Vanfleteren J, De Baere I, Jellie AM, Tate W, Trotman CNA. Commentary response. FEBS Lett 1993;320:284–287.
33. Honzatko RB, Hendrickson WA, Love WE. Refinement of a molecular model for lamprey hemoglobin from *Petromyzon marinus*. J Molec Biol 1985;184:147–164.
34. Steigemann W, Weber E. Structure of erythrocruorin in different ligand states refined at 1.4 Å resolution. J Molec Biol 1979;127:309–338.
35. Kao W-Y, Trewitt PM, Bergtrom G. Intron-containing globin genes in the insect *Chironomus thummi*. J Molec Evol 1994;38:241–249.
36. Kolatkar PR, Ernst SR, Hachert ML, Ogata CM, Hendrickson WA, Merritt EA, Phizackerley RP. Structure determination and refinement of homotetrameric hemoglobin from *Urechis caupo* at 2.5 Å resolution. Acta Cryst 1992;B48:191–199.
37. Jhiang SM, Riggs AF. The structure of the gene encoding chain *c* of the hemoglobin of the earthworm, *Lumbricus terrestris*. J Biol Chem 1989;264:19003–19008.
38. Dixon B, Walker B, Kimmins W, Pohajdak B. A nematode hemoglobin gene contains an intron previously thought to be unique to plants. J Molec Evol 1992;35:131–136.
39. Royer WE Jr, Hendrickson WA, Chiancone E. The 2.4-Å crystal structure of *Scapharca* dimeric hemoglobin. Cooperativity based on directly communicating hemes at a novel subunit interface. J Biol Chem 1989;264:21052–21061.
40. Titchen DA, Glenn WK, Nassif N, Thompson AR, Thompson EOP. A minor globin gene of the bivalve mollusc *Anadara trapezia*. Biochim Biophys Acta 1991;1089:61–67.
41. Bolognesi M, Onesti S, Gatti G, Coda A, Ascenzi P, Brunori M. *Aplysia limacina* myoglobin. Crystallographic analysis at 1.6 Å resolution. J Molec Biol 1989;205:529–544.
42. Arutyunyan ÉG, Kuranova IP, Vainshtein BK, Steigemann W. X-ray structural investigation of leghemoglobin. VI. Structure of acetate-ferrileghemoglobin at a resolution of 2.0 Å. Sov Phys Crystallogr 1980;25:43–58.
43. Christensen T, Dennis ES, Peacock JW, Landsmann J, Marcker KA. Hemoglobin genes in non-legumes: cloning and characterization of a *Casuarina glauca* hemoglobin gene. Plant Molec Biol 1991;16:339–344.

Tracing Biological Evolution in Protein and Gene Structures.
M. Gō and P. Schimmel, editors.

Module substitution in globins: preparation and association characteristics of chimeric hemoglobin subunits and myoglobin

Keisuke Wakasugi, Koichiro Ishimori and Isao Morishima
Division of Molecular Engineering, Graduate School of Engineering, Kyoto University, Kyoto, Japan

Abstract. The genes of the α- and β-subunits in native hemoglobin (Hb A) and of myoglobin (Mb) are made up of three exons interrupted by two introns. Gō demonstrated that there is a correlation between the subunit structure and exon pattern, and proposed a "module" structure as a compact protein structural unit, and showed that modules M1, M2 + M3, and M4 correspond to exon 1, 2, and 3, respectively, in the genes of α- and β-subunits [1]. The analysis of the correlation of hemoglobin function with its exon pattern showed that the residues associated with the defined function are concentrated in the specific exons encoding the "module". In this study, to investigate the functional and structural significance of the "modular structure" in Hb subunits and Mb, we prepared chimeric globins, in which the module M4 of the β-subunit or Mb was replaced by that of the α-subunit by use of mutagenesis. The spectroscopic data for the "chimera βα-subunit" have revealed the structure in the heme vicinity and ligand-binding properties were those of the β-subunit-like, and the association properties were those of the α-subunit-like [11]. The chimeric Mbα-subunit associates with the β-subunit, not with α or Mb (K. Wakasugi, K. Ishimori, I. Morishima, 1994;(to be submitted)). The results of our experiments indicate that the predominant role of the module M4 is the subunit association and suggests that the modules might correspond to structural and functional units to have the advantage of producing stable functional proteins.

Key words: chimeric globins, hemoglobin, module substitution, myoglobin, subunit association.

Introduction

The accumulation of information about protein and gene structures have suggested that exons correspond to discrete units of protein structure. Using a diagonal plot, Gō [1] found a correlation between the globin structure and the exon pattern and proposed that there are four "modules" which could perhaps be called compact units, where the exons divided off from one another with exception of the large central exon which compromises two compact units [2–5]. By the application of the same algorithm applied to the globins, the module structure were identified in chicken egg-white lysozyme [6] and many other proteins. Recently, however, Stoltzfus et al. [7] investigated the relationship between the intron positions and the distance of each C_α atom from the center of mass of the domain in which the atom resides and concluded that no significant correspondence between exons and units of protein structure was detected, suggesting that the putative correspondence does not exist and that the

Address for correspondence: Dr Isao Morishima, Division of Molecular Engineering, Graduate School of Engineering, Kyoto University, Kyoto 606, Japan. Tel.: +81-75-753-5921. Fax: +81-75-751-7611.

"module" theory is untenable.

This article is intended as a characterization of function and structure of "module"-substituted proteins that are constructed by a combination of modules from different kinds of proteins. It offers the key to an understanding of the roles of the "module" in protein function and structure. In the present study, we have chosen globins as a template protein since the globin structure is one of the first examples that gave the distinct modular structure [1].

The genes of the globins (myoglobin, α- and β-subunits in vertebrate hemoglobins) are made up of three exons, which correspond to three structural units, M1, M2 + M3, and M4. Although the identities of the amino acid sequences of the modules M1, M2 + M3, and M4 between the α- and β-subunits are only 46%, 40%, and 36%, respectively, in the mature protein coding region, the three-dimensional folding of the peptide chain is almost the same.

The analysis of hemoglobin function in terms of its modular structure showed that the amino acid residues associated with the heme contacts are concentrated in the central modules (M2 + M3) and the $\alpha1\beta1$ contact cluster is located in the module M4, as illustrated in Fig. 1 [8]. Craik et al. [9,10] isolated the module M2 + M3 coded by the central exon of the β-subunit using a protease and indicated that the fragment binds the heme stoichiometrically and tightly, generating a strong Soret absorption band as well as a characteristic absorption band in the visible spectrum. This confirmed a prediction that the product of the central exon, the module M2 + M3, is a complete functional domain that binds heme tightly and specifically. In contrast to the central exon encoding the module M2 + M3, there is no experimental evidence that would suggest the functional and structural significance of the module M4, although the module M4 is considered to play a crucial role in the subunit association of hemoglobin.

In the present study, to elucidate the correlation between the module M4 and the association properties of globins, we replaced the module M4 in a globin with that in the other globin by the mutagenesis. Our results also shed light on the aspects of the relationship between the modules and the structure and function of the globin chain.

Materials and Methods

Expression vector construction and protein preparation

The expression vector for module-substituted globins and purified proteins were prepared as previously reported by Wakasugi et al. [11]. We checked whether or not the desired mutation was correctly introduced and that no other mutation unexpectedly occurred in the engineered protein by means of fast-atom-bombardment mass spectrometry on tryptic digests [12].

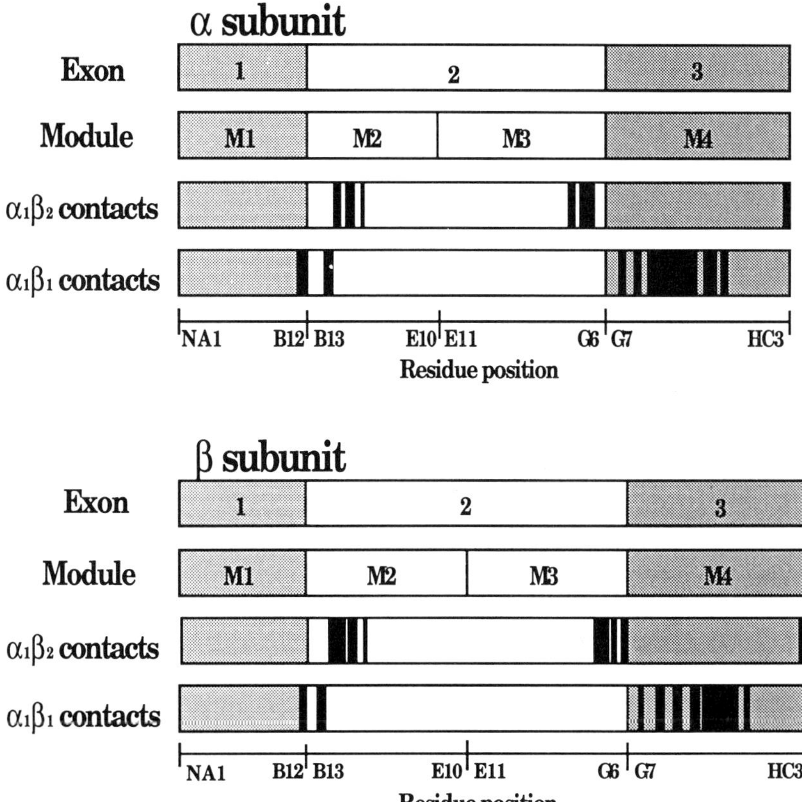

Fig. 1. Exon boundaries and residues with well defined functional roles in human hemoglobin proposed by Eaton [8].

Gel chromatography

Measurements were made by gel fraction on a Sephacryl S-200 HR column (0.8 × 65.5 cm) [13] as described by Wakasugi et al. [11]. The buffer used in all experiments was 50 mM Tris –0.1M Cl^- and 1 mM Na_2EDTA, pH 7.4 at 20°C and saturated with CO. The flow rates were 9.0 ml/h. Myoglobin (horse heart), and Blue Dextran were used for the calibration runs.

Resonance Raman and NMR spectra

Resonance Raman spectra were measured by using a model JEOL-400D Raman spectrometer equipped with a cooled RCA31034a photomultiplier as previously described [14,15]. The sample volume was 50 μl and the concentration was 50–100 μM/heme. Measurements were carried out at 20°C in the 50 mM Tris –0.1 M Cl^-, pH 7.4 buffer. ^{1}H-NMR spectra at 300 MHz were recorded on a Nicolet NT-300 spectrometer equipped with a 1280 computer system [16,17]. Protons shifted were referenced with respect to the water signal, which is 4.8 ppm downfield from the proton resonance of 4,4-dimethyl-4-silapentane-1-sulfonate (DSS) at 21°C.

Oxygen equilibrium experiments and carbon monoxide rebinding kinetics

Oxygen equilibrium curves were determined by using an improved version [18,19] of an automatic oxygenation apparatus [20]. The wavelength of the detection light was 560 nm. The temperature of the sample in the oxygenation cell was maintained constant within ±0.05°C. Simultaneous with measurement, the oxygenation data were acquired by a microcomputer (Model PC-98XA, Nippon Electric Co., Tokyo) that was interfaced to the oxygenation apparatus [21]. The resulting data analysis was performed using the same computer. The CO rebinding kinetic measurements were made by using the flash photolysis method as previously described [22].

Results

Gel chromatography of the chimera βα-subunit

Figure 2 illustrates the gel chromatographies of the carbonmonoxy derivatives of the α-, β-, and chimera βα-subunits and their mixtures. As shown in the chromatogram, the α- and β-subunits are completely or partially dissociated to the monomer at 10 μM (trace B and trace C), while the elution volume of the chimera βα-subunit was identical to that of the cross-linked Hb, indicating that the chimera βα-subunit forms homotetramer (chimera $\beta\alpha)_4$. Although the dissociation constant of the chimera βα-subunit was not estimable due to its instability, our preliminary experiments showed that the chimera βα-subunits must be tetrameric above 1 μM.

In the chromatogram of the mixture of the chimera βα- and α-subunits (trace E), two peaks were observed, which coincide with the peaks for the isolated chimera βα- and α-subunits (trace D and trace B). This observation implies that the chimera βα-subunit cannot bind to the native α-subunit. On the other hand, the mixture of the chimera βα-and β-subunits (trace F) exhibited only one peak, corresponding to that of the cross-linked Hb. This elution pattern clearly shows that the mixture is a tetrameric protein. The dissociation of the mixture to the β- and chimera βα-subunits were not observed on the chromatogram above 1 μM (figure not shown).

As shown in Table 1, the substitution of module M4 in Mb for that in the α-

Table 1. Association properties of the module-substituted globins.

Globin	Self-association	+ α-subunit	+ β-subunit
Chimera βα	Tetramer $[(\beta\alpha)_4]$	Tetramer $[(\beta\alpha a)_4]$	Tetramer $[(\beta\alpha)_2\beta_2]$
Chimera αMb	Dimer $[(\alpha Mb)_2]$	Dimer $[(\alpha Mb)_2]$	Dimer $[(\alpha Mb)\beta]$
Chimera Mbα	Polymer $[(Mb\alpha)_n]$	Polymer $[(Mb\alpha)_n]$	Dimer $[(Mb\alpha)\beta]$

Note: Concentration was 1 μM.

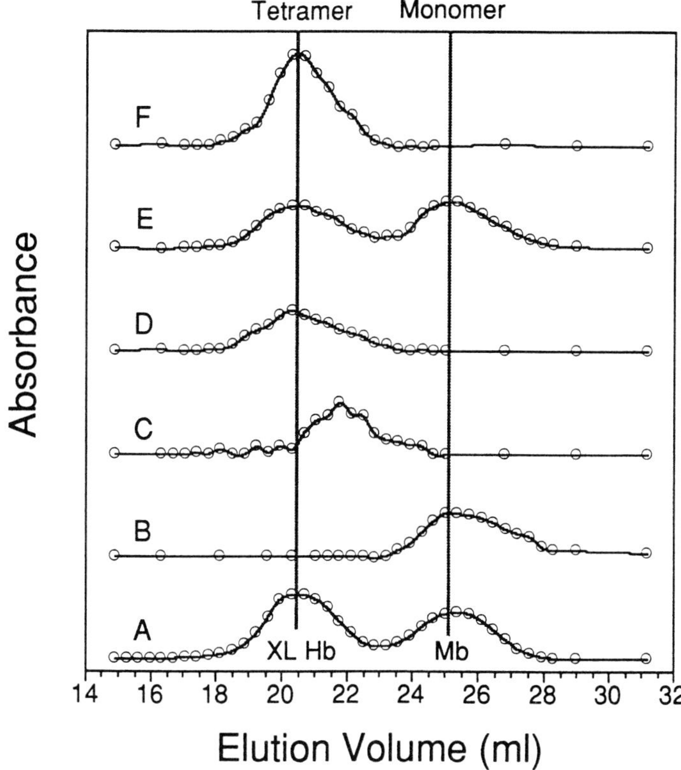

Fig. 2. Chromatography of carbonmonoxy forms of subunits on a Sephacryl S-200 HR column. The mixture of carbonmonoxy myoglobin (Mb) and cross-linked hemoglobin (XL Hb) (A), α-subunit (B), β-subunit (C), chimera βα-subunit (D), the mixture of the α- and chimera βα-subunits (E), and the mixture of the β- and chimera βα-subunits (F). Conditions: 50 mM Tris-0.1M Cl^-, 0.1 M NaCl, and 1 mM Na_2EDTA, pH 7.4, at 20°C. Monitoring wavelength was 420 nm and sample concentration was 10 μM.

subunit also drastically changed the association property of Mb. The module M4-substituted Mb, chimera Mbα-subunit, aggregates and does not dissociate above 0.1 μM (result not shown). However, in the presence of the β-subunit, the polymeric chimera Mbα-subunit dissociates to monomer and one molecule of the chimera Mbα-subunit preferentially binds to one molecule of the β-subunit to form a heterolytic dimer [(Mbα)β]. On the other hand, the association property of the counterpart chimera globin, chimera αMb-subunit, is quite different from that of chimera Mbα-subunit. Two molecules of chimera αMb-subunit form a homodimer $[(\alpha Mb)_2]$ and, by addition of the native β-subunit, the homodimer dissociates and chimera αMb-subunit binds to the β-subunit to form a heterodimer [(αMb)β] as found in the chimera Mbα-subunit. By addition of the α-subunit, no significant changes were induced in the elution pattern of the chimera αMb (figure not shown).

Resonance Raman spectra of the chimera βα-subunit

In Table 2 the position of the Fe-N_ε stretching mode for the α,` β, chimera βα-subunits and their mixtures are summarized. At pH 7.4, the resonance Raman line of the Fe-N_ε stretching in the isolated chimera βα-subunit was observed at 224 cm^{-1}, which is identical to that of the isolated α- and β-subunits [23]. The position of the Raman line of the chimera βα-subunit indicates that the strain is not imposed on the Fe-His bond in the chimera βα-subunit as found for the isolated subunits of hemoglobin. The Raman line at 215 cm^{-1} for Hb A is characteristic of the "T" structure [23,24], whereas the mixture of the chimera βα- and α-subunit exhibited the Raman line at 224 cm^{-1}, corresponding to the "R" structure in Hb A [23]. The Raman line of the Fe-N_ε stretching for the mixture of the chimera βα- and β-subunits was observed at 222 cm^{-1}, indicating that the Fe-N_ε bond is also less stretched than that of Hb A in the deoxy state. However, addition of 5 mM IHP caused a lower wave-number shift (4 cm^{-1}) for the mixture of the chimera βα- and β-subunits, implying that binding of IHP can induce some structural changes in the complex. On the other hand, the Raman line for the mixture of the chimera βα- and α-subunits was insensitive to addition of IHP.

NMR spectra of the chimera βα-subunit

For native Hb A, the ring current-shifted proton peak at –6.6 ppm has been assigned to the γ_1-methyl protons of the α- and β-Val E11, which serves as a marker for the tertiary structure in the heme vicinity [25]. As summarized in Table 3, the peaks of the γ_1-methyl proton in Val E11 for the isolated α- and β-subunits were observed at –6.6 and –6.8 ppm, respectively. The corresponding signal of the chimera βα-subunit appeared at –6.8 ppm, the same position as that of β-subunit. The two signals were observed for the mixture of the chimera βα- and α-subunits at –6.6 and –6.8 ppm, which correspond to the γ_1-methyl resonances of the isolated α- and chimera βα-subunits, respectively. Thus, we can conclude that the tertiary structures in the heme

Table 2. Resonance Raman line for Fe-N_ε (His-F8) bond of the isolated subunits, their mixtures, and Hb A [11].

	Raman shift (cm^{-1})	
	– IHP	+ IHP
α	223	223
β	224	224
chimera βα	224	224
chimera βα + α	224	224
chimera βα + β	222	218
Hb A	215	215

Experimental conditions: Hb concentration, 50–100 μM on heme basis; in 50 mM Tris –0.1 M Cl^-, pH 7.4 at 20°C; in the absence and presence of 5 mM IHP.

Table 3. Proton NMR shifts of isolated subunits, their mixtures, and Hb A [11].

	Resonance position (ppm from H_2O)							
	CO form					Deoxy form		
	Val	E11	Hydrogen bond			Proximal		His (NH)
α	–6.6			–			73.0	
β		–6.8		–				81.0
chimera βα		–6.8		–				81.0
chimera βα + α	–6.6	–6.8		–			73.0	81.0
chimera βα + β	–7.0	–7.9	6.4				73.0	81.0
Hb A	–6.6		8.2	7.4	5.9	59.5	72.0	

Experimental conditions: Hb concentration, 1 mM on heme basis; in 50 mM Tris –0.1M Cl^-, pH 7.4 at 20°C.

vicinity of the chimera βα-subunits are little perturbed by mixing with the α-subunit and also showed that the chimera βα-subunit cannot bind to the α-subunit.

On the other hand, the corresponding signals for the mixture of the chimera βα- and β-subunits were observed at –7.0 and –7.9 ppm. Since both of the isolated chimera βα- and β-subunits exhibit their NMR signals from the methyl group of Val E11 at –6.8 ppm, the shifts of the NMR signals in the mixture imply that the tertiary structural changes in the heme cavity of the chimera βα- and β-subunits are induced by the association with the β-subunit.

In the hydrogen-bonded proton region, the exchangeable proton signals were observed at 8.2, 7.4, and 5.9 ppm for carbonmonoxy native Hb A. They have been assigned to the intersubunit hydrogen bond between Asp-$\alpha_1$126(H9) and Tyr-$\beta_1$35(C1) [26], His-$\alpha_1$103(G10) and Asn-$\beta_1$108(G10) [26], Asp-$\alpha_1$94(G1) and Asn-$\beta_2$102(G4) [27], respectively. No exchangeable proton signals were observed for the carbonmonoxy form of the mixture of the chimera βα- and α-subunits, nor for the isolated α-, β-, and chimera βα-subunits. However, the NMR spectrum of the chimera βα- and β-subunits complex exhibited a new signal at 6.4 ppm in the hydrogen-bonded proton region, suggesting that some interactions through the new hydrogen bond were induced by the association of the subunits.

In the deoxygenated α- and β-subunits, an exchangeable proton resonance was observed at 73 and 81 ppm, respectively, which has been assigned to the $N_\delta H$ proton in the proximal histidine [28]. As shown in Table 3, the $N_\delta H$ resonance of the chimera βα-subunit appeared at 80 ppm, corresponding to that of the β-subunit.

NMR spectrum of the mixture of the chimera βα- and α-subunits (figure not shown) show two peaks at 81 and 73 ppm and they can be assignable to the proximal histidine $N_\delta H$ in the chimera βα- and α-subunits, respectively. No significant shifts and changes of their signal intensities in these two signals due to mixing the two subunits also implies that the chimera βα-subunit could not associate with the α-subunit, as indicated by the gel chromatography.

On the other hand, for the mixture of the chimera βα- and β-subunits, two peaks were observed at 73 and 80 ppm with the same signal intensities. Since the peak positions of the isolated chimera βα- and β-subunits are at around 80 ppm, the appearance of the peak at 73 ppm implies that the some structural changes in the heme vicinity were induced in either of the subunits by the association. It should also be noted that the peak at 73 ppm has the same signal intensity as that of the peak at 80 ppm. The same signal intensities of the two signals indicate that the ratio of the chimera βα- and β-subunits in the mixture is 1:1.

Oxygen equilibrium of the chimera βα-subunit

In Table 4, we summarized the oxygen binding properties of the chimera βα-subunit and its mixture with the β-subunit. The P_{50} value of the mixture of the chimera βα- and β-subunits was estimated to be 0.62 mmHg at pH 8.4, which indicates significantly lower oxygen affinity than that for the isolated β-chain. The n_{max} value of the mixture was almost unity. Unfortunately, the quantitative analysis of the chimera βα-globin has not yet been successful due to its rapid autoxidation. However, our preliminary results showed that the chimera βα-subunit has a very high oxygen affinity comparable to that for the native isolated chains.

Carbon monoxide rebinding kinetics in the chimera βα-subunit and their mixture of the native subunit

The binding of CO to the α-subunit appeared as a simple bimolecular association reaction on an ms time scale [29–31]. Under our experimental conditions, the bimolecular association rate constant amounted to 4.33 $\mu M^{-1}s^{-1}$ as listed in Table 5. In contrast to the α-subunit, photolysis of the carbonmonoxy β-subunit gave biphasic rebinding kinetics [32]. The bimolecular association rate constants could be estimated at 13.7 and 5.30 $\mu M^{-1}s^{-1}$. These rate constants are virtually consistent with those reported previously [33]. In the photolysis of the carbonmonoxy chimera βα-subunit, the biphasic rebinding was observed and their rate constants were 14.8 and 4.93 $\mu M^{-1}s^{-1}$, which are almost identical to those of the β-subunit as shown in Table 5.

It is quite interesting that the time course of the carbonmonoxide recombination for the mixture of the chimera βα- and β-subunits (figure not shown) experienced the two "fast" (k_{on} = 10.2 and 4.99 $\mu M^{-1}s^{-1}$) and one "slow" (k_{on} = 0.419 $\mu M^{-1}s^{-1}$)

Table 4. Oxygen equilibrium parameter values for the isolated β-subunit, the mixture of native β- and chimera βα-subunits and Hb A [11].

	P_{50}	n_{max}
β	0.37	0.97
chimera βα + β	0.62	1.02
HB A	2.3	2.99

Experimental conditions: Hb concentration, 60 μM on heme basis; in 50 mM Tris –0.1M Cl^-, pH 8.4 at 25°C.

Table 5. Kinetic constants for carbon monoxide rebinding of chimera βα Hbs [11].

	Rate constant/$\mu M^{-1}s^{-1}$		Slower rate constant/ $\mu M^{-1}s^{-1}$
	k_{fast}	k_{slow}	
α	–	4.33 ± 0.01	–
β	13.7 ± 0.1	5.30 ± 0.01	–
chimera βα	14.8 ± 1.0	4.93 ± 0.36	–
chimera βα + α	14.7 ± 1.1	5.33 ± 0.12	–
chimera βα + β	10.2 ± 0.5	4.99 ± 0.17	0.419 ± 0.008
Hb A	10.9 ± 0.1	5.22 ± 0.12	0.237 ± 0.006

Hb concentration: about 30 μM on the heme basis. In 50 mM Tris –0.1M Cl^- buffer, pH 7.0 at 20.0°C. Wavelength of observation: 436 nm. 10 mm path-length cells were used. Photolysis was accomplished with a 300-ns laser flash (590 nm).

components as listed in Table 5, whereas that for the isolated chimera βα- or β-subunits consisted of only two "fast" components. The rate constants for the mixture of the chimera βα- and β-subunits were almost the same as those of native Hb (10.9, 5.22, and 0.237 $\mu M^{-1}s^{-1}$), although the fraction of the slow phase in the mixture of the chimera βα- and β-subunits (~5%) was much smaller than that of Hb A (~30%). On the other hand, no kinetic alterations were observed by mixing of the chimera βα- and α-subunits.

Discussion

Association properties of the chimera globins

As shown in Fig. 2, the chimera βα-subunit more readily associates into tetramer than the native β-subunit. It is somewhat surprising that the association property of the chimera βα-subunit is *not* intermediate between that of the α- and β-subunits. Since the native β-subunits can assemble into a tetramer [13], it can form both $\beta_1\beta_1$ and $\beta_1\beta_2$ type contacts with neighboring β-subunits within a β_4 tetramer. If the module M4 is replaced by the corresponding module of the α-subunit, then $\beta_1\beta_1$ contacts would be changed to the $\alpha_1\alpha_1$ type. The α-subunit is predominantly monomeric [13], implying that the subunits cannot assemble into dimers or tetramers through the $\alpha_1\alpha_1$-type contacts. This is contrary to what was observed in the chimera βα-subunit. However, it should be noted that the residues involved in the $\alpha_1\beta_1$ contact, which is more appropriate for subunit association into dimer or tetramer than the $\alpha_1\alpha_1$, $\beta_1\beta_1$, $\beta_1\beta_2$, and $\alpha_1\beta_2$-type contact, are also located in the modules M1 and M2 + M3 as shown in Fig. 1. By the module substitution, the residues in the module βM1 and βM2 + M3 in the chimera βα-subunit can form contacts with the module αM4 in neighboring chimera βα-subunits. This implies that the tetrameric chimera βα-subunit has the $\alpha_1\beta_1$-type contact, although the $\alpha_1\beta_1$-type contact in the chimera βα-subunit is incomplete.

In the mixture of the chimera $\beta\alpha$- and β-subunits, their NMR signals (Table 3) and elution patterns (Fig. 2) show that the mixture is a heterotetramer, which consists of two chimera $\beta\alpha$-subunits and two β-subunits, (chimera $\beta\alpha)_2\beta_2$. Although the chimera $\beta\alpha$-subunits bind tightly with each other as mentioned above, the formation of (chimera $\beta\alpha)_2\beta_2$ means the dissociation of the tetrameric chimera $\beta\alpha$-subunit, (chimera $\beta\alpha)_4$, to the dimers or monomers. Since the module M4 in the α- and β-subunits includes the amino acid residues located in the $\alpha_1\beta_1$ intersubunit as shown in Fig. 1, the new $\alpha_1\beta_1$-type interactions should be formed in (chimera $\beta\alpha)_2\beta_2$ instead of the $\alpha_1\alpha_1$-type interactions in (chimera $\beta\alpha)_4$. The $\alpha_1\beta_1$-type interactions between the module M4 in the α- and β-subunits are much more potent to associate subunits than the $\alpha_1\alpha_1$-type interactions in (chimera $\beta\alpha)_4$, resulting in the dissociation of (chimera $\beta\alpha)_4$ and formation of (chimera $\beta\alpha)_2\beta_2$.

On the other hand, such $\alpha_1\beta_1$-type interactions cannot be formed in α_2(chimera $\beta\alpha)_2$ because both of the subunits have the module M4 of the α-subunit and the $\alpha_1\alpha_1$, $\alpha_1\alpha_2$, $\alpha_1\beta_2$-type interactions are predominant in α_2(chimera $\beta\alpha)_2$. These interactions are rather weak to associate the subunit and would not be enough to dissociate (chimera $\beta\alpha)_4$ to dimers and to form α_2(chimera $\beta\alpha)_2$.

As listed in Table 2, the substitution of the module M4 in Mb for that in the α-subunit induced polymerization of the chimeric protein, indicating that the structure of the chimera Mbα-subunit is quite different from that of Mb and α-subunits. Our preliminary circular dichroism measurement showed remarkable reduction of the negative ellipticity at 222 nm in the chimera Mbα-subunit. This reduction implies that the folding of the chimera Mbα-subunit is not completed and some hydrophobic amino acid residues might be exposed to the solvent, resulting in the nonspecific aggregation. In the "reversed" chimeric protein, chimera αMb-subunit, the reduction of the ellipticity was less drastic than that in the chimera Mbα-subunit. Although Mb is a monomeric protein and the α-subunit is also monomeric below 10 μM, the elution pattern of the chimera αMb showed that the two molecules of the chimeric proteins form a homodimer [$(\alpha Mb)_2$] above 1 μM. To elucidate the reasons for the dimerization of the chimera αMb-subunit, the structural analysis of the chimeric proteins is in progress.

Since chimera Mbα-subunits have the module M4 from the α-subunit in which the $\alpha_1\beta_1$-type interactions are localized, it would be quite natural that the chimera Mbα-subunit can preferentially bind to the β-subunit. However, it should be noted that the chimera αMb-subunit, which has the module M4 from Mb, can also bind to the β-subunit, not to the α-subunit. In the chimera $\beta\alpha$-subunit, we suggested that the amino acid residues which participate the $\alpha_1\beta_1$ interactions are also located in the modules M1 and M2 + M3. These observations imply that the modules M1 and M2 + M3 in the α-subunit play essential roles in the formation of the $\alpha_1\beta_1$-type interactions as well as the module M4.

Our preliminary results of the association properties of the module-substituted Mbs showed that the chimera αMb and Mbα-subunit binds to the β-subunit to form a heterodimer, (Mb$\alpha)\beta$ and (αMb)β, respectively, and cannot form a tetramer up to 100 μM. To form a tetrameric protein such as $(Mb\alpha)_2\beta_2$ or $(\alpha Mb)_2\beta_2$, it should be necessary to form the $\alpha_1\beta_2$-type interaction which localized in the modules M2 + M3. Since the modules M2 + M3 of the chimera Mbα originated from Mb, the $\alpha_1\beta_2$-type interactions are weak and the chimeric protein should remain in dimer. In the chimera αMb, however, it is still unclear why the chimeric protein forms dimer, not tetramer.

Structure and function of the chimera βα-subunit and their mixture of the native subunit

In the deoxygenated chimera βα-subunit, the Raman line of the Fe-N_ε stretching and the hyperfine-shifted resonance of the proximal histidyl $N_\delta H$ clearly show that the structure around the Fe-N_ε bond in the heme vicinity is almost conserved regardless of the substitution of the module M4 in the β-subunit. This is also shown by the resonance of methyl group of Val E11 in the carbonmonoxy form. Although the oxygen-binding properties of the chimera βα-subunit have not been clear due to its rapid autoxidation, the kinetics of carbonmonoxide for the chimera βα-subunit correspond to those for the isolated β-subunit. Therefore, the substitution of the module M4 in the β-subunit for that in the α-subunit could not seriously affect the structure in the heme vicinity.

Table 3 indicates that the NMR peak positions of the chimera βα- and α-subunits in the mixture are identical to those of the isolated chimera βα- and β-subunits, respectively. The rebinding kinetic data also indicate that no new components were formed in the mixture of the chimera βα- and α-subunits, showing that there are no interactions between the subunits.

In the NMR spectra of the mixture of the deoxy chimera βα- and β-subunits, however, the $N_\delta H$ resonance of one subunit shifted from 81 to 73 ppm by mixing the two subunits (Table 3). This upfield shift of the $N_\delta H$ resonance implies that a strain is imposed on the bond between the heme iron and the proximal histidine [34]. In the kinetics of (chimera $\beta\alpha)_2\beta_2$ with carbonmonoxide, a new component having a slower binding rate was also observed and the slow binding rate corresponds to that of the tetrameric T-state Hb A. These observations clearly indicate that the chimera βα-subunit interacts with the native β-subunit.

It is also quite interesting that the oxygen affinity of the mixture is significantly lower than that of the isolated chimera βα- or β-subunit as listed in Table 4. The lower oxygen affinity of the mixture suggests that the interactions between the chimera βα- and β-subunit in the mixture are significant in the formation of the functional Hb molecule, and the subunit interface in the mixture has some native Hb A characters. It is, therefore, concluded that by the substitution of the module M4 in the β-subunit for that in the α-subunit, the chimera βα-subunit can bind to the β-subunit with structural and functional changes that are also observed in the formation of the native tetrameric Hb A from the native α- and β-subunits.

Module as a structural and functional unit

As discussed above, the structures of the heme vicinity are mainly determined by the modules M1 and M2 + M3, and the module M4 has a predominant role in the subunit association. Using the arginine-specific protease clostripain, Craik et al. [9,10] showed that the central fragment of the β-subunit, which corresponds to the module M2+M3, bound heme stoichiometrically and tightly at micromolecular concentrations, generating a strong Soret absorption band as well as a characteristic absorption band in the visible spectrum. The Soret band appeared at the same wavelength and had the same shape as in hemoglobin. Sanctis et al. [35] also indicated that the minimyoglobin, which is a domain of 108 amino acid residues (32 to 139), obtained by digestion of horse heart apomyoglobin with clostripain, bound protoheme in a 1:1 molar ratio. Flash photolysis experiments of the reconstituted minimyoglobin showed that it was very similar to native myoglobin in the combination reaction with carbonmonoxide and with oxygen, and in the oxygen replacement reaction by carbon monoxide. On the basis of the detailed kinetic data [36–38], it was concluded that the protein fragment encoded by the central exon of the myoglobin gene is the domain responsible for ligand-linked conformational transitions.

In the present study, the experimental data clearly show that by the substitution of the module M4 the β-subunit gets the binding ability to the β-subunit instead of the α-subunit without the large structural changes around heme vicinity. In Mb, the substitution of the module M4 for that in the α-subunit induced the protein association and specific binding to the β-subunit. In other words, the module M4 plays a crucial role in the formation of the functional tetrameric molecule of hemoglobin. We can also conclude that the module M4 is a structural and functional unit and the modules encoded by the corresponding exons are functional units as well as structural units to have the advantage of producing stable functional proteins by combinations of such exons.

Conclusions

By the substitution of the module M4 in human myoglobin and hemoglobin subunits, we could prepare several novel globins, in which association properties are very unique. Our experiments also provided the first experimental evidence that module M4 is a structural and functional unit and suggested that the exchanges of the modules can be able to have advantages to produce novel stable functional proteins and study the evolution of structural and functional relationships of proteins. In chimera βα-subunit, we can substitute the module M4 without large structural changes in the other modules, suggesting that the function and structure of one module is almost independent of those of the other modules. However, it is too early to conclude that the module is an *independent* unit for protein structure and function since we also indicated the large structural changes in protein structure accompanied by the module substitution in Mb.

Acknowledgements

We are grateful to Prof Teizo Kitagawa (Institute for Molecular Science) for resonance Raman spectral measurements.

References

1. Gō M. Nature 1981;291:90–92.
2. Leder A, Miller HI, Hamer DH, Seidman JG, Norman B, Sullivan M, Leder P. Proc Natl Acad Sci USA 1978;75:6187–6191.
3. Nishioka Y, Leder P. Cell 1979;18:875–882.
4. Konkel DA, Maizel JV Jr, Leder P. Cell 1979;18:865–873.
5. Konkel DA, Tilghman SM, Leder P. Cell 1978;15:1125–1132.
6. Gō M. Proc Natl Acad Sci USA 1983;80:1964–1968.
7. Stoltzfus A, Spencer DF, Zucker M, Longsdon JM Jr, Doolittle WF. Science 1994;265:202–207.
8. Eaton WA. Nature 1980;284:183.
9. Craik CS, Buchman SR, Beychok S. Proc Natl Acad Sci USA 1980;77:1384–1388.
10. Craik CS, Buchman SR, Beychok S. Nature 1981;291:87–90.
11. Wakasugi K, Ishimori K, Imai K, Wada Y, Morishima I. J Biol Chem 1994;269:18750–18756.
12. Wada Y, Matsuo T, Sakurai T. Mass Spect Rev 1989;8:379–434.
13. Valdes, Ackers. J Biol Chem 1989;252:74–81.
14. Imai K, Fushitani K, Miyazaki G, Ishimori K, Kitagawa T, Wada Y, Morimoto H, Morishima I, Shih DT-b, Tame J. J Molec Biol 1991;218:769–778.
15. Ishimori K, Imai K, Miyazaki G, Kitagawa T, Wada Y, Morimoto H, Morishima I. Biochemistry 1992;31:3256–3264.
16. Ishimori K, Morishima I. Biochemistry 1986;25:4892–4898.
17. Ishimori K, Morishima I. Biochemistry 1988;27:4060–4066.
18. Imai K. Meth Enzymol 1981;76:438–449.
19. Imai K. Allosteric Effects in Haemoglobin. London: Cambridge University Press, 1982.
20. Imai K, Morimoto H, Kotani M, Watari H, Hitata W, Kuroda M. Biochim Biophys Acta 1970;200:189–196.
21. Imai K. Meth Enzymol 1994;232:559–576.
22. Unno M, Ishimori K, Morishima I. Biochemistry 1990;29:10199–10205.
23. Nagai K, Kitagawa T, Morimoto H. J Biol Chem. 1980;136:271–289.
24. Hori H, Kitagawa T. J Am Chem Soc 1980;102:3608–3613.
25. Lindstrome TR, Noren IBE, Charache S, Lehmann H, Ho C. Biochemistry 1972;11:1677–1681.
26. Russu IM, Ho NT, Ho C. Biochim Biophys Acta 1987;914:40–48.
27. Fung LW-M, Ho C. Biochemistry 1975;14:2526–2535.
28. Nagai K, La Mar GN, Jue T, Bunn HF. Biochemistry 1982;21:842–847.
29. Antonini E, Brunori M. Hemoglobin and Myoglobin in their Reaction with Ligands. Amsterdam: North-Holland Publishing Co., 1970.
30. Reisberg PI, Olson JS. J Biol Chem 1980;255:4151–4158.
31. Mims MP, Porras AG, Olson JS, Nobel RW, Peterson JA. J Biol Chem 1983;258:14219–14232.
32. Philo JS, Lary JW, Schuster TM. J Biol Chem 1988;263:682–689.
33. Unno M, Ishimori K, Morishima I. Biochemistry 1991;30:10679–10685.
34. La Mar GN. In: La Mar GN, Horrocks WD Jr, Holm RH (eds) NMR of Paramagnetic Molecules, Chapter 3. New York: Academic Press, 1973.
35. Sanctis GD, Falcioni G, Giardina B, Ascoli F, Brunori M. J Molec Biol 1986;188:73–76.
36. Sanctis GD, Falcioni G, Giardina B, Ascoli F, Brunori M. J Molec Biol 1988;200:725–733.
37. Sanctis GD, Falcioni G, Grelloni F, Desideri A, Polizio F, Giardina B, Ascoli F, Brunori M. J Molec Biol 1992;222:637–643.
38. Iorio EED, Yu W, Calonder C, Winterhalter KH, Sanctis GD, Falcioni G, Ascoli F, Giardina B, Brunori M. Proc Natl Acad Sci USA 1993;90:2025–2029.

Tracing Biological Evolution in Protein and Gene Structures.
M. Gō and P. Schimmel, editors.

Three-dimensional profiles: assessing the compatibility of an amino acid sequence with a three-dimensional structure

James U. Bowie[1], Roland Lüthy[2] and David Eisenberg[1]
[1]*Department of Chemistry and Biochemistry and DOE Laboratory of Structural Biology and Molecular Medicine, University of California, Los Angeles, California; and* [2]*Amgen, Thousand Oaks, California, USA*

Abstract. We have devised a method to test whether a given amino acid sequence is likely to adopt a particular three-dimensional structure. The approach has been used to find sequences in the sequence database that adopt structures that are already known; to assess the correctness of model structures; as part of an energy function for de novo protein folding; and for local conformation prediction [1–6]. The environments of every residue in the three-dimensional structure are first assigned to different environment classes based on the area of the side chain that is buried, the fraction of the side chain that is covered by polar atoms and the secondary structure. Scores, derived from a database of known protein structures, that reflect the probability of finding each of the amino acid types in the environment classes are then used to find the optimal alignment of the test amino acid sequence with the protein structure. The overall score is a measure of the compatibility of the amino acid sequence for the structure.

Introduction

One of the primary forces that favor the folded state of a protein is the hydrophobic effect [7]. This is manifest in the formation of an apolar interior that is shielded from solvent and a polar exterior in contact with solvent. Moreover, residues in the hydrophobic core of a protein structure are much more sensitive to the deleterious effects of mutations than residues on the surface [8–13]. As a consequence, any protein sequence that can adopt a particular fold, must have apolar residues at appropriate positions in its sequence so that they can be placed in the hydrophobic core in the folded structure. As a consequence of the necessity for forming a hydrophobic core, the pattern of apolar and polar residues is characteristic of any sequence that can adopt a particular fold [8,13,14]. This pattern provides a considerable amount of information about the suitability of a particular amino acid sequence for a given fold and is the primary information that is used in the three-dimensional profile method [2].

Inverse protein structure prediction

Typically when we think about predicting protein structure, we have the following question in mind. What structure does a given sequence fold into? This is an

Address for correspondence: Dr J.U. Bowie, Molecular Biology Institute, University of California at Los Angeles, 405 Hilgard Avenue, Los Angeles, CA 90024-1570, USA.

extremely difficult problem which has largely stymied workers in the field for over 30 years. Drexler and Pabo, thinking about protein design, suggested that finding a sequence that could fold into a particular structure is a much simpler problem than finding the structure a sequence adopts because all the tertiary interactions would already be known [15,16]. Pabo called this inverted protein folding. We have simplified Pabo's idea even further by limiting the problem to known folds and sequences. The goal then is to find known sequences which could adopt structures that we already know. We refer to this as inverse structure prediction.

One simple approach to inverted structure prediction is to find sequence similarity between a protein of known structure and a protein of unknown structure. If significant similarity can be found, the protein of known structure can then often be used as a model for the protein of unknown structure. However, sequences can adopt very similar folds yet show essentially no sequence similarity. For example, it came as a complete surprise that the structures of actin and the ATP-binding domain of heat shock cognate protein 70 (HSC-70) were quite similar, yet the level of sequence similarity is too low to be detected by standard sequence homology searching methods [17].

Is there some way that we could have determined in advance that the sequence of actin is compatible with the structure of HSC-70 or that the sequence of HSC-70 is compatible with the actin structure? To do this it is necessary to derive a set of rules that can relate a one-dimensional sequence to a three-dimensional structure. But clearly the set of rules must be quite flexible since there can be many significant structural differences between proteins that adopt recognizably similar folds. For example, the orientations of secondary structure elements may change somewhat and insertions and deletions can occur in the sequence.

Our approach to this problem is outlined in Fig. 1. We start with a known three-dimensional structure and determine three things about each residue's environment in the structure: 1) the area of the side chain that is buried, 2) the fraction of the surrounding atoms that are polar and 3) the secondary structure of the position. Based on these three parameters, each residue is classified into one of 18 different environment classes. In this manner we convert a complex three-dimensional structure into a simple one-dimensional string that represents the sequence of environments that the residues in the structure adopt. It is possible to look at a database of protein structures and their sequences and determine probabilities for seeing each of the 20 amino acid types in each of the 18 different environment classes. From these probabilities one can derive scores for pairing amino acid types with environment classes. We call these scores 3D-1D scores. With these scores, one can actually align an amino acid sequence to the environment class string for a structure using algorithms that are commonly used for sequence homology searches [18,19]. Indeed, one is simply aligning two sequences, one a sequence of amino acid types, the other a sequence of environment classes.

To facilitate alignments and database searching we construct a position-specific scoring table, called a profile, which was originally developed by Gribskov and Eisenberg [20,21] for sequence similarity searches. Since our profile tables are

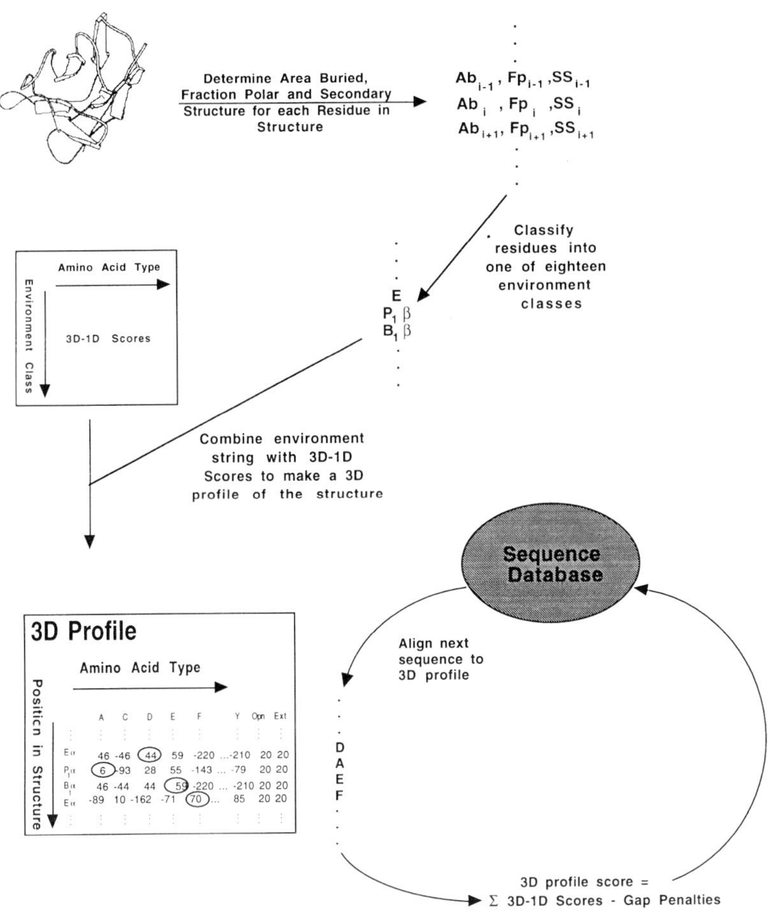

Fig. 1. Overview of the application of three-dimensional profiles to inverse structure prediction.

derived from three-dimensional structures rather than sequences, we call them three-dimensional profiles. Part of a three-dimensional profile for the sperm whale myoglobin structure is shown in Table 1. The first column of each row lists the position of the residue in the myoglobin structure. The second column of each row lists the environment class of the position. The next 20 columns in each row list the 3D-1D score for placing each of the 20 amino acids at that position in the alignment. For example, to put the charged residue Asp at position 1, a fully exposed position, would give a favorable score of 22, but to put Asp at position 2, a highly buried position, would give an unfavorable score of –128. In this manner, the profile embodies a general feature of sequences that could adopt a particular protein structure: polar residues should be excluded from the hydrophobic core. The last two rows give the gap opening and gap extension penalties which can vary based on position. In the example shown, gap penalties in coil regions are low while for

Table 1. Part of a three-dimensional profile.

Position in fold	Environment class	Amino acid type						Penalty	
		A	C	D	V	W	Y	Opn	Ext
1	E	12	–46	22	–91	–214	–94	2	0.02
2	B_2	66	–5	–128	60	102	112	2	0.02
3	Eα	46	–44	44	–110	–135	–210	200	200
4	$P_2\alpha$	6	–93	28	–48	–114	–79	200	200
5	Eα	46	–44	44	–110	–135	–210	200	200
6	$P_2\alpha$	6	–93	28	–48	–114	–79	200	200
7	$B_2\alpha$	–69	–10	–162	68	50	85	200	200
8	Eα	46	–44	44	–110	–135	–210	200	200
9	$P_2\alpha$	6	–93	28	–48	–114	–79	200	200
10	$B_1\alpha$	–66	–73	–197	66	100	18	200	200

Note: The first column lists the position in the protein structure and the second column lists the environment class of the position. The environment class definitions are given in Bowie et al. [2]. Briefly, the residues are first divided into three classes based on area buried to give a highly buried class B, partially buried class P and an exposed class, E. The highly buried class and the partially buried classes are further subdivided according to the fraction of their surfaces that are covered by polar atoms. This gives classes a total of six classes B_1, B_2, B_3 and P_1 and P_2 and E. Each of these classes are further subdivided into three more classes based on the secondary structure: α-helix, β-sheet or coil to give a total of 18 classes. The full length three-dimensional profile contains a row for each sequence position. The subsequent columns for each row list the 3D-1D scores for placing each of the 20 amino acid types at that position in the alignment to the environment class string. Many of the columns for the scores have been deleted for compactness. The last two columns list penalties for opening or extending a gap at each position in the alignment.

regions of secondary structure, the gap penalties are kept high. This is done because insertions and deletions in related protein structures usually occur within coil regions and less frequently in regions of secondary structure.

Once this profile is constructed, it is possible to align every sequence in a large sequence database to the three-dimensional profile as described by Gribskov and Eisenberg for sequence profiles [20]. Those sequences that score most highly with the profile are the most likely to adopt a fold similar to the structure used to create the three-dimensional profile. In other words, they are the most compatible with the starting fold. To distinguish this type of database search from homology searches, we have called them compatibility searches.

In a number of test cases, compatibility searches were able to find proteins that adopt similar structures yet have little or no apparent sequence similarity. An example of the results of a compatibility search using a profile made from the actin structure is shown in Table 2. Members of the HSC-70 family, which are now known to adopt similar structures to actin [17], score highly in this search, but do not score highly in a homology search. Another example is shown in Table 3 which lists the top-scoring proteins in a compatibility search using a three-dimensional profile made from the structure of the ligand-binding domain of the aspartate receptor from

Table 2. Compatibility search with a three-dimensional profile for actin.

Sequence name	ZScore
Top-scoring actin	88.11
Lowest scoring actin	9.29
68-kD Heat shock protein – mouse	8.12
70-kD Heat shock protein – frog	7.95
70-kD Major heat shock protein – fruit fly	7.03
70-kD Heat shock cognate protein – bovine	6.99
HNRNP complex, protein C – frog	6.74
70 kD Heat shock cognate protein – human	6.31

Note: These results are from Bowie et al. [2]. All sequences that received a ZScore greater than 6.0 are listed. A gap opening penalty of 5.0 and a gap extension penalty of 0.2 were used. The ZScore is the number of standard deviation units of the alignment score from the mean alignment score for other sequences of similar length.

Escherichia coli [22]. After the aspartate receptor sequences in the database, the next highest scoring proteins are apoliproteins. The structure of apolipoprotein E has been determined [23] and both apolipoprotein E and the aspartate receptor are long four-helix-bundle structures. There is, however, no apparent sequence similarity between apolipoproteins and the aspartate receptor. Thus, compatibility searching has the potential to find structural homologs that would be difficult to identify by standard sequence similarity methods. Nevertheless, not all distant homologs can be identified by compatibility searching and efforts are underway to improve the method. Wilmanns and Eisenberg have improved the method by incorporating information about residue neighbor preferences into the scoring of compatibility and Zhang and Eisenberg have introduced continuous profiles to smooth the scoring at boundaries between environment group divisions [5,6].

Table 3. Compatibility search with a three-dimensional profile for the aspartate receptor.

Sequence name	ZScore
Aspartate Receptor – *E. coli*	18.98
Aspartate Receptor – *S. typhymurium*	17.49
Apolipoprotein A-I precursor – chicken	7.88
Apoliprotein A-I precursor – bovine	7.73
Gas Vesicle Protein C – *Fremyella diplosiphon*	7.21
Apoliprotein A-I precursor – human	7.04
Glued protein – fruit fly	6.96
Apolipoprotein A-I precursor – baboon	6.82
Apolipoprotein A-I precursor – macaque	6.82
Apolipoprotein E precursor – guinea pig	6.77
Apolipoprotein E precursor – mouse	6.67
Apolipoprotein A-I precursor – pig	6.67

Note: All sequences that received a ZScore greater than 6.0 are listed. A gap opening penalty of 5.0 and a gap extension penalty of 0.05 were used. The ZScore is the number of standard deviation units of the alignment score from the mean alignment score for other sequences of similar length.

Three-dimensional profiles and model assessment

Methods for the assessing the correctness of model protein structures are useful for both structure prediction and for the verification of experimentally derived structures. Indeed, mistakes have occasionally been made in the solution of protein structures by X-ray crystallography. These mistakes can range from sequence frameshifts or exchange of β-strands, to inverting entire protein chains [24–27]. Three-dimensional profiles can be used to assess the correctness of model structures no matter how they were derived [3]. In this application, rather than ask whether a different sequence is compatible with a particular structure, we ask whether a structure is compatible with its own sequence.

The approach is illustrated in Fig. 2. Suppose we have two models for structures a sequence might adopt. Because the structural models are different, the environments of residues in the two models will not be the same. Consequently, the three-dimensional profiles will also be different. Because the environments of residues in the correct model will be more appropriate for residues in the sequence than the environments of the incorrect model, the three-dimensional profile score should be higher for the correct structure than the incorrect structure. For all tests so far conducted where both a correct and incorrect model are known, the three-dimensional profile score is indeed higher for the correct model [3].

For grossly misfolded structures, the score alone can be sufficient to distinguish a correct structure from an incorrect structure. When only part of the structure is incorrect, however, the score alone may not always be low enough to identify an incorrect structure. In such cases, the errors can often be revealed by three-dimensional profile window plots. The three-dimensional profile window plot is a graph of the average score in consecutive 21 residue segments (or windows) of a protein sequence. An example of three-dimensional profile window plots for the initial and final models of the small subunit of Rubisco and HIV protease are shown in Fig. 3. In the case of Rubisco, the small subunit was essentially traced backwards in the original model. Accordingly, the three-dimensional profile window plot for the incorrect structure is lower throughout the sequence than the plot for the corrected structure and dips below zero in two regions. None of the high-resolution, well refined structures tested thus far ever dip below zero in a three-dimensional profile window plot [3]. Thus, the profile window plot for the incorrect structure of Rubisco is clearly indicative of problems. In the case of HIV protease, the electron density was initially misinterpreted at the C-terminus. As would be expected, the scores in the three-dimensional profile window plot dip down precipitously in this area. In this manner, the three-dimensional profile window plot can be used to highlight areas of a structure that warrant further attention.

Three-dimensional profiles and protein folding

If we are to fold proteins on a computer, it will be necessary to develop an energy

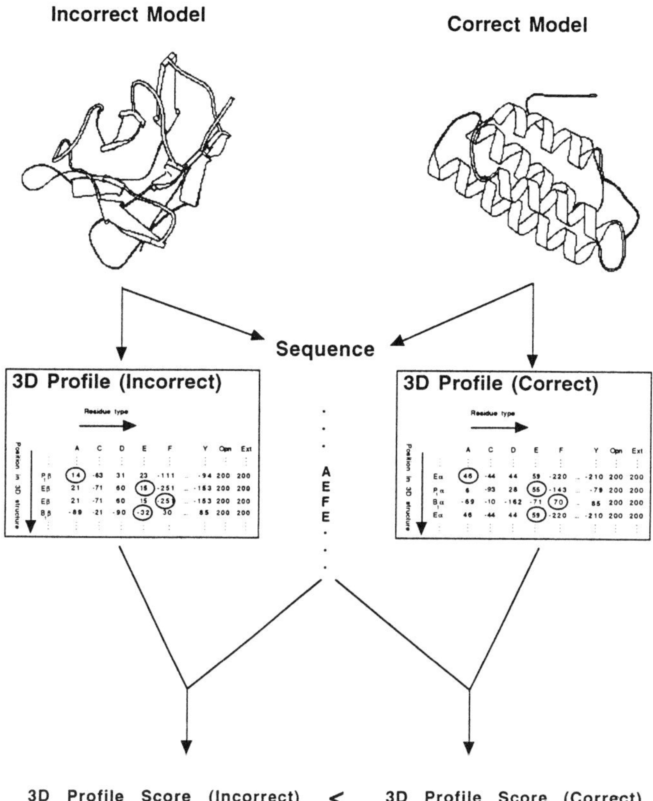

Fig. 2. Overview of structure verification with three-dimensional profiles.

function that can distinguish a correct structure from an incorrect structure. As shown in the previous section, three-dimensional profiles can be used for this purpose. Certainly, the three-dimensional profile score alone cannot distinguish a correct structure from an incorrect structure during a folding simulation since three-dimensional profiles contain no information about geometrical features such as steric overlaps. We have explored the utility of incorporating the three-dimensional profile score into an energy function for use in folding simulations [1].

There are a number of extremely difficult obstacles that need to be overcome if we are to fold proteins on a computer. First of all, we need an energy function for which the correct structure or a conformation near the correct structure is at the global minimum. Second, we need a method to minimize the energy function that is capable of passing over the large number of local minima to get to the global minimum. Finally, because it is possible to search only a tiny fraction of all possible conformations, we need some way to direct the search of conformation space toward the correct structure.

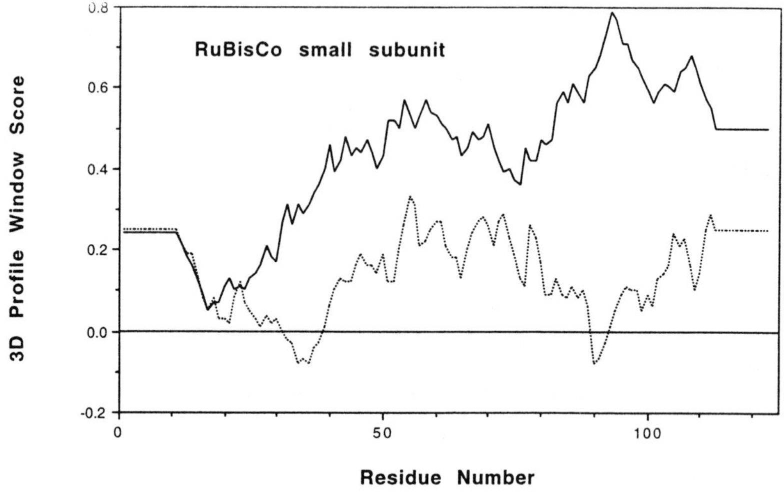

B)

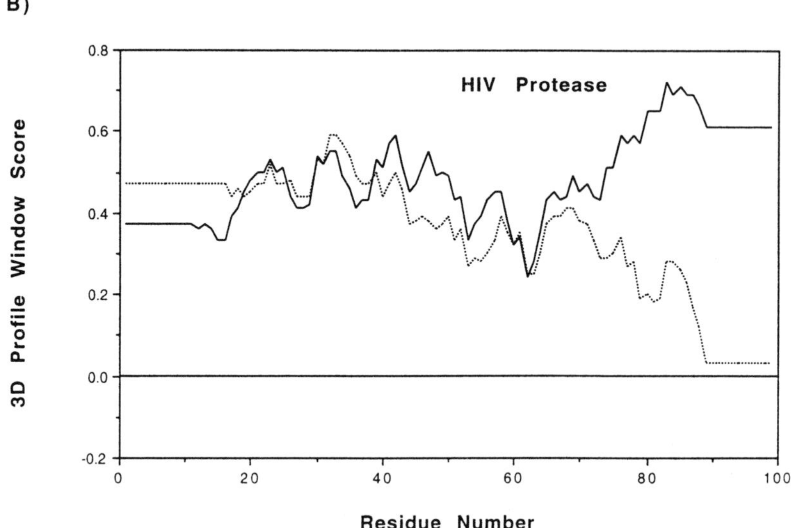

Fig. 3. Profile window plots. The average three-dimensional profile score in 21-residue windows along the sequence is plotted for the small subunit of Rubisco (A) and HIV protease (B). The three-dimensional profile window plots for the corrected structures are show as solid lines and for the initial structures as dotted lines.

A guiding energy function

The energy function we developed includes the following parameters: 1) the three-dimensional profile score; 2) the total accessible surface area of the conformation as a measure of compactness; 3) the number of atoms that are within 80% of the sum of their van der Waals radii; 4) the number of atoms that are within 50% of the sum of their van der Waals radii as measures of steric overlaps; 5) the hydrophobicity contrast function for the entire conformation to measure the segregation of nonpolar atoms inside and polar atoms outside; and 6) the ratio of the longest to the shortest axis of the inertial ellipsoid for the conformation to measure globularity.

To combine these very different parameters into an overall score, S, for a conformation, we used the following function:

$$S = \sum_{i=1}^{N} W_i \left| \frac{(R_i - 1.0)}{\sigma_i} \right|^{X_i}$$

where the summation is over the parameters. W_i is the weight for that parameter and X_i is the exponent for that parameter. R_i is the ratio of the parameter value for the conformation and its target value. The target values are determined by measuring these parameters for proteins of known structure. The standard deviation of the ratio R_i for proteins of known structure is σ_i. If value of parameter i for a conformation exactly equals its target value, R_i will equal one and that parameter will contribute nothing to the overall score. However, if the value of parameter i deviates from its target value, its contribution to the overall score will depend on how precisely that parameter can be defined for known protein structures, reflected in σ_i, as well as the weight, W_i, and the exponent, X_i, given to that parameter.

What are the best weights W_i and exponents X_i for this function? One might choose weights and exponents such that the correct conformation is at the global minimum energy and the minimum energy is well separated from the energy of all other conformations. This criterion alone, however, is not sufficient. An extreme example is an energy function that has a constant, high value for every conformation but the correct one. Because every conformation would have the same energy except the correct one, the energy would not provide any information about whether a particular conformation was close or far from the correct one. Consequently, this hypothetical flat energy function would require an exhaustive search of conformation space. An energy function that maintains a gradient in the direction of the correct structure could in essence direct the search of conformation space toward the correct structure. In this sense, the ideal energy function would be one that mimics root mean square deviation or distance matrix error (DME) from the correct structure. Indeed if we use DME as our energy function, we are able to fold proteins to conformations very near the correct one. Thus, we sought weights and exponents for our energy function that would mimic DME as closely as possible. We did this by generating thousands of structures with the sequence of 434 repressor, a 63-residue DNA-binding protein, and adjusting the weights and exponents of the energy function to produce the optimal correlation between the conformation score S and DME.

Local conformation enrichment

To further restrict our search of conformation space, we used the three-dimensional profile method to find fragments of structure that were more likely to be correct than random fragments. Two fragment selection procedures were performed: 1) short fragments, nine residues in length, were selected for each nonoverlapping segment of sequence and 2) long fragments ranging in length from 15 to 25 residues were selected.

To select short fragments, nine-residue segments of sequence were extracted from the sequence to be folded and matched against all possible environment patterns in a database of unrelated protein structures. The 15 top-scoring fragment conformations for each segment of sequence were stored and became the conformation library for that segment of sequence. This procedure indeed enriches for fragments that are more likely to be correct. The fraction of random nine-residue fragments in a database of 71 nonhomologous proteins that were found to be within 1.75 Å DME on C^{α} atoms was 0.17. The fraction of nine-residue fragments in the enriched pools within 1.75 Å DME on C^{α} atoms was 0.31. Thus, about one-third of randomly chosen fragments from these enriched fragment pools will be close to the correct conformation.

It is possible to measure, to some extent, the likelihood that a conformation selected from the enriched pools would be roughly correct. There are three types of fragment pools that one can imagine: 1) a pool in which virtually all the conformations are correct and therefore very similar (the enrichment procedure performed well), 2) a pool in which few of the conformations are correct and they are all very dissimilar (the enrichment procedure was incapable of selecting a particular conformation), and 3) a pool in which few of the conformations are correct, but they are all very similar (the enrichment procedure reliably selected the same wrong fragment conformation). If the third possibility is rare, the diversity of the population can be used as a measure of probability that a fragment selected from the pool will be correct. We measure diversity, D, of the pool as follows:

$$D = \frac{\left(\sum_{i}^{n} \sum_{j \neq i}^{n} DME_{ij} \right)}{n(n-1)},$$

where DME_{ij} is the distance matrix error on Ca positions between fragments i and j and n is the number of fragments in the pool. Figure 4 shows a histogram of diversity versus fraction of the conformations in an enriched pool that are within 1.75 Å DME of the correct structure. There is clearly a very strong correlation. This information was used to weight mutation frequencies within a particular fragment during the evolutionary optimization (see below). The higher the diversity, the higher the mutation rate. We note that this approach might be useful in weighting secondary structure predictions as well. While we did not attempt to predict the local structure specifically, Yi and Lander have applied a modification of three-dimensional profiles for the prediction of secondary structure [4].

Longer fragments were selected by alignment without gaps of the sequence to be folded with three-dimensional profiles of unrelated protein structures. The conformations of the top-scoring aligned regions between 15 and 25 residues long were saved in a long fragment library. Structure fragments longer than 25 residues were not found to be useful. Again, this procedure was found to enrich for fragments that resemble the true structure [1].

Energy minimization using an evolutionary algorithm

To minimize the energy function for a particular structure, we used an evolutionary algorithm. Initially, a population of conformations was built up by randomly selecting fragment conformations from the appropriate fragment conformation library. These conformations were encoded as dihedral angles. Next, new conformations were generated from the initial conformations by first copying the initial population of conformations and then altering the new conformations by mutation and recombination events. Mutation events were simply alterations of the dihedral angles and recombination events were swaps of a randomly selected regions of a structure with another randomly selected conformation in the population. The new population was thereby doubled. The score, S, was then determined for each conformation and each conformation was made to compete with 10 randomly chosen conformations in the population. The winner of each competition was simply the one with the better score. The winning half of conformations from the new population were kept and the remainder discarded. By this procedure, a new, more fit population was generated. The cycle was then repeated until the fittest conformation no longer improved for a preset number of generations.

We tested the ability of the algorithm to fold three small helical proteins: 434 repressor, the engrailed homeodomain and the B domain of protein A. The 434 repressor test is biased by knowledge of the crystal structure since this protein was

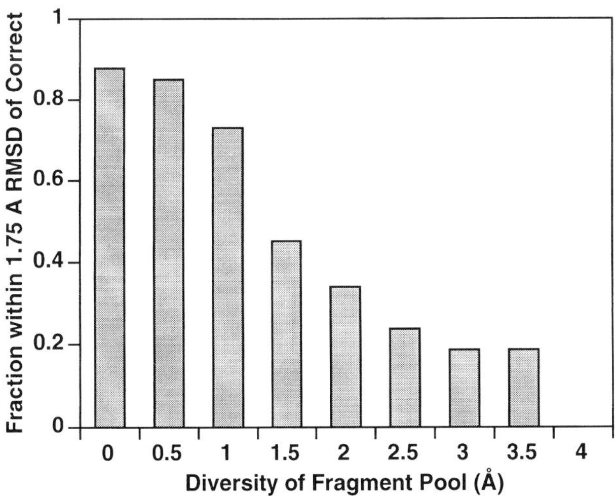

Fig. 4. Diversity of fragment pool predicts the fraction of close conformations in the enriched fragment libraries.

used to optimize the energy function. The other two proteins represent objective test cases, except for the fact that they were chosen because they are small α-helical proteins and are therefore more likely to be constrained by similar parameters as is 434 repressor. There are no parameters in our energy function that would specifically favor β-sheet formation. For each of the three proteins, the evolutionary minimization was performed 3 separate times using 400 conformation populations. For both 434 repressor and the B-domain of protein A, conformations with folds similar to the correct fold were found among the 400 conformations in the final population for two out of the three miminizations. For engrailed, the correct fold was obtained in each of the three attempts. The best DME values obtained for all the tests were 3.0 Å, 2.3 Å and 3.0 Å for 434 repressor, the engrailed homeodomain and protein A respectively. For none of the minimizations, however, was the best conformation the top-scoring one and frequently, the top-scoring conformation was far from the correct structure. Thus, the energy function clearly has significant energy minima far from the correct structure. However, while the correct conformation is not at the global minimum, the energy function appears to be able to direct the search to the correct region of conformation space. For example, in the attempts to fold 434 repressor, we found that one-third of the structures in the final populations had structures within 4.0 Å DME. Given the huge number of conformations a 63-residue protein could adopt, this represents an enormous enrichment. Thus, while much work is clearly needed to improve both the selectivity of the energy function and to generalize it to other protein types, our approach to weighting the energy function to maintain a gradient in the direction of the correct structure should prove useful.

Conclusion

The three-dimensional profile method embodies a general requirement of any foldable sequence: the ability to create a suitable hydrophobic core. The generality and simplicity of the three-dimensional profile method, has facilitated its incorporation into diverse problems in protein folding and it seems likely that the three-dimensional profile approach will see continued use in its original or altered forms in the future.

Acknowledgements

We thank NIH, NSF and DOE for support.

References

1. Bowie J, Eisenberg D. Proc Natl Acad Sci USA 1994;91:4436–4440.
2. Bowie JU, Lüthy R, Eisenberg D. Science 1991;253:164–170.
3. Lüthy R, Bowie JU, Eisenberg D. Nature 1992;356:83–85.
4. Yi T-M, Lander E. J Molec Biol 1993;232:1117–1129.

5. Zhang K, Eisenberg D. Prot Sci 1994;3:687–695.
6. Wilmanns M, Eisenberg D. Proc Natl Acad Sci USA 1993;90:1379–1383.
7. Dill K. Biochemistry 1990;29:7133–7155.
8. Bashford D, Chothia C, Lesk AM. J Molec Biol 1987;196:199–216.
9. Lesk AM, Chothia C. J Molec Biol 1980;136:225.
10. Lesk AM, Chothia C. J Molec Biol 1982;160:325.
11. Lesk AM, Chothia C. J Molec Biol 1982;160:325–342.
12. Reidhaar-Olson JF, Sauer RT. Science 1988;241:53.
13. Bowie JU, Reidhaar-Olson JF, Lim WA, Sauer RT. Science 1990;247:1306–1310.
14. Sweet R, Eisenberg D. J Molec Biol 1983;171:479–488.
15. Drexler KE. Proc Natl Acad Sci USA 1981;78:5275–5278.
16. Pabo C. Nature 1983;301:200.
17. Flaherty KM, McKay DB, Kabsch W, Holmes KC. Proc Natl Acad Sci USA 1991;88:5041–5045.
18. Needleman SB, Wunsch CD. J Molec Biol 1970;48:443–453.
19. Smith TF, Waterman MS. Adv Appl Math 1981;2:482–489.
20. Gribskov M, McLachlan AD, Eisenberg D. Proc Natl Acad Sci USA 1987;84:4355.
21. Gribskov M, Lüthy R, Eisenberg D. Meth Enzymol 1990;183:146–159.
22. Bowie J, Pakula A, Simon M. Acta Cryst 1995;D51:145–154.
23. Wilson C, Wardell M, Weisgraber K, Mahley R, Agard D. Science 1991;252:1817–1822.
24. Chapman M, Suh S, Curmi P, Cascio D, Smith W, Eisenberg D. Science 1988;241:71–74.
25. Ghosh D, O'Donnell S, Furey WJ, Robbins A, Stout C. J Molec Biol 1982;158:73–109.
26. de Vos A, Tong L, Milburn M, Matias P, Jancarik J, Noguchi S, Nishimura S, Miura K, Ohtsuka E, SH K. Science 1988;239:888–893.
27. McClarin J, Frederick C, Wang B, Greene P, Boyer H, Grable J, Rosenberg J. Science 1986;234: 1526–1541.

Author index

Keyword index